Modulation Spectroscopy

SOLID STATE PHYSICS

Advances in
Research and Applications

Editors

Frederick Seitz
Rockefeller University, New York, New York

David Turnbull
*Division of Engineering and Applied Sciences, Harvard University
Cambridge, Massachusetts*

Henry Ehrenreich
*Division of Engineering and Applied Physics, Harvard University
Cambridge, Massachusetts*

The following monographs are published within the framework of the series:

1. T. P. DAS and E. L. HAHN, *Nuclear Quadrupole Resonance Spectroscopy,* 1958
2. WILLIAM LOW, *Paramagnetic Resonance in Solids,* 1960
3. A. A. MARADUDIN, E. W. MONTROLL, and G. H. WEISS, *Theory of Lattice Dynamics in the Harmonic Approximation,* 1963
4. ALBERT C. BEER, *Galvanomagnetic Effects in Semiconductors,* 1963
5. ROBERT S. KNOX, *Theory of Excitons,* 1963
6. S. AMELINCKX, *The Direct Observation of Dislocations,* 1964
7. JAMES W. CORBETT, *Electron Radiation Damage in Semiconductors and Metals,* 1966
8. JORDAN J. MARKHAM, *F-Centers in Alkali Halides,* 1966
9. ESTHER M. CONWELL, *High Field Transport in Semiconductors,* 1967
10. C. B. DUKE, *Tunneling in Solids,* 1969
11. M. CARDONA, *Modulation Spectroscopy,* 1969

MODULATION SPECTROSCOPY

Manuel Cardona

PHYSICS DEPARTMENT, BROWN UNIVERSITY
PROVIDENCE, RHODE ISLAND

1969

ACADEMIC PRESS   *New York and London*

COPYRIGHT © 1969, BY ACADEMIC PRESS, INC.
ALL RIGHTS RESERVED
NO PART OF THIS BOOK MAY BE REPRODUCED IN ANY FORM,
BY PHOTOSTAT, MICROFILM, RETRIEVAL SYSTEM, OR ANY
OTHER MEANS, WITHOUT WRITTEN PERMISSION FROM
THE PUBLISHERS.

ACADEMIC PRESS, INC.
111 Fifth Avenue, New York, New York 10003

United Kingdom Edition published by
ACADEMIC PRESS, INC. (LONDON) LTD.
Berkeley Square House, London W1

LIBRARY OF CONGRESS CATALOG CARD NUMBER: 55-12299

PRINTED IN THE UNITED STATES OF AMERICA

Foreword

The sharp structure in the optical constants of semiconductors at photon energies considerably greater than the band gap was observed a little more than ten years ago. Its subsequent interpretation led to the development of an important new experimental means to obtain detailed information concerning the band structures of these solids as well as metals and insulators. Indeed, because such experimental results permit comparison with band calculations of many materials over a sufficiently broad energy range, that their reliability could be ascertained more completely than before. In the present volume Cardona, one of the outstanding contributors to this field, reviews the ideas and techniques that have led to these developments, surveys the experimental results, and relates them to the theory. The author emphasizes the more recently developed modulation techniques in which, for example, the variation of the reflectance with respect to a time varying external disturbance such as an electric field, temperature, or stress is measured rather than the reflectance itself. The derivative of the reflectance obtained in this way yields very sharp spectra and as a result correspondingly more precise theoretical information. Cardona rounds out this more specialized discussion with a succinct but quite general survey of the theoretical framework underlying the description of the optical properties of solids. This book thus forms a complement to the earlier articles by Stern and Phillips that appeared respectively in Volumes 15 and 18 of this series.

F. SEITZ
D. TURNBULL
H. EHRENREICH

Preface

The study of the electronic optical properties of solids, in particular semiconductors and insulators, regained considerable attention during the 1950's. This was due, in part, to the advances in materials preparation technology associated with the invention of the transistor, and to theoretical work leading to the detailed knowledge of the band structure of germanium and related materials. In the mid 1950's optical studies were usually confined to the vicinity of the lowest (fundamental) absorption edge. These studies were extended in the late 1950's and early 1960's to photon energies well above the fundamental edge, with the normal incidence reflection technique. By the mid 1960's the electronic structure in the optical constants of a large number of materials had been investigated and tentatively interpreted. Activity in the field began to taper off, probably as a result of having reached the limit of possibilities of the existing experimental methods, until the electroreflectance technique of Seraphin and Hess appeared (1965). Since then the vast amount of activity in electroreflectance and other optical modulation methods has brought new vigor to the field of electronic optical properties of solids. A number of very ingenious experimental techniques have been developed; their applicability extends beyond solid state spectroscopy.

In this volume we present the theoretical background necessary for the understanding of the electronic optical properties of solids and their dependence on external perturbations. The main techniques used for modulation spectroscopy are described and typical experimental results are presented and analyzed.

While writing this book the author has benefited greatly from his association with a number of colleagues at Brown University. Thanks are due, in particular, to D. E. Aspnes, F. Cerdeira, R. A. Forman, A. Gavini, E. Matatagui, F. H. Pollak, J. E. Rowe, and K. L. Shaklee, who were responsible for the development of some of the techniques discussed here. Thanks are also due to a large number of workers for granting permission to reproduce drawings and to Miss Susan Desilets and Miss Sharon Perlow for skillfully typing the manuscript. The author is particularly indebted to his wife, Inge, for help in editing and proofreading and for her understanding and patience while the book was being written.

The work of the Brown Group on modulation spectroscopy has been supported by the National Science Foundation, the Army Research Office, Durham, and the Advanced Research Projects Agency. The author has been, while writing this book, the recipient of an Alfred P. Sloan Foundation Fellowship.

Providence, Rhode Island   MANUEL CARDONA
May, 1969

Contents

Foreword v
Preface vii

I. **Introduction and Historical Survey** 1

II. **Optical Properties of Electrons in Solids**
    1. Introduction 9
    2. One-Electron Model 10
    3. Intraband Effects 14
    4. Critical Points and Direct Transitions 15
    5. Indirect Transitions 23
    6. Excitons 25
    7. Broadening 47
    8. Experimental Techniques 55
    9. Assignment of Optical Structure to Interband Critical Points .. 65

III. **Modulation Techniques**
    10. Fundamentals 89
    11. General Techniques 97
    12. Modulation Techniques and Band Structure Dependence on Static Parameters 102
    13. Kramers–Kronig Analysis 103

IV. **Wavelength Modulation**
    14. Experimental Techniques 105
    15. Line Shapes 109
    16. Results 112

V. **Temperature Modulation**
    17. General Considerations 117
    18. Experimental Techniques 123
    19. Results 125

VI. Stress Modulation

20. General Considerations 137
21. Experimental Techniques: Piezoabsorption and Piezoreflectance .. 148
22. Results 150

VII. Electric Field Modulation

23. Introduction 165
24. Theory 166
25. Experimental Techniques 202
26. Experimental Results 224

VIII. Modulation Techniques and Dependence of Band Structure on Static Parameters

27. Introduction: Effects of Temperature and Doping 277
28. Electroreflectance in Binary and Pseudobinary Alloys 280
29. Uniaxial Stress Measurements 284
30. Modulation Techniques in the Presence of a Steady Magnetic Field 300

Appendix I. A Few Relationships Involving Airy Functions .. 325

Appendix II. Convolution Expression for $\epsilon(\omega\,\mathscr{E})$; Experimental Absorption Edges 329

Appendix III. k · p Perturbation Theory, Effective Masses, and Effective g-Factors of Semiconductors 331

References 335

Author Index 345

Subject Index 353

I. Introduction and Historical Survey

The lines observed in the absorption and emission spectra of nearly isolated atoms and ions are extremely sharp. As a result, their wavelengths can be determined with great accuracy. The large amount of extremely precise spectroscopic data for atoms and ions accumulated during the last century and the early part of this century has been responsible for the initial development and many subsequent refinements of quantum theory. Molecular spectra, while usually less sharp than atomic spectra, are also sharp. Positions of spectral lines can be determined with enough accuracy to check quantum mechanical calculations of the electronic structure of the molecules. By the same token one concludes that a study and understanding of the optical properties of a solid would contribute to the knowledge of its electronic structure. Experimental information about such structure is particularly interesting since it involves many-body interactions among a much larger number of particles than present in either atoms or molecules.

The high particle density of solids, however, makes their optical spectra rather broad and hence often uninteresting from the experimental point of view. The large degeneracy of the atomic levels is split by the interatomic interactions into quasi-continuous bands. As a result of these splittings and of the high electronic densities in solids, the penetration depths for electromagnetic radiation are of the order of 500 Å through most of the optical spectrum. Such small penetration depths make absorption spectroscopy with bulk single crystal materials extremely difficult except in the immediate vicinity of an energy gap (semiconductors and insulators).

From the point of view of a theoretical understanding of the optical spectra of solids it is reasonable to start with those solids whose spectra bear a close relationship to those of the isolated atoms, ions, or molecules. No such relationship is immediately obvious for materials with either covalent, metallic, or, to some extent, ionic bonding. This explains why the interpretation of the optical properties of these materials had to await the fairly sophisticated picture of their band structures which became available in the late 1950's. The electronic spectra of molecular crystals, on the contrary, should bear a close relationship to those of the isolated molecules. An interpretation based on the spectra of isolated molecules perturbed by the relatively weak intermolecular interactions should be possible.

Early studies of the absorption of molecular crystals and their similarity with those of the isolated molecules were described by Becquerel.[1] Extensive studies of optical properties of molecular crystals were carried out in the light of the contemporary advances in quantum theory in Germany[2] and in the Soviet Union[3] in the 1930's. We shall not concern ourselves with this subject since optical modulation techniques have not yet been applied to molecular crystals. For a survey of the field of optical properties of molecular crystals we refer the reader to the review articles of McClure[4] and Wolf.[5]

In view of the short penetration depths mentioned above, the requirement of samples thin enough for transmission measurements suggests the use of evaporated thin films. Covalent crystals are difficult to obtain by vacuum deposition as opposed to ionic crystals which usually evaporate in molecular form. This fact enabled Hilsch and Pohl in the late 1920's[6] to obtain absorption spectra for evaporated films of alkali halides which differ little from those obtained three decades later from the reflectivity spectrum of bulk materials.[7] They attributed the peaks observed at the absorption edge to excitations of electrons from the halogen ion to the surrounding alkali ions. To this date, this explanation remains generally accepted with slight modification. Hilsch and Pohl noticed[6] that in many alkali halides two absorption peaks appear near the absorption edge; the energy splitting between these two peaks is mostly determined by the halogen ion. Franck et al.[8] pointed out that the splitting of the doublet mentioned above is essentially the same as the splitting of the P_1–P_2 spectral terms of the halogen atoms. This atomic splitting was at that time already known to be due to spin orbit interaction. The observation of spin orbit splittings in the optical spectra of solids remains, to this date (see Section 9), one of the most powerful tools for identifying optical transitions.

The quantum theory of light absorption by crystals was developed in the 1930's. Frenkel[9] and Peierls[10] described the excited state of the crystal as a properly symmetrized combination of highly localized atomic or quasi-molecular excitations. This picture was appropriate, at least qualitatively, for the alkali halides. Wannier[11] chose to represent the excited state by using the one-electron band model as a basis. The excited electron would be bound to the hole left behind by Coulomb attraction. This picture, valid for the case of a weak Coulomb interaction, i.e., high dielectric constant and small effective mass, leads to a series of sharp hydrogenic lines in the gap below the band-to-band absorption continuum. The peaks in the alkali halides reported by Hilsch and Pohl[6] were broad, as opposed to those predicted by the exciton theories. The first observation of very sharp excitonic lines was probably due to Hayashi and Katzuki[12] for Cu_2O at low temperatures. This work was followed by that of Gross and co-workers[13] and Nikitine and co-workers.[14] The work of Eby et al.[15] on the alkali halides at low temperature sharpened the exciton lines and discovered new structure not seen in the room temperature measurements of Hilsch and Pohl.

Bardeen and co-workers discussed in 1954 the theory of the line shape of direct and indirect band-to-band transitions in the vicinity of an absorption edge.[16] They proposed the now well-known $(\omega - \omega_g)^{1/2}$ and $(\omega - \bar{\omega}_g)^2$ line shapes of the absorption coefficient for direct and indirect transitions, respectively. In 1957, Elliott[17] reported calculations of the effect of Wannier excitons on the band-to-band line shapes of Bardeen et al.[16] He showed that the $(\omega - \bar{\omega}_g)^2$ dependence of indirect band-to-band transitions should become $(\omega - \bar{\omega}_g)^{1/2}$ in the presence of excitonic interaction.

In spite of the early work of Bloch, Brillouin, Slater, Wigner, and Wilson,[18] which laid the foundations of the one-electron band theory of solids, very little work was done towards relating actual band structure calculations to optical properties until the mid-1950's. Herman's calculations,[19] together with the results of cyclotron resonance measurements,[20] led to the conclusion that the lowest absorption edge of germanium and silicon was indirect[21]: the photon energy dependence of the absorption coefficient was $(\omega - \bar{\omega}_g)^2$, as predicted by Bardeen et al.[16] Later, upon close examination of the absorption edge of germanium with high resolution, Macfarlane et al.[22] found deviations from the $(\omega - \bar{\omega}_g)^2$ behavior and attributed them to the formation of indirect excitons.[17]

From that moment, optical experiments and band structure calculations for germanium and related materials, e.g., zincblende and wurtzite, went hand in hand while the understanding of the optical spectra of other families of materials lagged behind owing to insufficiencies in the knowledge of their band structure.

Until the mid-1950's most of the optical measurements with single crystals were confined to the wavelength region in the vicinity and below the lowest absorption edge; measurements on evaporated thin films were not very reproducible and thus they were considered unreliable. At that time a number of authors began to measure optical constants of single crystals above the edge by means of reflection techniques. Such techniques involved reflectivity with polarized light at oblique incidence,[23] ellipsometry,[24] and normal incidence reflectance accompanied by the use of dispersion relations.[25,26] Because of its simplicity, the latter of these techniques has been the most productive to date. Recent advances in epitaxial crystal growth have made possible transmission work on high-quality single crystal thin films.[27]

The theoretical work of Phillips[28] and of Roth and Lax[29] in 1959 produced a breakthrough in the interpretation of optical spectra of solids above the fundamental edge. These authors assigned, on the basis of the known band structure of germanium, the peak observed by Philipp and Taft[26] and by Archer[24] for this material around 2 eV to direct transitions at critical points in the [111] and equivalent directions of **k**-space. While many identifications of optical structure have had to be changed in view of subsequent work, this explanation of the 2 eV peak of germanium is still generally accepted today.[30]† Striking confirmation of this interpretation was obtained by Tauc and Antončik[31] as they observed the spin orbit splitting of the 2 eV peak of germanium. This doublet shall be referred to as E_1, $E_1 + \Delta_1$. The remarkable similarity of the optical spectra of all germanium and zincblende-type materials soon became apparent.[32,33] A review of the large amount of data obtained by the normal incidence reflection technique can be seen in the review articles of Phillips[34] and Cardona.[35]

We should mention at this point the parallel advances in the understanding of the optical constants of metals. The theoretically simplest of all metals, the alkali metals, are extremely difficult to work with experimentally. Most of the early work was concerned with the deter-

---

† However, the critical points, believed to be at the edge of the zone by Phillips[28] and Roth and Lax,[29] have been moved to the inside of the zone.

mination of their plasma frequencies,[36] largely independent of band structure details. Stimulated by Segall's band calculations for copper,[37] Ehrenreich and Philipp[38] determined the optical constants of Cu, Ag, and Au by the normal incidence reflection technique and obtained a reasonable picture of the transitions contributing to the optical structure. Most of the observed structure, however, was quite broad and an analysis as detailed as that performed for the germanium–zincblende materials was not possible. The work of Mayer and El Naby for the alkali metals[39] came as a great surprise: peaks were found in the optical constants which could not be explained in terms of the one-electron picture. This work generated considerable theoretical speculation[40,41]; the observed structure, however, has yet to be confirmed by other experimenters.[42]

As a result of the accumulation of optical data it became clear that energy band calculations based on first principles, with no adjustable parameters, were not accurate enough to interpret those data quantitatively. Calculations in which a number of parameters is left adjustable so as to fit *some* of the existing data soon became quite popular and successful for correlating experimental results. Among these calculations we mention the pseudopotential method of Phillips and Kleinman[30,43] and Herman's method[44] of orthogonalized plane waves (OPW) with a number of Fourier coefficients of the potential left adjustable. Pseudopotential calculations with high-speed computers are fast enough to permit a sampling of the band structure at a large number of points. With such sampling, an actual calculation of the optical constants of the solid can be performed, as first done by Brust and co-workers.[30,45] Such calculation produces rather definitive assignments of the observed optical structure. As an example, we mention the discovery by Brust *et al.*[30] of the M_1 nature (saddle point) of the critical points involved in the 2 eV peaks of germanium.

It has become clear, as a result of the work mentioned above, that the gross features of the observed optical spectra of a number of solids *above the fundamental edge* can be accounted for in terms of the one-electron band picture (with the possible exception of the Mayer–El Naby structure of the alkali metals). The question then naturally arises of whether there are any recognizable many-body effects, such as exciton effects, in the optical properties above the fundamental edge. Cardona and Harbeke[46] suggested that exciton effects were necessary to account for the sharpness of the $E_1, E_1 + \Delta_1$ peaks of zincblende materials at low temperatures. Phillips[47] discussed a theoretical model for such excitons which emphasized the hyperbolic nature of the corresponding critical points M_1. Recent interest in this subject is exemplified by the work of Velický and Sak,[48] Duke and Segall,[49] Toyozawa et al.,[50] and Hermanson.[51]

Considerable progress in our understanding of experimental structure has been due to the measurements by the Harvard group of the effect of hydrostatic pressure on the optical properties.[52,53] Measurements of optical spectra in the presence of a strong magnetic field have also contributed to our knowledge of the energy band parameters, in particular, effective masses and g-factors. Especially noteworthy is the work of Gross and Zakharchenia,[54] on the Zeeman effect of excitons in a magnetic field for Cu_2O and of Hopfield and Thomas[55] and Dimmock and Wheeler[56] for wurtzite-type materials. The original work of Burstein and co-workers[57] and of Lax and co-workers (see Zwerdling et al.[58]) on magnetooptical effects in interband transitions should also be mentioned. Photoemission measurements, mainly performed by Spicer[59] and by the Bell group[60] have also contributed to our understanding of optical transitions in solids; energy distribution measurements of the photoemitted electrons yield, in particular, the energy of the final states with respect to the vacuum level.

While considerable progress in studying transitions above the lowest edge has been made by means of reflectivity measurements, the possibilities of these measurements are limited. The structure observed is relatively broad and it is usually superimposed on a large structureless background. Accurate reflectivity measurements are somewhat cumbersome, especially at low temperatures. Seraphin's original work on electroreflectance in germanium,[61] opened a wide field of possibilities for studying optical structure. It became clear as a result of this work that fairly sharp structure could be obtained by measuring instead of the reflectivity R, the derivative of R with respect to some external parameter, such as an applied electric field. The simplicity and sensitivity of such measurements could be greatly enhanced by making full use of the advantages of phase sensitive detection. The techniques of piezoreflectance[62,63] and thermoreflectance[64] appeared in quick succession. Modulation techniques have also been helpful for transmission measurements.[65-67] The large number of possible variations and modifications of these modulation methods will constitute the main subject of this book.

Special mention should be made, at this point, of the applications of these methods to study deformation potentials under static uniaxial stress,[67,68] and magnetooptical phenomena.[69,70]

Electric field modulation (electroreflectance) has probably become the most popular of all modulation techniques, due mainly to the simplicity of the electrolytic method.[71] The theoretical groundwork for this type of modulation was done by Franz[72] and Keldysh[73] (hence the name Franz–Keldysh effect) and extended by Callaway,[74] Tharmalingham,[75] and Aspnes.[76] Early experimental work on the subject was done by transmission and hence was limited to energies below or near the lowest gap.[77,78] The theoretical work on the effect of an electric field on the optical constants described above neglects exciton interaction between electron and hole. The electro-optic effect of hydrogenic excitons has been treated by Duke and Alferieff.[79]

The purpose of this book is to give the background needed for the study of optical modulation techniques as applied to solid state spectroscopy and to discuss the experimental techniques and results obtained from the standpoint of the information they contain about energy band structures. In Chapter II we discuss the theory of the electronic optical properties of solids, illustrated with a few typical experimental results. This discussion is, of necessity, incomplete; its purpose is only to acquaint the reader with the main principles and facts. Many derivations have thus been condensed to the minimum judged necessary to convey the spirit of the method. Original references and references to review articles are given whenever appropriate. Chapter III discusses the general principles involved in modulation spectroscopy and the details common to all techniques. Chapter IV discusses the technique of wavelength modulation, sometimes called an *external* modulation method in contrast to the *internal* methods in which the properties of the sample are modulated. In spite of the theoretical simplicity of the wavelength technique, the few results available illustrate its inherent experimental difficulties. Chapter V discusses temperature modulation, a technique in many cases equivalent to wavelength modulation. Chapter VI describes uniaxial stress modulation and Chapter VII electric field modulation. The techniques in Chapters VI and VII have the common feature that the modulation may lower the symmetry of the sample; as a result these methods can be used to investigate the symmetry of optical transitions. Chapter VIII discusses the use of modulation techniques in the presence of a steady perturbation, such as uniaxial stress, or a large magnetic field.

At the time of writing this book the rate at which work in the field is being published does not show any signs of tapering off. We chose, arbitrarily, December 1967 as the cutoff point for literature references, except for a few available to us in preprint form. Even so, some pertinent references may have been inadvertently omitted, victims of the recent explosion in the field.

We would like to call the reader's attention to a number of excellent review articles and books on the electronic optical properties of solids, such as the recent book on the "Optical Properties of the III–V Compounds,"[80] the articles by Phillips,[34] and Tauc,[81] the Proceedings of the 1966 Varenna Summer School,[53] the Proceedings of the 1965 Paris Conference on the electronic structure of metals and alloys,[42] the Proceedings of the 1962 St. Andrews Conference on Polarons and Excitons,[82] the treatises on excitons by Knox[83] and by Dexter and Knox,[84] the recent book by Greenaway and Harbeke,[85] and the book by Moss.[86] In addition we also recommend the Proceedings of the past eight International Conferences on the Physics of Semiconductors, those of the Schenectady conference on Semiconducting Compounds,[87] and those of the Providence Conference on II–VI Compounds.[88]

II. Optical Properties of Electrons in Solids

1. Introduction

The linear response of a nonmagnetic medium to transverse electromagnetic radiation ($\mathbf{V} \cdot \mathbf{D} = 0$, $\mathbf{V} \cdot \mathbf{B} = 0$) is completely described by the frequency and wavevector dependent dielectric tensor $\boldsymbol{\varepsilon}$ or by the conductivity tensor $\boldsymbol{\sigma} = -(i\omega/4\pi)(\boldsymbol{\varepsilon} - 1)$ where ω is the angular frequency. These tensors depend, in general, on the frequency and wavevector of the radiation. In the phenomena to be discussed here, the radiation field varies little over the characteristic atomic dimensions, and hence, the wavevector dependence of $\boldsymbol{\varepsilon}$ and $\boldsymbol{\sigma}$ (spatial dispersion) will be neglected. The components of the tensors $\boldsymbol{\sigma}$ and $\boldsymbol{\varepsilon}$ are not all independent. Thermodynamics requires these tensors to be symmetric in the absence of a magnetic field (Onsager relations).[89] The causal nature of the response to an electromagnetic field imposes a relationship between the real and the imaginary parts of $\boldsymbol{\varepsilon}$: no response to an applied field can appear *before* the field is applied.[90-92] This relationship is, in the absence of magnetic fields:

$$\varepsilon_r(\omega) - 1 = \frac{2}{\pi} P \int_0^\infty \frac{\omega' \varepsilon_i(\omega')}{\omega'^2 - \omega^2} d\omega', \tag{1.1}$$

and its inverse:

$$\varepsilon_i(\omega) = \frac{2\omega}{\pi} P \int_0^\infty \frac{\varepsilon_r(\omega')}{\omega^2 - \omega'^2} d\omega', \tag{1.2}$$

where $P\int$ designates the Cauchy principal part of the integral. Equations (1.1) and (1.2) are called the Kramers–Kronig relations.

For isotropic and cubic materials $\boldsymbol{\varepsilon}$ and $\boldsymbol{\sigma}$ reduce to scalars. The propagation of a plane electromagnetic wave is isotropic and is determined by the scalar complex refractive index $n = n_r + i n_i = \varepsilon^{1/2}$.

The real and imaginary parts of n are related to those of ε by:

$$n_r^2 - n_i^2 = \varepsilon_r$$
$$2n_r n_i = \varepsilon_i \qquad (1.3)$$

The electric field **E** of a plane wave propagating along the x direction has the form:

$$\mathbf{E} = \mathbf{E}_0 \exp(-i\omega[t - (xn_r/c)]) \cdot \exp(-\tfrac{1}{2}\alpha x), \qquad (1.4)$$

where $\alpha = 2\omega n_i/c$ is the absorption coefficient and c is the speed of light in vacuum. The real and the imaginary parts of n fulfill dispersion relations analogous in form to those for ε_r and ε_i. They can be considered as the result of the requirement of relativistic causality: no signal can propagate at a speed higher than c.

2. ONE-ELECTRON MODEL[93-96]

We shall treat the problem of the response of electrons to an electromagnetic field in a perfect solid by using the random phase approximation.[95-96] In this approximation the effect of the external fields on the many-electron wave functions is obtained by calculating the response of the one-electron wave functions in a self-consistent potential. We shall neglect local field corrections,[97] i.e., the dependence of the self-consistent potential on the external field. Atomic units will be used throughout this paper ($\hbar = 1$, $e = 1$, $m = 1$) unless otherwise specified.

We shall assume that the real and imaginary parts of the dielectric constant tensor can be diagonalized with respect to the same set of axes and we shall choose these axes as axes of coordinates. This restriction holds for all crystal systems other than monoclinic and triclinic. Let us calculate the linear response of Bloch electrons in a solid to an electric field $\mathbf{E}_0 e^{-i\omega t} = (i\omega/c)\mathbf{A}_0 e^{-i\omega t}$, where $\mathbf{A}_0 e^{-i\omega t}$ is the vector potential (we consider only electric dipole transitions). In the Schrödinger representation, the one-electron Hamiltonian is (neglecting spin orbit effects):

$$H = H_0 + (1/c)\mathbf{p} \cdot \mathbf{A}_0 e^{-i\omega t} = H_0 + H', \qquad (2.1)$$

to first order in $\mathbf{A}_0$. H_0 is the Hamiltonian of the crystal in the absence of radiation. The effect of the perturbation Hamiltonian of

2. ONE-ELECTRON MODEL

the radiation H' on the electron wave functions can be easily calculated in the interaction representation[94,98]

$$\Psi_I(t) = e^{iH_0(t-t_0)}\Psi(t)$$
$$H_I'(t) = e^{iH_0(t-t_0)}H'e^{-iH_0(t-t_0)}. \quad (2.2)$$

By integrating the equation of motion of Ψ_I:

$$i\,\partial\Psi_I/\partial t = H_I'\Psi_I, \quad (2.3)$$

we obtain, to first order in H':

$$\Psi_I(t) = \Psi(t_0) - i\int_{t_0}^{t} H_I'(t')\Psi(t_0)\,dt'. \quad (2.4a)$$

The wave function $\Psi(t)$ in the Schrödinger representation is obtained from Eqs. (2.4a,b):

$$\Psi(t) = e^{-iH_0(t-t_0)}\,\Psi_I(t). \quad (2.4b)$$

We want to apply the perturbation Hamiltonian H' in an adiabatic manner and hence we shall replace ω in Eq. (2.1) by $\omega + i\delta$, with δ an infinitesimal positive number. For $t_0 \to -\infty$, $\Psi(t_0)$ becomes in Eq. (2.4a) the time independent Bloch function. For this reason we shall take the lower limit of integration in Eq. (2.4a) to be minus infinity.

In order to calculate the diagonal components of the conductivity tensor σ_{jj} we must calculate the jth component of the current density J_j at the point $\mathbf{r}$:

$$J_j = -\tfrac{1}{2}\sum_{\text{occupied } k}[\Psi_k^*(t)p_j\Psi_k(t) - \Psi_k(t)p_j\Psi_k^*(t) + (2/c)A_j\Psi_k^*\Psi_k], \quad (2.5)$$

and find its average $\langle J_j \rangle$ over a volume V of large dimensions compared with atomic dimensions but small compared to the wavelength and penetration depth $(1/\alpha)$ of the light. The wave functions Ψ_k are taken normalized to the volume V. The conductivity tensor is extracted from the obtained linear relationship between $\langle \mathbf{J} \rangle$ and the external electric field $\mathbf{E}$: $\langle J_j \rangle = \sigma_{jj} E_j$.

When we substitute Eqs. (2.4a,b) into Eq. (2.5) and average the current, the unperturbed wave functions $\Psi_k(t_0)$ for $t_0 \to -\infty$ do not contribute to the first two terms in (2.5). The contribution from any Bloch function Ψ_k is cancelled by that from its time reversed Ψ_{-k}. The unperturbed wave functions give, however, the only first-order contribution to the third term in Eq. (2.5). The jth component of the average current is, to first order in E_j [95]:

$$\langle J_j \rangle = -\frac{E_j}{i\omega V} \left(\sum_{\text{occupied } k} \langle k|p_j|l\rangle\langle l|p_j|k\rangle \right.$$

$$\left. \times \left[\frac{1}{\omega_{lk} + \omega + i\eta} + \frac{1}{\omega_{lk} - \omega - i\eta} \right] \right) - \frac{1}{i\omega} E_j N, \quad (2.6)$$

where N is the electron density and $\omega_{lk} = \omega_l - \omega_k$. Equation (2.6) is obtained by replacing Eqs. (2.4a,b) into Eq. (2.5), averaging over the volume, and using the completeness relation:

$$\sum_l \Psi_l^*(r)\Psi_l(r') = \delta(r, r'). \quad (2.7)$$

We shall take the volume of normalization $V = 1$.

Equation (2.6) gives the dielectric constant tensor:

$$\varepsilon_{jj}(\omega) = 1 - \frac{4\pi}{\omega^2} \sum_{k \text{ occupied}} \left(\langle k|p_j|l\rangle\langle l|p_j|k\rangle \frac{2\omega_{lk}}{\omega_{lk}^2 - (\omega + i\eta)^2} \right) - \frac{4\pi}{\omega^2} N \quad (2.8)$$

Equation (2.8) can be transformed with the help of the $\mathbf{k} \cdot \mathbf{p}$ sum rule[93,95]:

$$-1 = -\frac{\partial^2 \omega_k}{(\partial k_j)^2} - \sum_{l \neq k} \frac{2\langle k|p_j|l\rangle\langle l|p_j|k\rangle}{\omega_{lk}}, \quad \omega_{lk} = \omega_l - \omega_k \quad (2.9)$$

into an expression exhibiting explicitly the difference between interband and intraband terms in ε_{jj}. Multiplying Eq. (2.9) by $4\pi/\omega^2$, and summing over all occupied states, we obtain an expression for $(4\pi/\omega^2)N$. Substituting it into the right-hand side of Eq. (2.8) we obtain:

$$\varepsilon_{jj}(\omega) = 1 + 4\pi \sum_{\substack{k \text{ occupied} \\ l \text{ empty}}} \frac{F_{jj}^{lk}}{\omega_{lk}^2 - (\omega + i\eta)^2} - \frac{4\pi}{\omega^2} \sum_{k \text{ occupied}} \frac{\partial^2 \omega_k}{\partial k_j^2} \quad (2.10)$$

The "oscillator strength" F_{jj}^{lk} is equal to $2\langle k|p_j|l\rangle\langle l|p_j|k\rangle \cdot \omega_{lk}^{-1}$. The sum in Eq. (2.10) has been restricted to k occupied and l empty since for both states occupied the kl terms are cancelled by the lk terms (hence we do not have to worry about the exclusion principle). The last sum in Eq. (2.10) represents the intraband or "free electron" contribution $\boldsymbol{\varepsilon}^f$ to the dielectric constant. The first sum $\boldsymbol{\varepsilon}^b$ involves

2. ONE-ELECTRON MODEL

only transitions between two different bands (interband). The convergence parameter η must be made to tend to zero if no scattering is present. Scattering can be taken into account phenomenologically by replacing η in Eq. (2.10) by a collision frequency $\omega_\tau \neq 0$. Similarly ω^2 can be replaced by $\omega(\omega + i\omega_\tau')$ in the intraband terms.†

Let us assume $\omega_\tau = 0$. The sums in Eq. (2.10) can be easily transformed into integrals over **k**-space. In Eq. (2.10) we have assumed we are dealing with a perfect solid, and have neglected the effect of phonons (to be treated later); **p** only couples states of the same **k**-vector (in the reduced zone scheme), hence the sum over l in Eq. (2.10) must only be extended to initial and final states such that $\mathbf{k}_l = \mathbf{k}_k$, and therefore it remains a sum over the discrete set of all possible transitions whose final state has the same **k** as the initial state (direct transitions). The volume integrals can be written as integrals over a constant energy surface, followed by an integration over the energy ω:

$$\varepsilon_{jj}(\omega) = 1 + \frac{1}{\pi^2} \int_{\omega_{lk}} \int_{S_{lk}} \frac{F_{jj}^{lk}}{\omega_{lk}^2 - (\omega + i\eta)^2} f(\omega_k)[1 - f(\omega_l)] \frac{dS_{lk}\, d\omega_{lk}}{|\mathbf{V_k}\omega_{lk}|}$$

$$- \frac{1}{\pi^2\omega^2} \int_{\omega_k} \int_{S_{\omega_k}} \frac{\partial^2 \omega_k}{(\partial k_j)^2} f(\omega_k) \frac{dS_{\omega_k}\, d\omega_k}{|\mathbf{V_k}\omega_k|}. \qquad (2.11)$$

We shall assume completely degenerate statistics. In this case the energy integrals have as limits for the energy of the initial state the lowest energy ω_m and the Fermi energy ω_F. The real and imaginary parts of the interband contribution to $\boldsymbol{\varepsilon}$ in Eq. (2.11) can be separated by using the equation:

$$\int_{\eta \to +0} \frac{F(x)}{x + i\eta}\, dx = \mathrm{P} \int \frac{F(x)}{x}\, dx - i\pi\, \delta(x). \qquad (2.12)$$

We obtain:

$$\varepsilon_{jj}^b = \frac{1}{\pi^2} \mathrm{P} \int_{\omega_{lk}} \int_{S_{lk}} \frac{F_{jj}^{lk}}{\omega_{lk}^2 - \omega^2} \frac{dS_{lk}\, d\omega_{lk}}{|\mathbf{V_k}\omega_{lk}|} + \frac{i}{2\pi\omega} \int_{S_{\omega_{lk}=\omega}} \frac{F_{jj}^{lk}}{|\mathbf{V_k}\omega_{lk}|} dS_{lk}, \qquad (2.13)$$

where the states k and l are supposed to be filled and empty respectively. For cubic materials, $\boldsymbol{\varepsilon}$ reduces to a scalar: The tensor oscillator

† For a detailed discussion of the effect of collisions on the optical constants see Ehrenreich.[94]

strength F_{jj}^{lk} can then be replaced by the symmetrized scalar oscillator strength:

$$F^{lk} = \tfrac{1}{3}\sum_j F_{jj}^{lk} = \frac{2|\langle k|\,\boldsymbol{p}\,|l\rangle|^2}{3\omega_{lk}}. \tag{2.14}$$

3. Intraband Effects

The intraband contribution to $\boldsymbol{\varepsilon}$ may be large in the region where the interband contribution is nondispersive, e.g., for ω much smaller than the smallest ω_{lk}. This situation occurs in monovalent metals and in heavily doped semiconductors. It is then convenient to replace the interband contribution in Eq. (2.11) by a frequency independent real term $\varepsilon_{0,jj}$:

$$\varepsilon_{jj} = \varepsilon_{0,jj} - \frac{1}{\pi^2 \omega^2}\int_{\omega_m}^{\omega_F}\int_{S_{\omega_k}} \frac{\partial^2 \omega_k}{\partial k_j^2}\frac{dS_k\, d\omega_k}{|\boldsymbol{\nabla}_{\mathbf{k}}\omega_k|}; \tag{3.1}$$

ω_m represents the bottom of the conduction band and ω_F the Fermi level. For cubic materials with only one set of equivalent parabolic conduction band valleys (not necessarily isotropic), Eq. (3.1) becomes[99] (even at finite temperature):

$$\varepsilon = \varepsilon_0 - \frac{4\pi N_c}{m_c^* \omega^2} \tag{3.2}$$

where N_c is the conduction electron density and m_c^* the so-called conductivity or optical effective mass:

$$\frac{1}{m_c^*} = \frac{1}{3}\left[\frac{1}{m_x} + \frac{1}{m_y} + \frac{1}{m_z}\right], \tag{3.3}$$

where m_x, m_y, m_z are the three principal components of the effective mass tensor. Equation (3.2) has been extensively used for effective mass determination in semiconductors[99] and metals[36]: if we determine $\boldsymbol{\varepsilon}$ from the reflectivity of the material and N_c from its Hall effect, Eq. (3.2) can be used to obtain ε_0 and m_c^*. Lifetime broadening effects† can be phenomenologically introduced in Eq. (3.2) by replacing ω^2 by $\omega(\omega + i\omega_\tau')$ where ω_τ', the phenomenological collision

---

† For a detailed discussion of the effect of collisions on the optical constants see Ehrenreich.[94]

frequency, is not the same as that for interband transitions defined above. We obtain:

$$\varepsilon = \varepsilon_0 - \frac{4\pi N_c}{m_c^* \omega(\omega + i\omega_\tau')}. \tag{3.4}$$

4. Critical Points and Direct Transitions[34,100-102]

As already mentioned, the interband contribution to ε is produced by direct transitions if imperfections (phonons, defects, etc.) are neglected. The oscillator strength F^{lk} is subject to the selection rules for the matrix elements of **p** (see Eq. (2.14)). The simplest and perhaps the most common of the selection rules is that based on parity. If a crystal has inversion symmetry, parity is a good quantum number for the Bloch functions at $\mathbf{k} = 0$. Hence at the center of the Brillouin zone, F^{lk} is zero if the k and l states have the same parity. Parity is destroyed for the Bloch functions of states with $\mathbf{k} \neq 0$ inside the Brillouin zone but is recovered again at some high symmetry points on the surface of the Brillouin zone. In any case, F^{lk} is usually a smooth function[103,104] of **k** and no structure or sharp features in the optical constants are likely to arise from structure in F^{lk}.† If for a given pair of bands F^{lk} is zero at the center and at several high-symmetry points of the Brillouin zone boundary, it is not likely to reach high values anywhere inside the Brillouin zone and the transitions between the bands of that pair are nearly forbidden everywhere. Very often the energy bands of a solid are obtained by applying a small perturbation to the free electron bands for the same lattice.[34] The matrix elements of **p** between states which do not arise from degenerate free electron states are nearly zero, since the corresponding free electron matrix element is zero. If the states arise from degenerate free electron states, they both contain a mixture of the same plane waves and, if the transition is not forbidden by symmetry, the matrix element of **p** is of the order of the magnitude of the **k**-vector (in the extended zone) of the corresponding free electron states. These results are very useful for estimating the relative intensities of interband transitions. For a given pair of bands, F_{jj}^{lk} can sometimes be

† Optical structure owing to structure in oscillator strength can sometimes be observed in low symmetry crystals.[104]

taken out of the surface integral in Eq. (2.13), since it does not vary much over the constant energy surface of integration.

Equation (2.13) suggests[28] that a singularity in the optical constants, and hence sharp structure, may occur whenever $|\mathbf{V_k}\omega_{lk}|_{\omega_{lk}=\omega} = 0$ or equivalently when:

$$[\mathbf{V_k}\omega_l - \mathbf{V_k}\omega_k]_{\omega_{lk}=\omega} = 0. \tag{4.1}$$

The points in **k**-space for which Eq. (4.1) is fulfilled are called critical points or Van Hove singularities.[101,102] These critical points occur by symmetry at $\mathbf{k} = 0$ or at high symmetry points on the Brillouin zone boundary: in this case $\mathbf{V_k}\omega_k = \mathbf{V_k}\omega_l = 0$. They also may occur along high symmetry lines, e.g., along axes of three- or four-fold symmetry. In this case the components of the individual gradients $\mathbf{V_k}\omega_k$ and $\mathbf{V_k}\omega_l$ perpendicular to the symmetry axis are zero. The gradients along the symmetry axis are not zero but they may be accidentally equal at one particular point, and hence Eq. (4.1) may be satisfied at this point. Symmetry critical points of the type described above are more likely to occur than those at arbitrary points of **k**-space. Critical points along symmetry lines, for instance, are determined by the vanishing of only one component of Eq. (4.1) since the vanishing of the other two components is secured by symmetry; general critical points require the simultaneous vanishing of three components of the gradient in Eq. (4.1).

An expansion of ω_{lk} in power series of k_x, k_y, and k_z around a critical point gives:

$$\omega_{lk} = \omega_g + \frac{k_x^2}{2m_x} + \frac{k_y^2}{2m_y} + \frac{k_z^2}{2m_z} + \cdots, \tag{4.2}$$

where **k** is referred to the axes of the effective mass tensor. Equation (4.2) enables us to classify the critical points into four categories according to the signs of the principal effective masses. The critical points are labelled M_s where s is the number of negative masses in Eq. (4.2) ($s = 0, 1, 2, 3$). M_0 corresponds to a minimum, M_3 to a maximum, and M_1 and M_2 to saddle points in ω_{lk}. It is also of interest to consider one- and two-dimensional critical points which occur when two or one of the effective masses in Eq. (4.2) are very large. One-dimensional critical points are also important in the treatment of interband transitions in the presence of a magnetic field (see Chapter VIII). There are three types of two-dimensional (maximum,

4. CRITICAL POINTS AND DIRECT TRANSITIONS

minimum, and saddle point) and two types of one dimensional (maximum and minimum) critical points.

The behavior of the interband dielectric constant near a critical point can be found by replacing Eq. (4.2) into Eq. (2.13). The imaginary part of ε becomes (we shall omit the superscript b):

$$\varepsilon_i = \frac{F}{2\pi\omega} \int_{\substack{S \\ \omega_{lk}=\omega}} \frac{dS_\omega}{|\nabla_\mathbf{k}\omega_{lk}|} = \frac{2\pi^2 F}{\omega} N_d(\omega_{lk} = \omega) + \text{smooth background}, \quad (4.3)$$

where N_d is the joint or combined density of states. We have assumed that the oscillator strength is independent of $\mathbf{k}$. Such an assumption does not affect the singular behavior of ε_i provided F does not vanish at the critical point (allowed transitions). The integration of Eq. (4.3) can be readily performed for the various types of critical points. Three-dimensional maxima (M_3) and minima (M_0) yield the well-known density of states:

$$\begin{aligned} N_d(\omega) &\propto \text{Re}(\omega - \omega_g)^{1/2} + C; \quad \text{for an } M_0 \text{ critical point} \\ N_d(\omega) &\propto \text{Re}(\omega_g - \omega)^{1/2} + C; \quad \text{for an } M_3 \text{ critical point,} \end{aligned} \quad (4.4)$$

while the density of states around two- and one-dimensional maxima and minima is:

$$\begin{aligned} N_d(\omega) &= C_1; \quad \omega < \omega_g; \quad C_1 > C_2 \quad \text{for a two dimensional maximum} \\ N_d(\omega) &= C_2; \quad \omega > \omega_g; \quad C_1 < C_2 \quad \text{for a two dimensional minimum} \end{aligned} \quad (4.5)$$

$$\begin{aligned} N_d(\omega) &\propto \text{Re}(\omega - \omega_g)^{-1/2} + C; \quad \text{for a one dimensional minimum} \\ N_d(\omega) &\propto \text{Re}(\omega_g - \omega)^{-1/2} + C; \quad \text{for a one dimensional maximum.} \end{aligned} \quad (4.6)$$

The constant energy surfaces around three-dimensional saddle points M_1 and M_2 are hyperboloids and hence, in these cases, the surfaces of integration in Eq. (4.3) extend to infinity. Such infinite extension is artificial; it comes from having assumed perfectly quadratic bands. Actually, deviations from the quadratic behavior must take place as one moves away from the critical point and approaches the boundaries of the Brillouin zone. We can simulate the existence of a Brillouin zone boundary, and thus remove the divergence which results in the density of states if hyperbolic bands are extended to infinity, by introducing a cutoff in $|\mathbf{k}|$ of the order of the size of the Brillouin zone.[100,101] We then find for three-dimensional saddle points:

$$\left.\begin{array}{ll} N_d \propto C - (\omega_g - \omega)^{1/2}; & \text{for } \omega < \omega_g \\ N_d \propto C; & \text{for } \omega > \omega_g \end{array}\right\} \text{ near an } M_1 \text{ critical point.}$$

$$\left.\begin{array}{ll} N_d \propto C; & \text{for } \omega < \omega_g \\ N_d \propto C - (\omega - \omega_g)^{1/2}; & \text{for } \omega > \omega_g \end{array}\right\} \text{ near an } M_2 \text{ critical point.}$$

(4.7)

Equations (4.7) show that the density of states around any three-dimensional critical point has the form:

$$N_d \propto C \pm \text{Re}[\pm(\omega - \omega_g)]^{1/2}. \tag{4.8}$$

The four types of critical points are obtained for the four possible combinations of signs in Eq. (4.8).

The density of states around a two-dimensional saddle point is:

$$\begin{array}{ll} N_d \propto -\ln(\omega_g - \omega); & \text{for } \omega < \omega_g \\ N_d \propto -\ln(\omega - \omega_g); & \text{for } \omega > \omega_g. \end{array} \tag{4.9}$$

The shapes of the density of states (or ε_i for allowed transitions) around the various types of critical points described above are illustrated in Fig. 1. In order to obtain ε_i one must multiply N_d by $2\pi^2 F^{lk}/\omega = (4\pi^2/3)|\langle k|\mathbf{p}|l\rangle|^2 \times \omega^{-2}$, with $|\langle k|\mathbf{p}|l\rangle|^2$ practically independent of frequency. Nonparabolicity effects have to be included whenever the frequency range is wide enough to make the inclusion of the ω^{-2} term in ε_i [Eq. (4.3)] necessary. In this case the assumption of constant oscillator strength must also be reexamined. These effects, however, do not affect the singular behavior of ε_i in the vicinity of the critical points.

The real part of the dielectric constant near a critical point can be also calculated with Eq. (2.13). Let us assume an M_0 critical point and perfectly parabolic bands extending to infinite energies. We obtain:

$$[\varepsilon_r(\omega) - 1] \propto \int_0^\infty \frac{(\omega' - \omega)^{1/2}}{(\omega'^2 - \omega^2)} F(\omega_{lk} = \omega') \, d\omega'. \tag{4.10a}$$

While the integral of Eq. (4.10a) depends explicitly on the functional relationship between F and ω, the singular behavior does not, provided the integral converges. Under the assumption of constant matrix elements of p, i.e., $F \propto \omega^{-1}$, we obtain:

$$[\varepsilon_r(\omega) - 1] \propto \int_0^\infty \frac{(\omega' - \omega_g)^{1/2}}{\omega'(\omega'^2 - \omega^2)} d\omega'. \qquad (4.10b)$$

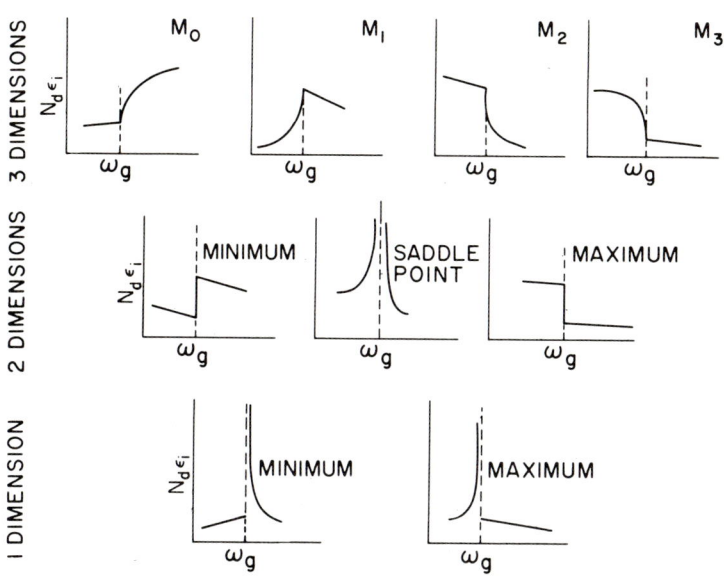

FIG. 1. Singular behavior of the density of states, i.e., the imaginary part of the dielectric constant ε_i for allowed transitions, in the neighborhood of the various types of one-, two-, and three-dimensional critical points.

Eq. (4.10b) can be evaluated as a contour integral with the contour shown in Fig. 2. We find[92,105]:

$$\varepsilon_r(\omega) - 1 \propto \begin{cases} \omega^{-2}[2\omega_g^{1/2} - (\omega + \omega_g)^{1/2}]; & \text{for } \omega > \omega_g \\ \omega^{-2}[2\omega_g^{1/2} - (\omega_g + \omega)^{1/2} - (\omega_g - \omega)^{1/2}]; & \text{for } \omega < \omega_g \end{cases}$$

(4.11a)

We thus see that the infinite slope $(\omega - \omega_g)^{1/2}$ which appeared in ε_i above ω_g appears in ε_r below ω_g. Deviations from parabolicity and contributions from other bands nonsingular at ω_g will not change this feature of ε_i at the critical point. As an illustration, we draw in Fig. 3 the dielectric constant ε_r below the absorption edge ω_g of cadmium telluride measured by Marple[106] and by Cardona.[107] The solid curve is:

$$\varepsilon_r = 6.52[1 + (\omega/5.3)^2] + 5.9\omega^{-2}\omega_g^{1/2}[2 - (1 + \omega/\omega_g)^{1/2} - (1 - \omega/\omega_g)^{1/2}]$$

(4.11b)

and represents a fit with Eq. (4.11a) using the proportionality constant as the only adjustable parameter. Allowance has been made for the existence of another strong edge at 5.3 eV, which produces a small

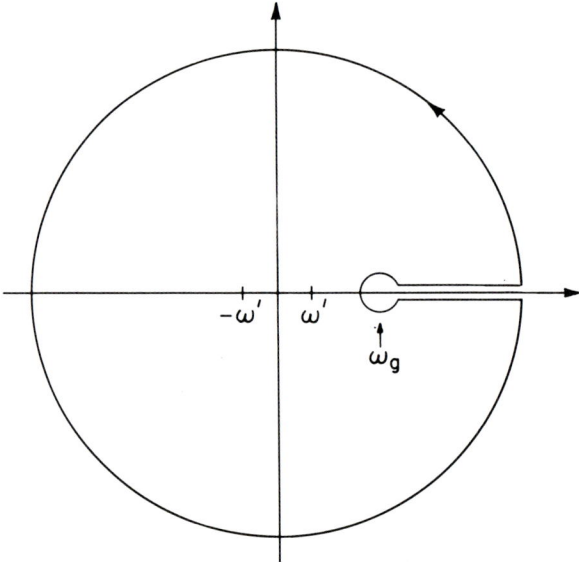

FIG. 2. Contour of integration used to calculate the integral of Eq. (4.10b).

residual dispersion below ω_g. This edge is assumed to be responsible for all the remaining polarizability not due to the M_0 edge at 1.49 eV. Its dispersion[107] has been represented by a single frequency term:

$$\Delta\varepsilon \propto \frac{1}{1 - (\omega/5.3)^2} \approx 1 + (\omega/5.3)^2 \qquad (4.11c)$$

A similar result is obtained for an M_3 critical point. In this case the two expressions in Eq. (4.11a) for $\omega > \omega_g$ and $\omega < \omega_g$ are interchanged and a change in sign is introduced by the sign reversal in the energy denominator of Eqs. (4.10). The shape of the resulting ε_r

singularity is sketched in Fig. 4. The shapes of ε_r around M_1 and M_2 critical points are obtained by the same procedure. A square root singularity arises in all cases since the density of states always has a square root singularity. The singularity in ε_r occurs below ω_g when that of ε_i is above ω_g and vice versa. The results obtained for the

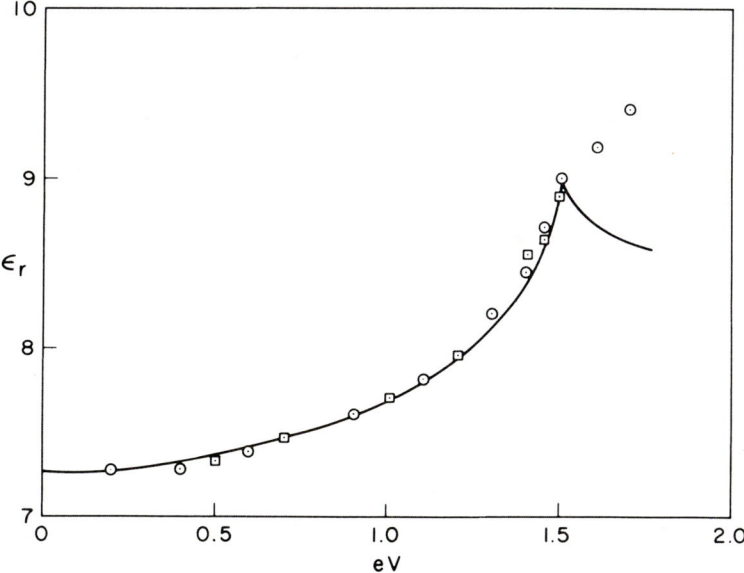

FIG. 3. Real part of the interband dielectric constant of CdTe below the lowest absorption edge (direct) ω_g as determined by Marple[106] □ and by Cardona[107] ○. The solid line is a theoretical fit to the experimental points [(see Eq. (4.11c)].

singular part of the dielectric constant near any type of allowed three-dimensional critical point can be represented by the expression:

$$\varepsilon \propto i^{r+1}(\omega - \omega_g)^{1/2}, \qquad (4.11d)$$

where r is the subindex which designates the type of critical point.

The shape of ε_r around a two-dimensional minimum can be obtained from the equation:

$$[\varepsilon_r(\omega) - 1] \propto \int_\omega^\infty \frac{d\omega'}{\omega'(\omega'^2 - \omega^2)} = -\frac{1}{\omega^2}\ln\left|1 - \frac{\omega^2}{\omega_g^2}\right|. \qquad (4.12)$$

The shapes of $\varepsilon_r(\omega)$ near two- and one-dimensional critical points are sketched in Fig. 4. For a one-dimensional minimum, for instance, we have:

$$[\varepsilon_r(\omega) - 1] \propto \begin{cases} \omega^{-2}[2\omega_g^{-1/2} - (\omega + \omega_g)^{-1/2}]; & \text{for } \omega > \omega_g \\ \omega^{-2}[2\omega_g^{-1/2} - (\omega - \omega_g)^{-1/2} - (\omega + \omega_g)^{-1/2}]; & \text{for } \omega < \omega_g. \end{cases} \quad (4.13)$$

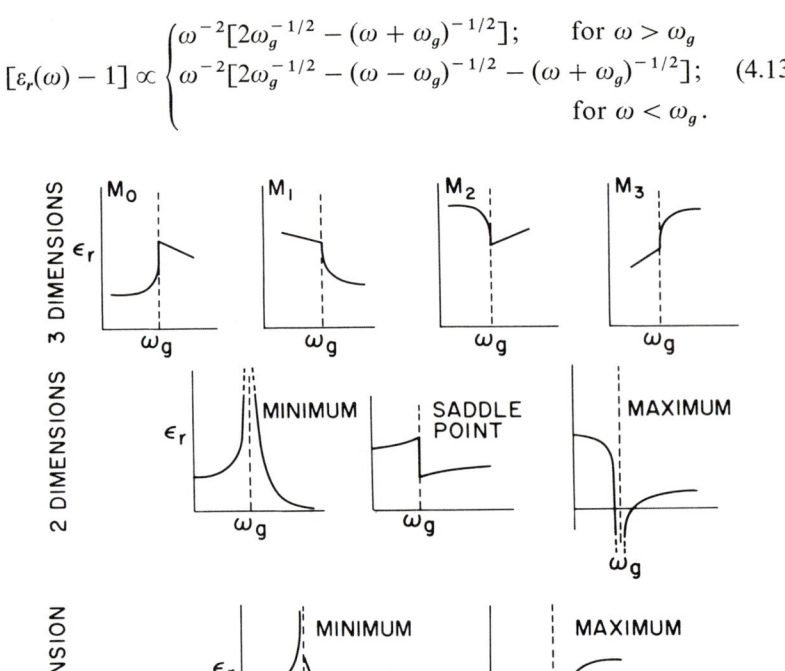

FIG. 4. Real part of the dielectric constant ε_r near a Van Hove singularity in one, two, and three dimensions.

It is also of interest to consider the behavior near a critical point of a transition forbidden at the critical point which becomes allowed as one moves away from it. The matrix elements of **p** are, to first order, linear functions of the components of **k**. As a result, an extra factor $|\omega - \omega_g|$ must be introduced in the integrand of Eq. (4.3). The $\frac{1}{2}$ exponents in Eq. (4.11a) must be replaced by $\frac{3}{2}$ and therefore ε_i and ε_r do not have an infinite singularity in their derivatives at ω_g for three-dimensional critical points; the square roots in Eqs. (4.4) and (4.7) must be replaced by $(\omega - \omega_g)^{3/2}$. Therefore, forbidden transitions at three-dimensional critical points will be difficult to observe in solids if other stronger transitions are present at the frequency at which they occur. A $(\omega - \omega_g)^{3/2}$ singularity is very difficult to detect when it occurs superimposed on a frequency dependent background.

Square root singularities are also found for forbidden transitions in the neighborhood of one-dimensional critical points. Eqs. (4.4) and (4.11a) remain valid for forbidden transitions in one dimension.

The two-dimensional case presents some curious features which are sketched in Fig. 5. The shape of ε_i around two-dimensional maxima

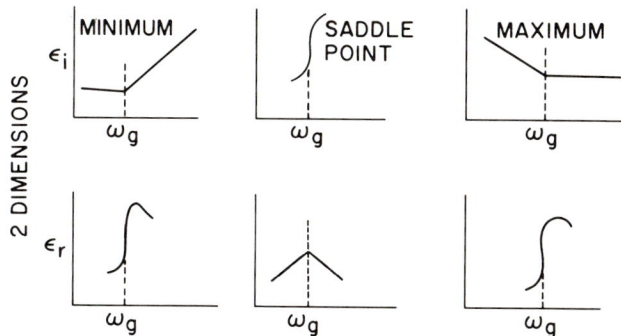

FIG. 5. Singularities in ε_r and ε_i for forbidden direct transitions in two dimensions.

and minima can be readily understood. It arises from the multiplication of Eq. (4.5) by $\omega - \omega_g$. A similar procedure yields for ε_i at a saddle point [see Eq. (4.9)]:

$$\varepsilon_i \propto (\omega - \omega_g) \ln(\omega_g - \omega); \quad \text{for } \omega < \omega_g$$
$$\varepsilon_i \propto (\omega - \omega_g) \ln(\omega - \omega_g); \quad \text{for } \omega > \omega_g. \quad (4.14)$$

Thus a logarithmic singularity (cusp point) is obtained for ε_i in the case of forbidden transitions at two-dimensional saddle points. The shapes of ε_r which correspond to forbidden transitions near two-dimensional critical points are also given in Fig. 5.

5. Indirect Transitions

The condition of **k**-conservation must be relaxed when one considers second-order transitions involving not only the electromagnetic field but also the electron–phonon coupling Hamiltonian. Transitions between an initial state $|0\rangle$ and a final state $|f\rangle$ of different **k** ($\mathbf{k}_0 \neq \mathbf{k}_f$) can take place provided a phonon of crystal momentum $\mathbf{k} = \pm(\mathbf{k}_f - \mathbf{k}_0) + \mathbf{B}_l$ is destroyed (+) or created (−). $\mathbf{B}_l$ is a reciprocal lattice vector. A relaxation of the **k** selection rule can also be produced by the presence of defects (impurities, dislocations, surfaces, etc.).

An expression[16] for the imaginary part of the dielectric constant produced by indirect phonon-aided transitions can be easily obtained by the method of Section 2, keeping the second-order terms in Eqs. (2.4). The electromagnetic interaction takes the electron from the initial state to a virtual intermediate state of the same crystal momentum ($\mathbf{k}_0 = \mathbf{k}_i$), without conserving energy, while the phonon absorption or emission takes it from the intermediate state to the final state. Energy must be conserved in the total process and hence $\omega = \omega_f - \omega_0 \pm \omega_{\text{phon}}$ with the (+) sign corresponding to phonon emission and the (−) sign to phonon absorption. For transitions from the neighborhood of a valence band maximum to that of a conduction band minimum (three-dimensional bands are considered here) the density of states for the transitions from a given initial energy to a given final energy should be proportional to the product of the densities of initial and final states, assuming a broad phonon distribution. Hence, for this density of states, we find the product of two square root singularities of the type given in Eqs. (4.4). For a given photon energy ω, more than one initial state energy is possible since **k** is not conserved. By integrating the density of states ($\omega^{1/2} \cdot \omega^{1/2} \approx \omega$) over all possible initial states, one obtains the following energy dependence for ε_i[16]:

$$\varepsilon_i \propto |M|^2 \frac{(\omega - \omega_g \pm \omega_{\text{phon}})^2}{\omega^2(\omega_f - \omega_i)^2}; \quad \text{for } \omega > \omega_g \pm \omega_{\text{phon}}$$

$$\varepsilon_i = 0; \quad \text{for } \omega < \omega_g \pm \omega_{\text{phon}}.$$

(5.1)

M is the matrix element of the electron–phonon interaction, which contains the temperature dependent phonon occupation numbers. $|M|^2$ is therefore proportional to the Bose–Einstein function $f_B(\omega_{\text{phon}}/T)$ for phonon absorption and to $f_B + 1$ for phonon emission. Equations (5.1) describe a rather smooth variation of ε_i near ω_g with a singularity only in the second derivative of ε_i with respect to ω. This fact, and the fact that indirect transitions are weak second-order processes, makes observation of these transitions only possible at frequencies at which no other stronger processes (direct allowed transitions) occur. Indirect transitions, modified by exciton effects (see

Section 6d), have been seen and identified in a number of semiconductors for which the lowest energy gap is indirect, e.g., Ge, GaP, Si. In these materials phonon-aided indirect transitions are observed in the absorption spectrum at frequencies below the lowest direct edge. Because of the small strength of the indirect transitions, their contribution to n_r is very small and hence the absorption coefficient α is proportional to $\omega\varepsilon_i$ [see Eqs. (1.3) and (1.4)]. Over the reduced photon energy region of the measurements, a plot of α^2 versus ω yields straight lines, according to Eqs. (5.1), from which ω_{phon} and ω_g can be obtained.[21] The identification of phonon-aided indirect transitions is completed by observing that their intensity varies with temperature in the manner required by the phonon occupation numbers.

6. Excitons[82, 84, 108]

a. Introduction[17, 51]

Among the various types of many-body interactions neglected in Section 2, the Coulomb interaction between the excited electron and the hole left behind in the valence band is known to play an important role. We shall call any modifications in the one-electron spectra produced by this interaction exciton effects.

Let us consider these effects for semiconductors and insulators. Exciton effects in metals have been recently considered by Mahan[109] but will not be discussed here. Equation (2.13) can be easily generalized to take into account the exciton interaction. For optically isotropic, i.e., cubic, materials, and at 0°K, the interband contribution to ε becomes:

$$\varepsilon = 1 + 4\pi \sum_j \frac{F^j}{\omega_j^2 - (\omega + i\eta)^2}. \tag{6.1}$$

Where ω_j represents the energy of any excited electron hole pair with respect to the ground state $|0\rangle$ and the summation is extended to all possible excited pair states $|j\rangle$. The symmetrized oscillator strength F^j for transitions from the ground state to an excited pair state is:

$$F^j = \frac{2 |\langle j| \sum_i \mathbf{p}^i |0\rangle|^2}{3\omega_j}, \tag{6.2}$$

with $\mathbf{p}^i$ the linear momentum operator which operates on the ith electron.

26   II. OPTICAL PROPERTIES OF ELECTRONS IN SOLIDS

The imaginary part ε_i of Eq. (6.1) can be rewritten[51] in a convenient form by making use of the relationship given in Eq. (2.12). We obtain:

$$\varepsilon_i(\omega) = \frac{4\pi^2}{3\omega^2} \sum_j \left| \langle j | \sum_i \mathbf{p}^i | 0 \rangle \right|^2 \delta(\omega - \omega_j)$$

$$= -\operatorname{Im} \frac{4\pi}{3\omega^2} \sum_{jj'} \langle 0 | \sum_i \mathbf{p}^i | j \rangle \langle j | \frac{1}{z - H} | j' \rangle \langle j' | \sum_i \mathbf{p}^i | 0 \rangle$$

$$= -\operatorname{Im} \frac{4\pi}{3\omega^2} \langle 0 | \sum_i \mathbf{p}^i G \sum_i \mathbf{p}^i | 0 \rangle \qquad (6.3)$$

where:

$$G = 1/(z - H), \quad \text{and} \quad z = \omega + i\eta \quad (\eta \to +0).$$

The exciton eigenstates $|j\rangle$ must be eigenstates of the translation operator and hence they are characterized by a "center of mass" linear momentum $\mathbf{K}$. Only states with $\mathbf{K} = 0$ are optically active for direct transitions in the dipole approximation. A set of symmetrized $\mathbf{K} = 0$ electron hole wave functions can be easily written in terms of the electron Wannier functions of the conduction band $W_e^c(r_e)$ and the hole Wannier function of the valence band $W_h^v(r_h)$:

$$\Psi_R(r_e, r_h) = \frac{1}{N^{1/2}} \sum_{Rj} W_e^c(r_e - R_j - R) W_h^v(r_h - R_j), \qquad (6.4)$$

where N is the number of unit cells per unit volume. The electron and hole Wannier functions are:

$$W_{e(h)}^{c(v)} = \frac{1}{N^{1/2}} \sum_k u_{e(h)}^{c(v)}(r) e^{i\mathbf{k} \cdot \mathbf{r}}.$$

The symbols $u_{e(h)}^{c(v)}$ represent electron and hole Bloch functions of the valence and conduction band, according to sub- and superscripts. We assume that only one valence and one conduction band are involved in the formation of our pairs. The exciton wave functions can then be written as a linear combination of the $\Psi_R(r_e, r_h)$ functions:

$$\Psi(r_e, r_h) = \frac{1}{N^{1/2}} \sum_R \phi(R) \Psi_R(r_e, r_h). \qquad (6.5)$$

The matrix element of $\mathbf{p}$ which appears in Eq. (6.3) becomes, in the $\mathbf{k}_e$, $\mathbf{k}_h$ representation:

$$\langle \mathbf{k}_e^c, \mathbf{k}_h^v | \sum_i \mathbf{p}^i | 0 \rangle = \langle \mathbf{k}_e^c | \mathbf{p} | \mathbf{k}_e^v \rangle \cdot \delta(\mathbf{k}_e^c, \mathbf{k}_e^v). \qquad (6.6)$$

If $\langle \mathbf{k}_e^c | \mathbf{p} | \mathbf{k}_e^v \rangle$ is not zero at any **k** for a given pair of bands it is usually not very strongly **k**-dependent. It is then customary to set $\langle \mathbf{k}_e^c | \mathbf{p} | \mathbf{k}_e^v \rangle = \text{constant} = \langle c | \mathbf{p} | v \rangle$ for this set of bands. Under this assumption the optical matrix element becomes, in the Wannier representation:

$$\langle \Psi_{R_1} | \sum_i \mathbf{p}^i | 0 \rangle = \frac{1}{N} \sum_{\mathbf{k}_1} \langle c | \mathbf{p} | v \rangle e^{-i\mathbf{k}_1 \mathbf{R}_1}, \tag{6.7}$$

and the optical matrix element between the ground state and the pair eigenstate $\langle \Psi |$ becomes:

$$\langle \Psi | \sum_i \mathbf{p}^i | 0 \rangle = \phi(0) \langle c | \mathbf{p} | v \rangle. \tag{6.8}$$

According to Eq. (6.8), only the *even* component of the envelope function $\phi(0)$ gives allowed transitions. This conclusion only holds when the matrix element of **p** between Bloch states can be considered to be independent of **k**.

b. Hydrogenic (Wannier) Excitons[11,17]

If the exciton interaction potential $V(\mathbf{r})$ is a slowly varying function of the relative coordinate **r**, we can obtain possible envelope functions $\phi(\mathbf{r})$ for $\mathbf{k} = 0$ by solving effective mass equations associated with M_0 (minimum) and M_3 (maximum) Van Hove singularities in the vertical ($\Delta \mathbf{k} = 0$) energy difference between the conduction and the valence band (the case of M_1 and M_2 singularities will be considered later). For non degenerate bands these equations are of the form:

$$-\frac{1}{2}\frac{1}{\mathbf{m}} \cdot \nabla^2 \phi(\mathbf{r}) + V(\mathbf{r})\phi(\mathbf{r}) = W_{\text{ex}} \phi(\mathbf{r}) \tag{6.9}$$

where $1/\mathbf{m}$ is the inverse effective mass tensor and W_{ex} the exciton binding energy. The Coulomb potential $V(\mathbf{r})$ must be screened, at large distances, by the static dielectric constant of the crystal $\varepsilon_0 [V(\mathbf{r}) = -(1/\varepsilon_0 r)]$. As the distance decreases, a different dielectric constant should be used: the ionic contribution to ε_0 relaxes at smaller values of r and one should use the long wavelength infrared dielectric constant ε_{ir}. At still shorter distances (one lattice constant), ε_{ir} relaxes out and one obtains the unscreened Coulomb potential. Near the atomic cores, pseudopotential orthogonalization corrections cancel the Coulomb potential $V(\mathbf{r})$. The use of an "**r**-independent" dielectric constant is justified whenever the "Bohr radius" of the solution of Eq. (6.9) is very large compared with the lattice constant.

Without significant loss in generality, we can restrict ourselves to the treatment of Eq. (6.9) for spherical constant energy surfaces (scalar effective mass m^*). Equation (6.9) has, for $m^* > 0$ (M_0 critical point), a series of hydrogenic bound states below ω_g. Their energies are[17]:

$$\omega_l = \tilde{\omega}_g + W_{\text{ex},l} = \omega_g - \tfrac{1}{2}(m^*/\varepsilon_0^2 l^2) \quad \text{with } l = 1, 2, 3, \ldots, \quad (6.10)$$

where m^* is the reduced electron-hole† mass[110]:

$$\frac{1}{m^*} = \frac{1}{m_e^*} + \frac{1}{m_h^*}. \quad (6.11)$$

If we assume $\langle \mathbf{k}_e^c | \mathbf{p} | \mathbf{k}_e^v \rangle = \text{constant} \cdot \delta(\mathbf{k}_e^c, \mathbf{k}_h^v)$ (see Eq. (6.8)), the probability for transitions to these bound states is proportional to:

$$|\phi_l^s(0)|^2 = \frac{m^{*3}}{\pi l^3 \varepsilon_0^3}, \quad (6.12)$$

where $\phi_l^s(0)$ are the s-like solutions of Eq. (6.9). The ε_i spectrum is composed of a collection of δ-function peaks which cluster together near ω_g. The integrated intensity in this quasi-continuum below ω_g is[17]:

$$\varepsilon_i = \frac{2\pi}{\omega} F(2m^*)^{3/2} |W_{\text{ex},1}|^{1/2}, \quad (6.13)$$

where F is the oscillator strength for interband transitions. The real part of the dielectric constant ε_r has the form of the harmonic oscillator dispersion near each bound state energy:

$$\varepsilon_r \propto \frac{1}{\omega_l^2 - \omega^2} \quad (6.14)$$

The interband absorption continuum, given by Eq. (4.3) for $V(\mathbf{r}) = 0$, is modified by the presence of the exciton interaction. Making use of the wave functions $\phi^s(\mathbf{r})$ of the continuum of the hydrogen atom we obtain[17]:

---

† If the valence band is degenerate at the point under consideration, as is often the case, m_h is some average of the heavy and light hole masses. The exact calculation of this average is not easy, since Eq. (6.9) must be replaced by a set of four coupled differential equations.

6. EXCITONS

$$|\phi^s(0)|^2 = \frac{\gamma e^\gamma}{\sinh \gamma}, \quad \text{with} \quad \gamma = \pi\left(\frac{|W_{\text{ex},1}|}{\omega - \omega_g}\right)^{1/2}, \quad (6.15)$$

where $|W_{\text{ex},1}| = \frac{1}{2}(m^*/\varepsilon_0^2)$ is the binding energy of the exciton ground state [Eq. (6.10)]. The imaginary part of the dielectric constant ε_i becomes, in the interband continuum:

$$\varepsilon_i = \varepsilon_i^0 |\phi^s(0)|^2, \quad (6.16)$$

where ε_i^0 is the dielectric constant in the absence of exciton interaction. For $\omega - \omega_g \gg |W_{\text{ex},1}|$, $\phi(0)$ tends to 1 and ε_i becomes equal to ε_i^0. Hence the exciton interaction only modifies the optical properties in the neighborhood of the critical point ($\omega - \omega_g \lesssim |W_{\text{ex},1}|$). Very near the critical point

$$|\phi^s(0)|^2 \simeq 2\pi\left(\frac{|W_{\text{ex},1}|}{\omega - \omega_g}\right)^{1/2}$$

and ε_i becomes:

$$\varepsilon_i = \frac{2\pi}{\omega_g} F(2m^*)^{3/2} |W_{\text{ex},1}|^{1/2} \quad (6.17)$$

equal to the average value at the bound state quasi-continuum below the gap. Hence, if we disregard the discrete states below ω_g, a step singularity similar to that near a two-dimensional minimum [Eq. (4.5)] results. However, the fact that the quasi-continuum below joins the continuum above is going to modify this singularity and shift it to lower energies. Since the quasi-continuum is transformed into a continuum by broadening, a quantitative knowledge of the shape of the resulting singularity requires an analysis of the broadening of the exciton lines. The ε_i calculated from the model discussed above for the lowest direct edge of GaAs is sketched in Fig. 6. The calculations were performed using $W_{\text{ex},1}$ the ground state exciton binding energy, as an adjustable parameter. The best fit to the experimental data of Sturge[112] above ω_g was found for $|W_{\text{ex},1}| = 0.0034$ eV, which corresponds, for $\varepsilon_0 = 11$,[111] to $m^* = 0.031$. This mass is in reasonable agreement with the value obtained from Eq. (6.11) using $m_e^* = 0.065$ and the *light hole* mass $m_h^* = 0.078$.[113]

Excitons associated with conduction and valence band extrema at high-symmetry points such that $\langle \mathbf{k}_e^c | \mathbf{p} | \mathbf{k}_e^v \rangle = 0$ (due to a symmetry selection rule) are sometimes observed. The simplest case is that of Cu_2O: the lowest direct gap occurs at $\mathbf{k} = 0$ between states of the same parity.[114] In this case we can use the expansion $\langle \mathbf{k}_e^c | \mathbf{p} | \mathbf{k}_e^v \rangle = \mathbf{A} \cdot \mathbf{k}$ with $\mathbf{A}$ a k-independent tensor. The same steps which led to (6.8) lead to[17]:

$$\langle \Psi | \sum \mathbf{p}^i | 0 \rangle = \mathbf{A} \cdot \mathbf{\nabla}_\mathbf{r} \phi(0), \tag{6.18}$$

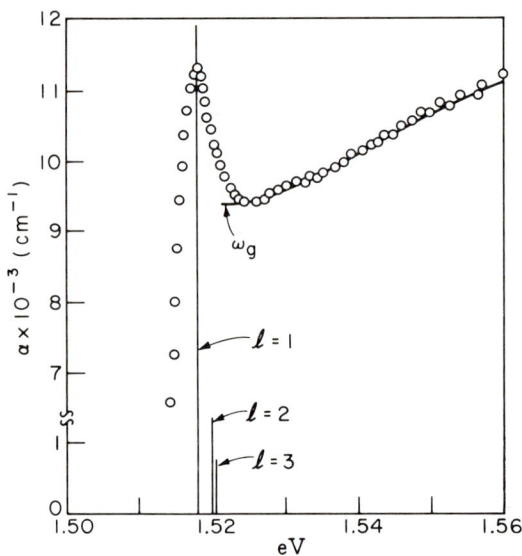

FIG. 6. Experimental and calculated exciton spectrum near the direct edge of GaAs at 21°K. The vertical axis represents the absorption coefficient α, approximately proportional to ε_i.

and, hence, allowed dipole transitions occur only to p-states ϕ_l^p of the exciton. Their intensities are proportional to[17]:

$$|\mathbf{\nabla}_\mathbf{r} \phi_l^p(0)|^2 \sim \frac{l^2 - 1}{l^5}. \tag{6.19}$$

Also in this case the exciton interaction modifies the continuum only near ω_g. As may be expected from the discussion of the allowed exciton, ε_i above ω_g joins smoothly the value for the exciton quasi-continuum below ω_g. The behavior of ε_i near ω_g is:

$$\varepsilon_i \propto \left[1 + \frac{\omega - \omega_g}{W_{\text{ex}, 1}} \right]. \tag{6.20}$$

The effective mass approximation can also be used to calculate the effect of the exciton interaction near an M_3 (maximum) singularity. In this case the principal effective masses are negative and the Coulomb interaction behaves as a repulsive interaction with positive masses; no bound states appear. The dielectric constant is modified, however, near the critical point in the following manner[48]:

$$\varepsilon_i = \varepsilon_i^0 \frac{\gamma e^\gamma}{\sinh \gamma}, \tag{6.21}$$

with

$$\gamma = -\pi \left(\frac{|W_{\text{ex},1}|}{\omega_g - \omega} \right)^{1/2}.$$

Again $\varepsilon_i = \varepsilon_i^0$ for $\omega_g - \omega \gg \omega_0$. Very near $\omega_g (\gamma \to -\infty)$, ε_i becomes[48]:

$$\varepsilon_i = \frac{2\pi}{\omega} (2m^*)^{3/2} F |W_{\text{ex},1}|^{1/2} e^{2\gamma} \tag{6.22}$$

The resulting line shape is shown in Fig. 7. According to Eq. (6.22), the exciton interaction destroys the singularity in the slope of ε_i at the

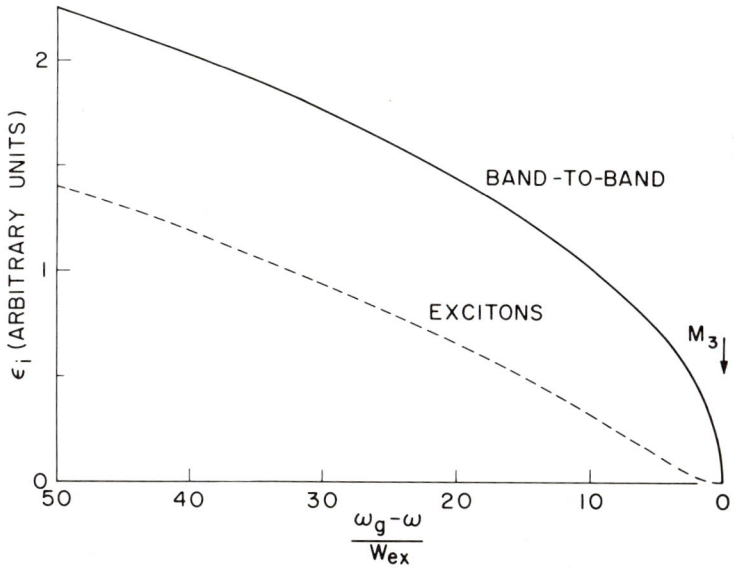

Fig. 7. Effect of the exciton interaction on direct allowed transitions near an M_3 (maximum) critical point [from B. Velický and J. Sak, *Phys. Status Solidi* **16**, 147 (1966)].

M_3 critical point. The dielectric constant ε_i tapers off very smoothly since all derivatives of ε_i with respect to ω are zero at M_3. Hence M_3 transitions are not likely to be observable: no such transition has been conclusively identified.

The effective mass approximation used above can be justified whenever the potential $V(\mathbf{r})$ does not vary much within a lattice constant. This is obviously not the case for a Coulomb potential near $\mathbf{r} = 0$. As mentioned above, pseudopotential orthogonalization corrections and the finite extent of the Wannier functions are going to remove the singularity in $V(\mathbf{r})$ for $\mathbf{r} \to 0$. If the Bohr radius of the solutions of Eq. (6.9) is large ($\varepsilon_0/m^* \gg a_0$, the lattice constant) for an M_0 critical point, the region around $\mathbf{r} = 0$ contributes little to the energy eigenvalues and eigenfunctions and hence the exact knowledge of $V(\mathbf{r})$ near $\mathbf{r} = 0$ is not required. The solutions for a pure Coulomb potential are a good approximation to the solutions for the true potential. The same result holds in the continuum of an M_0 or an M_3 critical point provided $\varepsilon_0/m^* \gg a_0$. The condition is fulfilled for most semiconductors and small gap insulators but does not hold for large gap insulators such as the alkali halides.

We want to examine now the possibility of applying the effective mass approximation to calculate exciton effects near saddle point singularities (M_1, M_2) for semiconductors and small band gap insulators. If we assume, as is implicitly done in the effective mass approximation, bands extending to infinity in $\mathbf{k}$-space, the wave functions can localize themselves enormously at no expense in kinetic energy since there are now states of energy ω_g with $\mathbf{k}$ extending all the way to infinity. As a result of the exciton localization in the region of strong variation of $V(\mathbf{r})$ (near $\mathbf{r} = 0$), not only a knowledge of the exact shape of $V(\mathbf{r})$ near $\mathbf{r} = 0$ may be required to calculate the eigenvalues but also the effective mass approximation may break down. If we disregard the question of the validity of the effective mass approximation and try to solve an effective mass equation of the form (6.9) with a negative mass along one or two directions, we would find no bound states for a pure Coulomb potential since any state can lower its potential energy by localizing itself more and more around $\mathbf{r} = 0$ without loss in kinetic energy. While an exact solution of the effective mass equation is not possible, Duke and Segall[49] have, in fact, shown that no bound states exist for several separable approximations to the Coulomb potential. This conclusion does not have much physical significance since real constant energy surfaces around

6. EXCITONS

M_1 and M_2 critical points *do not* extend to infinity: a cutoff at $|\mathbf{k}|$ of the order of the dimensions of the Brillouin zone must be introduced in the quadratic expansion of $\omega(\mathbf{k})$ as a function of k_i. Such a cutoff prevents localization of the exciton to a volume smaller than a unit cell and hence the effective mass equation may be expected to remain approximately valid provided $V(\mathbf{r})$ is made constant for $\mathbf{r} < a_0$ (see Fig. 8). As a result of this cutoff, bound states may reappear for an M_1 critical point.[49]

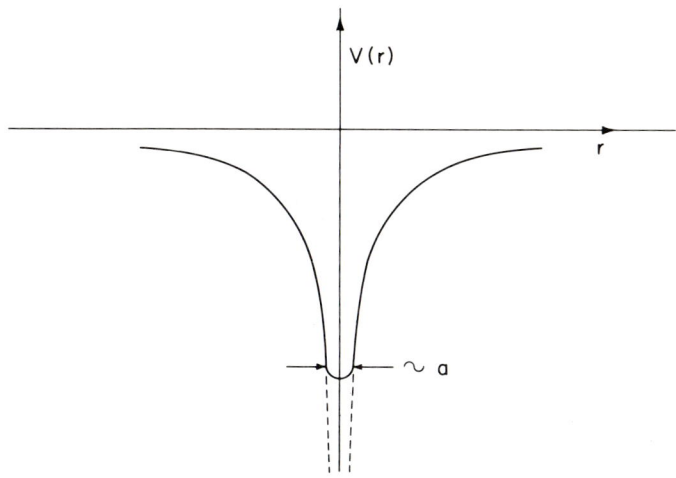

FIG. 8. Coulomb potential with cutoff appropriate for the qualitative treatment of hyperbolic critical points.

The solution of Schrödinger's equation with effective masses of different signs for different principal directions cannot be obtained exactly even in the case of cylindrical symmetry:

$$\left[-\frac{1}{2m_\perp}\left(\frac{\partial^2}{\partial x^2} + \frac{\partial^2}{\partial y^2}\right) - \frac{1}{2m_\|}\frac{\partial^2}{\partial z^2} + V(\mathbf{r}) \right]\phi(r) = W\phi(\mathbf{r}). \quad (6.23)$$

For an M_1 edge, $m_\perp > 0$, $m_\| < 0$; for an M_2 edge $m_\perp < 0$, $m_\| > 0$. Equation (6.23) can be treated for cases in which $|m_\|| \gg m_\perp$ with the aid of the adiabatic approximation.[48] Such cases are actually of great practical importance since most of the materials of the zinc blende–diamond family have an M_1 critical point in the [111] direction of **k**-space (see Fig. 9) which produces prominent structure in the optical spectra above the lowest direct gap (generally at $\mathbf{k} = 0$). At these critical points, and owing to a selection rule for the matrix elements of **p**, $|m_\parallel| \gtrsim 10 m_\perp$.[113] In order to illustrate these critical points usually labeled E_1 and $E_1 + \Delta_1$, we show in Fig. 9 the band structure of InSb obtained by the **k·p** method.[113]

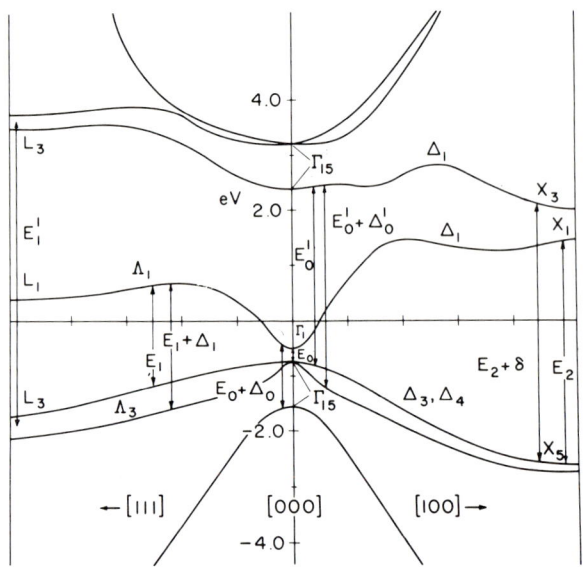

FIG. 9. Band structure of InSb showing the M_1 critical points $(E_1, E_1 + \Delta_1)$ in the [111] direction [see F. H. Pollak, C. W. Higginbotham, and M. Cardona, *J. Phys. Soc. Japan Suppl.* **21**, 20 (1966); *Proc. Intern. Conf. Phys. Semiconductors*, Moscow 1968 (to be published)]. The location along [111] of the $E_1, E_1 + \Delta_1$ critical points may not be the one indicated here since it depends rather critically on the parameters of the calculation.

The adiabatic wave function of the system is $\phi(x, y; z) \times \Psi(z)$ where $\phi(x, y; z)$ is the solution of Eq. (6.23) with z treated as a constant parameter ($m_\parallel = \infty$) and $\Psi(z)$ the solution of:

$$\left[-\frac{1}{2m_\parallel} \frac{\partial^2}{\partial z^2} + W_l(z) - W \right] \Psi(z) = 0, \quad (6.24)$$

where $W_l(z)$ is the eigenvalue of Eq. (6.23) with z treated as a parameter. The imaginary part of ε is proportional to $|\phi(0, 0; 0) \times \Psi(0)|^2$. For $z = 0$, Eq. (6.23) is the equation of the two-dimensional hydrogen atom which has bound states below ω_g at energies[48,115]:

$$W_l(0) = -\frac{2m_\perp}{\varepsilon_0^2 (2l-1)^2} \qquad l = 1, 2, \ldots. \tag{6.25}$$

The values of $|\phi_l^s(0)|^2$ for these states are:

$$|\phi_l^s(0)|^2 = \frac{16 m_\perp^3}{\pi \varepsilon^3 (2l-1)^3}, \tag{6.26}$$

And for the continuum above ω_g:

$$|\phi^s(0)|^2 = \frac{e^{-\gamma'}}{\cosh \gamma'} \quad \text{with } \gamma' = -\frac{\pi |W_1(0)|^{1/2}}{2(\omega - \omega_g)^{1/2}}.$$

Notice that the $l=1$ exciton is now considerably more dominant (27 times the intensity of the $l=2$ exciton) than in the three-dimensional case (8 times the intensity of the $l=2$ exciton). The value of $|\phi(0)|^2$ for $\omega = \omega_g$ is twice that for $\omega \to \infty$. The dielectric constant ε_i which would be obtained for $m_\parallel = \infty$ is sketched in Fig. 10.

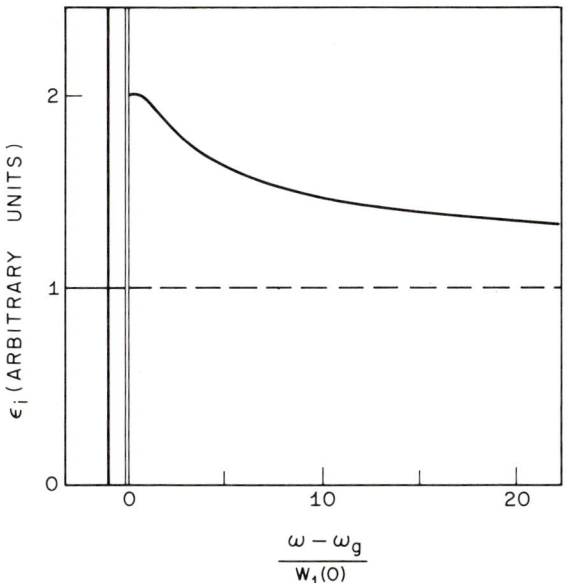

FIG. 10. Dielectric constant ε_i near a two-dimensional minimum critical point including exciton interaction [B. Velický and J. Sak, *Phys. Status Solidi* **16**, 147 (1966)]. This dielectric constant approximates that near an M_1 critical point provided the negative mass has a much larger magnitude than the positive masses.

When $m_\parallel$ is finite, but large, the modification to Fig. 10 due to $\Psi(z \neq 0)$ can be calculated by solving Eq. (6.24). This requires a knowledge of $W_l(z)$, which can be obtained by applying first-order perturbation theory to Eq. (6.23). Equation (6.24) can then be treated with the *WKB* approximation (the cutoff in $V(\mathbf{r})$ shown in Fig. 8 should eliminate any singularity in $W_l(z)$ for $z = 0$). The resulting spectrum around the $l = 1$ exciton as calculated by this method[48] for $m_\parallel = -50 m_\perp$ is shown in Fig. 11. The δ-function peak is broadened

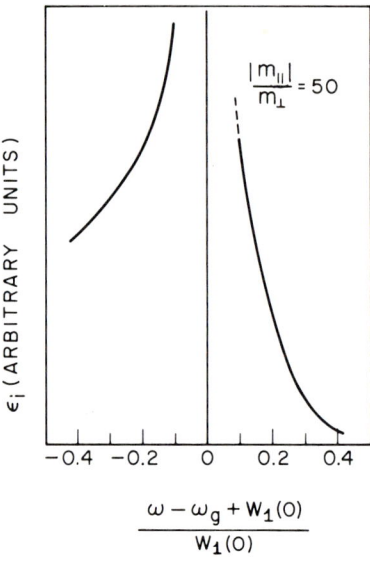

FIG. 11. Exciton resonance near an M_1 critical point for $|m_\parallel| = 50 m_\perp$ [from B. Velický and J. Sak, *Phys. Status Solidi* **16**, 147 (1966)].

by the interaction with the overlapping continuum produced by the negative $m_\parallel$. As $|m_\parallel/m_\perp|$ decreases, the broadening increases. The *WKB* approximation breaks down when $|m_\parallel| \approx m_\perp$.

As indicated earlier, exciton effects of the variety just discussed may be expected to exist near the E_1 (and its spin-orbit-split mate $E_1 + \Delta_1$) edge of germanium- and zincblende-type semiconductors. Sharp peaks have indeed been observed at these edges for a number of materials of this family[35,46] especially at low temperatures. As an example we show in Fig. 12 the optical density (log I_0/I, approximately proportional to the absorption coefficient) of an InSb thin film[46] at room temperature and at 77°K. The E_1, $E_1 + \Delta_1$ peaks of Fig. 12 sharpen considerably at low temperatures and it is tempting to interpret them

6. EXCITONS

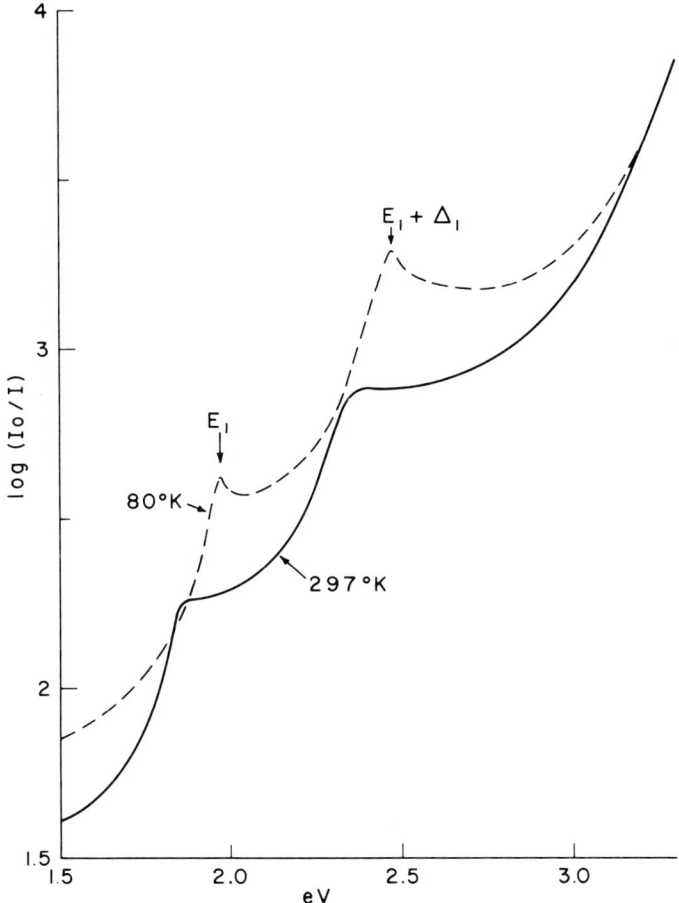

FIG. 12. Optical density ($\log(I_0/I)$) of a vacuum deposited InSb thin films showing the E_1 and $E_1 + \Delta_1$ peaks at room temperature and at 77°K [unpublished data, see also M. Cardona and G. Harbeke, *J. Appl. Phys.* **34**, 813 (1963)].

as excitons of the M_1 variety shown in Fig. 11. The transverse reduced mass of InSb at the [111] M_1 critical points[113] (see Fig. 9) is $m_\perp \approx 0.04$ while the longitudinal mass is $m_\parallel \approx -1$ and hence the adiabatic approximation should apply. With these parameters and the infrared dielectric constant $\varepsilon_0 = 15.5$, the binding energy of the corresponding two-dimensional exciton W_1 [Eq. (6.25)] would be 0.01 eV, which is of the order of the line widths of the peaks in Fig. 12. Hence one would not expect to see any excited states near the band gap singularity since the broadened ground state ($l = 1$) of the exciton overlaps the band edge. A similar situation is shown even more clearly in Fig. 13 for CdTe. The shape of the E_1 and $E_1 + \Delta_1$ peaks is obviously

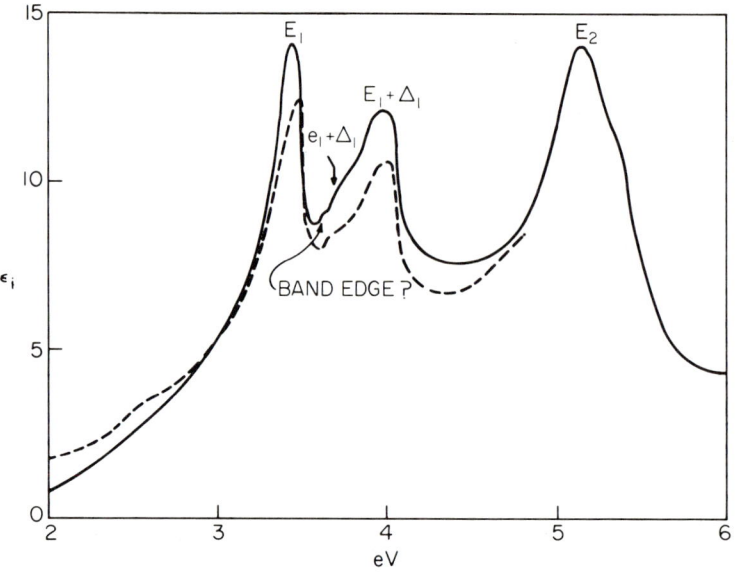

FIG. 13. Imaginary part of the dielectric constant of CdTe at 77°K (—) [see M. Cardona, *J. Appl. Phys.* **36**, 2181 (1965)], and at 20°K (- - -) [see D. T. F. Marple and H. Ehrenreich, *Phys. Rev. Letters* **8**, 87 (1962)].

incompatible with that of an M_1 critical point (see Fig. 1), especially in view of the fact that the high energy side of each one of these peaks is steeper than the low energy side. The binding energy of the possible two-dimensional excitons would be, for $m_\perp = 0.1$[116] and $\varepsilon_0 = 7.2$, 0.1 eV. This energy is also close to the line width of the E_1 peak and the band edge may be undistinguishable from the exciton peak. The E_1 band edge, however, could also give the weak additional structure suggested with a question mark in Fig. 13. This conclusion has to be regarded as speculative since the structure in question is not well resolved, especially in view of its closeness to the $e_1 + \Delta_1$ structure, attributed earlier to an M_0 edge at the L point (edge of the Brillouin zone in the [111] direction).

6. EXCITONS

The sharp structure of Figs. 12 and 13 could also be explained,[118] without invoking exciton effects, as due to strong deviations of the energy bands from parabolicity near the critical points, in particular, to an accidental near degeneracy of an M_1 and M_2 critical point. Such explanation is not compatible with our present knowledge of the band structure of zincblende- and diamond-type materials.[113,119]

c. Nonhydrogenic Excitons: Koster–Slater Interaction[48,50,120]

The effective mass approximation, and hence the theory of hydrogenic excitons, should not hold for crystals with small infrared dielectric constants ($\lesssim 3$) and large effective masses ($m^* \gtrsim 1$) such as the alkali halides and the solid rare gases.[121]† As indicated in Section 6b, it may also not hold for M_1 and M_2 critical points. Several schemes have been proposed to treat the electron hole interaction in these cases.[50,51] The common feature of these methods is the truncation of the interaction beyond a certain value of the electron hole separation **r**. The simplest of these models is that of a Koster–Slater[120] contact interaction potential $V(\mathbf{r}) = \delta(\mathbf{r})g$, equal to zero except when the Wannier electron and hole are on the same unit cell. In a similar manner as that used in the derivation of Eq. (6.8), it is easy to see that such an interaction becomes, in the **k**-representation:

$$\langle \mathbf{k} | V | \mathbf{k}' \rangle = \text{constant} = N^{-1} g, \tag{6.27}$$

for all states belonging to a given set of valence and conduction bands. N is the number of unit cells per unit volume and g is a constant (negative for attractive potentials) which gives the strength of the interaction.

The resolvent G for this problem can be calculated from the expression[51]:

$$G = (1 - G^0 V)^{-1} G^0, \tag{6.28}$$

where G^0 is the resolvent for $g = 0$, which in the Bloch representation is:

$$\langle \mathbf{k} | G^0 | \mathbf{k}' \rangle = \frac{1}{\omega + i\eta - \omega_{\mathbf{k}}} \delta(\mathbf{k}, \mathbf{k}'), \tag{6.29}$$

† The hydrogenic approximation still seems to hold reasonably well for the excitons associated with the lowest gap (at $k = 0$) of solid xenon.

where ω_k is the energy of the pair formed with a conduction electron and a missing valence electron of crystal momentum **k** for $g = 0$. The resolvent G for $g \neq 0$ can be easily evaluated by expanding (6.28) in power series and noticing, according to the rules of matrix multiplication, that:

$$G^0 V G^0 V = \left(N^{-1} g \sum_\mathbf{k} \frac{1}{\omega + i\eta - \omega_k} \right) G^0 V = -g F(\omega) G^0 V, \qquad (6.30)$$

where $F(\omega)$ can be found from the combined density of states $N_d(\omega)$:

$$F(\omega) = -N^{-1} \int_{-\infty}^{+\infty} \frac{1}{\omega - \omega' + i\eta} N_d(\omega') \, d\omega'. \qquad (6.31)$$

The operator $[1 - G^0 V]^{-1}$ thus becomes:

$$[1 - G^0 V]^{-1} = 1 + G^0 V [1 + g F(\omega)]^{-1}, \qquad (6.32)$$

and the resolvent G is [see Eq. (6.28)]:

$$G = \{1 + G^0 V [1 + g F(\omega)]^{-1}\} G^0. \qquad (6.33)$$

By substituting Eq. (6.33) into Eq. (6.3) we find:

$$\omega^2 \varepsilon_i \propto -\mathrm{Im}\, F(\omega) \frac{1}{1 + g F(\omega)} = -\mathrm{Im}\, F(\omega) \frac{1 + g F^*(\omega)}{|1 + g F(\omega)|^2}$$

$$= -|1 + g F(\omega)|^{-2} \, \mathrm{Im}\, F(\omega). \qquad (6.34)$$

If we make use of Eqs. (6.31) and (2.12) we finally obtain[48]:

$$\varepsilon_i(\omega) = |1 + g F(\omega)|^{-2} \cdot \varepsilon_i^0(\omega), \qquad (6.35)$$

where $\varepsilon_i^0(\omega)$ is the imaginary part of the dielectric constant for $g = 0$.

The imaginary part of $F(\omega)$ is, according to Eqs. (6.31) and (2.12), equal to $\pi N^{-1} N_d(\omega)$ while its real part is the Hilbert transform of $N_d(\omega)$:

$$\mathrm{Re}\, F(\omega) = N^{-1} \int_{-\infty}^{+\infty} \frac{d\omega'}{\omega' - \omega} N_d(\omega'). \qquad (6.36)$$

The integrand of Eq. (6.36) has, near a critical point of N_d, a behavior very similar to that of Eq. (1.1) and hence the singularities in $\mathrm{Re}\, F(\omega)$ near a critical point have the same form as those of $\varepsilon_r(\omega)$. However, it is impossible to integrate Eq. (6.36) for two- and three-dimensional critical points with nonphysical parabolic bands extending to infinity as we did for Eq. (1.1), since the integral in Eq. (6.36) diverges. This was not the case for Eq. (1.1) since $\varepsilon_i \approx \omega^{-2} N_d(\omega)$ and the extra

ω^{-2} factor makes the integral converge. The introduction of a convergence factor ω^{-2} in the integrand of Eq. (6.36) should not change the form of the singularities which is thus given by:

$$\text{Re } F(\omega) \simeq \omega^2 \int_{-\infty}^{+\infty} \frac{d\omega'}{\omega'^2(\omega' - \omega)} N_d(\omega'). \tag{6.37}$$

Hence the effect of the exciton interaction on ε_i in the Slater–Koster model is to mix at each critical point singularities of the ε_r-type (Fig. 3) with the ε_i singularities (Fig. 1). The result can be a sharpening or a blurring of the singularity. In order to examine this point we assume $gF(\omega) \ll 1$ and expand Eq. (6.35) in power series of g, keeping only the zero- and the first-order terms:

$$\varepsilon_i(\omega) = [1 - 2g \text{ Re } F(\omega)]\varepsilon_i^0(\omega). \tag{6.38}$$

The result obtained from Eq. (6.38) for the four kinds of three-dimensional critical points is sketched in Fig. 14. The real part of the dielectric constant ε_r, which corresponds to Eq. (6.38) can be easily obtained by compounding the effects of the square root singularities above and below ω_g, using Figs. 1 and 3. The results are also sketched in Fig. 14. Sharp peaks are obtained for ε_r at an M_0 singularity and for

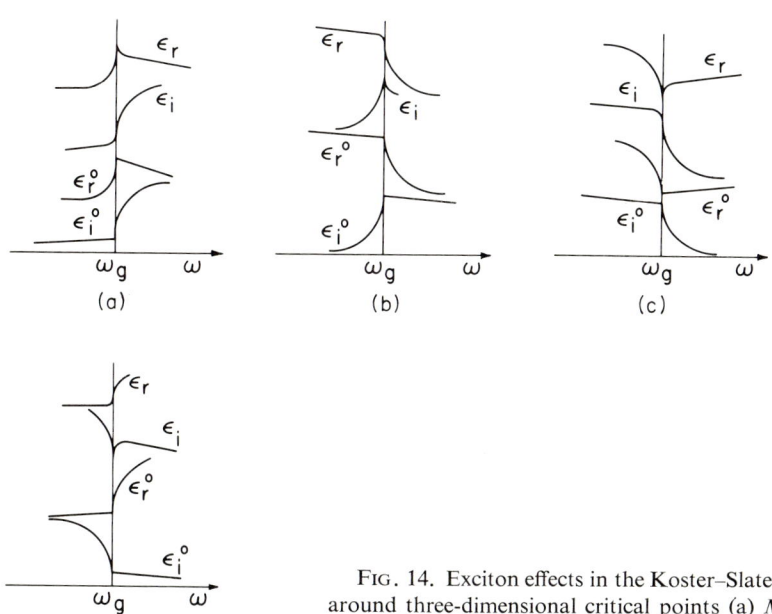

FIG. 14. Exciton effects in the Koster–Slater model around three-dimensional critical points (a) M_0, (b) M_1, (c) M_2, and (d) M_3 for small g [see Eq. (6.38)].

ε_i at an M_1 singularity. M_2 and M_3 singularities give sharp dips for ε_i and ε_r, respectively while inflexion points with an infinite slope are obtained in all other cases. Such inflexion points may be difficult to observe if broadened by phonons, defects, and electron correlations.

The singularity in the complex dielectric constant ε near a three-dimensional critical point can, according to the discussion above, be represented by

$$\varepsilon \propto b(\omega - \omega_g)^{1/2} + \text{constant}, \qquad (6.39)$$

where b is a complex number. For $g = 0$ (no exciton interaction), b is either real or pure imaginary as shown in Eq. (4.11c), $b = i^{r+1}$ near an M_r critical point. The complex number b is rotated counterclockwise in the complex plane for $g < 0$ (attractive interaction) as g is increased. The results of the exciton interaction in the Koster–Slater model are best understood with the help of Fig. 15. This figure shows ε_r and ε_i in the absence of exciton interaction for each type of three-dimensional critical point near the corresponding value of b in the complex plane. The attractive exciton interaction rotates b counterclockwise on a circle of radius one. The line shapes for a rotation of approximately 45° are also given.

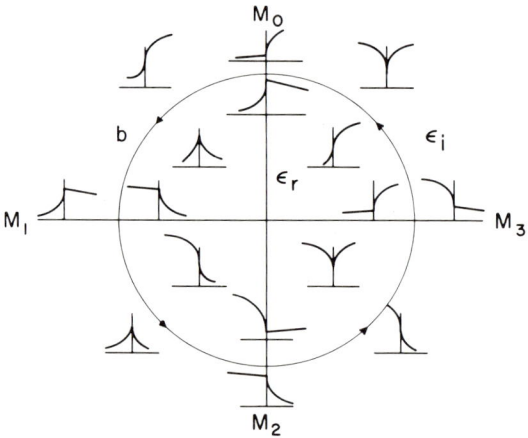

FIG. 15. Real (ε_r, inside circle) and imaginary (ε_i, outside circle) parts of the dielectric constant near three-dimensional Van Hove singularities. The effect of a small exciton interaction is obtained by mixing the M_i singularity with the $M_{i+1}(M_5 = M_0)$. From Y. Toyozawa, M. Inoue, T. Inui, M. Okazaki, and E. Hanamura, *J. Phys. Soc. Japan Suppl.* **21**, 133 (1967).

6. EXCITONS

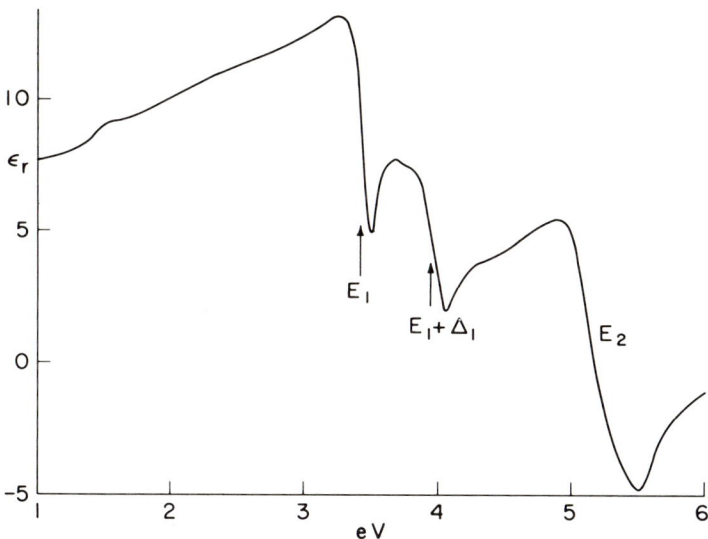

FIG. 16. Real part of the dielectric constant ε_r of CdTe at 77°K obtained by Kramers–Kronig analysis of reflectivity data. The corresponding imaginary part is given in Fig. 13. From M. Cardona, *J. Appl. Phys.* **36**, 2181 (1965).

Figure 16 shows ε_r for CdTe (ε_i shown in Fig. 13) as obtained from normal incidence reflection data. At the E_1 and $E_1 + \Delta_1$ energies the shape of ε_r looks qualitatively like that of Fig. 15 with $g < 0$ near an M_1 singularity. This confirms our earlier contention that exciton effects contribute to this line shape.

For large values of g, ($|gF(\omega)| \gtrsim 1$), a study of the line shape of Eq. (6.34) requires an exact knowledge of $F(\omega)$. Changes, including sign reversals, of the singularities of Fig. 14, e.g., a dip instead of a peak, may result for ε_i near M_1.[122]

For exciton interactions extending farther than the Koster–Slater model discussed above, calculations of the optical properties can be performed under simplifying assumptions. Figure 17 shows $\omega^2 \varepsilon_i$ for solid xenon as calculated by Hermanson[51] assuming tight binding energy bands and various ranges of the exciton interaction. The Koster–Slater approximation (dotted line) gives results which are far away from convergence. Good qualitative convergence is obtained for an interaction range larger than $2^{1/2}d$ (d is the nearest neighbor distance); the shape of the spectrum does not change substantially for longer ranges. Beside the bound states below the M_0 singularity,

Fig. 17 shows peaks associated with $L(M_1)$ and X critical points. The singularity at the M_3 critical point has disappeared and the spectra become rather smooth near the high energy threshold, in agreement with conclusions reached for the hydrogenic model in Section 6b.

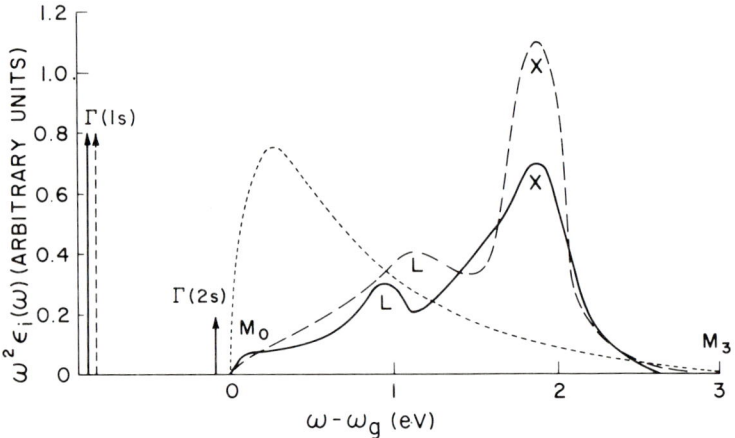

FIG. 17. $\omega^2 \varepsilon_i$ (in arbitrary units) for solid xenon calculated by Hermanson with inclusion of exciton effects for energy bands obtained by the tight binding method. The range of the exciton interaction has been assumed to be: zero (Koster-Slater interaction) (- - -), $2^{1/2}$ times the nearest neighbor distance d (- -), and $3d$ (—). From J. Hermanson, *Phys. Rev.* **150**, 660 (1966); **166**, 893 (1968).

d. *Indirect Excitons*[17]

Throughout Sections 6b and 6c, we have assumed that the absorption of a photon produces only electron hole pairs of zero center of mass momentum **K**. These excitons are composed of an electron in a conduction band with crystal momentum k_e^c and a hole in a valence band of crystal momentum $\mathbf{k}_h^v = -\mathbf{k}_e^v = -\mathbf{k}_e^c$. Hence the conduction band minimum and the valence band maximum from which we set up the effective mass Hamiltonian of Eq. (6.9) must occur at the same point of **k**-space if the exciton transitions are to be optically active. This restriction breaks down when we consider second-order transitions involving the simultaneous absorption of a photon and the absorption (or emission) of a phonon, similar to those discussed in Section 5. In general, these transitions are only going to be observable when they are not superimposed on stronger direct allowed transitions, e.g., when the lowest thermal energy gap is indirect. For this reason, we shall only consider indirect excitons derived from a valence band maximum and a conduction band minimum. Indirect transitions between other types of extrema occur normally superimposed on a direct allowed background and thus are not observable. Since phonons of all possible wave vectors $\mathbf{k}_{ph}$ can be emitted at all temperatures (and absorbed at high temperatures), the $\mathbf{K}=0$ selection rule breaks down and transitions to exciton states formed from valence and conduction valleys of different k-vectors become possible. These states cannot be considered as discrete states since $\mathbf{K}$ is not unique (all $\mathbf{k}_{ph}$ are available); an exciton band is obtained as a function of $\mathbf{K}$, the ground state occurs at $\mathbf{K}_0 = \mathbf{k}_{e,0}^c - \mathbf{k}_{e,0}^v$ where $\mathbf{k}_{e,0}^{c,v}$ is the k-vector of the corresponding band extremum. For $K \neq K_0$, the energy in the effective mass approximation is that of the $K = K_0$ state of the exciton W_l, obtained from the effective mass equation [Eq. (6.9)], plus the kinetic energy of the center of mass of the electron hole pair[17]:

$$W_l(K) = \frac{(K - K_0)^2}{2(m_e^* + m_h^*)} + W_l(K_0). \qquad (6.40)$$

Equation (6.40) represents a series of three-dimensional bands. The density of states is given by an equation similar to Eq. (4.4). Under the customary assumption that the optical and electron–phonon matrix elements are independent of k (the optical matrix element is, however, only allowed for direct transitions to the intermediate state), it is easy to see (Sections 4a and 4b) that the contribution to ε_i of transitions to the l-th exciton band of Eq. (6.40) is proportional to the density of states for such transitions[17]:

$$\varepsilon_i \propto N_d \propto [\omega - \omega_g + W_l \pm \omega_{\text{phon}}]^{1/2} = [\omega - \bar{\omega}_g]^{1/2}, \qquad (6.41)$$

where the term with $-\omega_{\text{phon}}$ corresponds to phonon emission and that with $+\omega_{\text{phon}}$ to phonon absorption. The apparent gap $\bar{\omega}_g$ of Eq. (6.41) is:

$$\bar{\omega}_g = \omega_g - W_l \pm \omega_{\text{phon}}.$$

The contribution of allowed indirect exciton transitions to ε_i is also proportional to $|\phi_l(0)|^2$, the square of the exciton envelope function at $r = 0$ [see Eq. (6.15)]. Hence Eq. (5.1), which did not exhibit a singularity in the slope of ε_i at $\omega = \omega_g \pm \omega_{\text{phon}}$, is modified by the exciton interaction so as to exhibit a singularity at $\bar{\omega}_g$ similar to that obtained for direct transitions between three-dimensional bands without exciton interaction. Because of the strong decrease in $|\phi_l(0)|$ with increasing l [Eq. (6.15)], we would normally expect to see only one or at most two ($l = 1$ and $l = 2$) edges of the type described by Eq. (6.41). At low temperatures, only phonon emission edges will be observable.

Figure 18 shows data obtained by Gershenzon et al.[123] for GaP, a

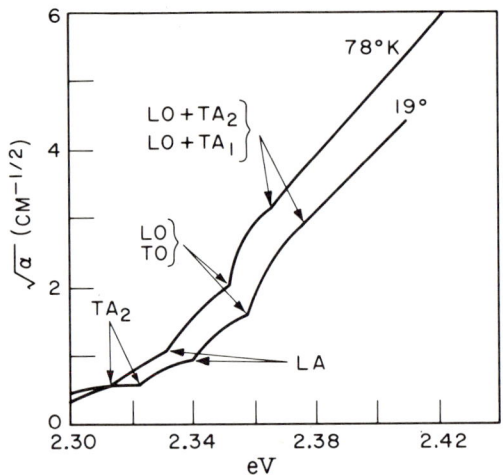

FIG. 18. Square root of the absorption coefficient α of GaP in the indirect absorption region showing exciton singularities. The nature of the contributing phonons is given at each singularity. [From M. Gershenzon, D. G. Thomas, and R. E. Dietz, *Proc. Intern. Conf. Phys. Semiconductors*, *Exeter* 1962, p. 752. Inst. Phys. Physical Soc., London, 1962).]

material with an indirect thermal gap. The square root of the absorption coefficient α (α is approximately proportional to ε_i near the critical point) has been plotted in order to accentuate the exciton effects: this plot would give straight lines (see Section 5) if such effects were absent. Instead we see the sharp singularities at certain energies which correspond to the singularities of Eq. (6.41). For this material, $k_e^v = 0$ and $k_e^c \simeq (2\pi/a_0)[100]$ (a_0 is the lattice constant). The study of the indirect exciton spectrum yields information about the phonon spectrum at or near the $(2\pi/a_0)[100]$ point of the Brillouin zone. Beside the one-phonon transitions discussed above, Gershenzon et al.

have identified in Fig. 18 third-order transitions corresponding to the simultaneous emission of two phonons (LO + TA). The identification is simplified considerably by the consideration of the group theoretical selection rules governing the various processes.[124]

7. Broadening

a. Band-to-Band Transitions

With the exception of the indirect transitions, considered in Section 5, the presence of phonons has been disregarded in our previous treatment of the optical properties of solids. While the effects of impurities, dislocations, and other defects can, at least in principle, be reduced to a minimum in pure and perfect enough crystals, phonons constitute "intrinsic" defects when the solid is pictured as a gas of electrons moving in the potential of periodically placed fixed ions; phonon absorption processes can certainly be eliminated by lowering the temperature but phonon emission processes always remain. Surfaces are also "intrinsic" defects. In the region of low absorption (indirect transitions) a thick crystal can be studied and hence surface effects minimized. However, the absorption coefficients for direct allowed transitions are between 10^4 and 10^6 cm^{-1}. Hence, the close proximity of surfaces is inevitable in the direct absorption region and deviations from the infinite solid theory described above would be expected. Very little is known about the magnitude of such deviations but, because of the agreement between experimental observations and calculations of the optical spectra which do not include surface effects, it is generally believed that the surface effects are small, provided that the surfaces are "good" (for the meaning of "good surfaces" see Section 8).

To first order, the effects of the phonons can be described as a broadening of the initial conduction—and valence band states by phonon emission and absorption and a violation of the **k**-selection rule due to the emission and absorption of phonons during the photon absorption process (indirect transitions). The effect of indirect transitions has already been considered. We have seen in Section 5 how the breakdown in **k**-conservation leads to the weakening of the singularity at a Van Hove critical point. In particular, we saw how the square root singularity near a three-dimensional critical point becomes a much weaker ω^2 singularity for indirect transitions. Indirect transitions will, in general, make Van Hove singularities less sharp.

Due to the linear nature of the dispersion relation for acoustical phonons near $\mathbf{k} = 0$ and the smallness of phonon energies as compared with electronic energies, there is, in general, always a double infinity of solutions to the equations:

$$\omega - \omega_g = \omega_{\text{phon}}$$
$$\mathbf{k}(\omega) - \mathbf{k}(\omega_g) = \mathbf{k}_{\text{phon}} \tag{7.1}$$

for such phonons (see Fig. 19). Hence, at the bottom of a conduction

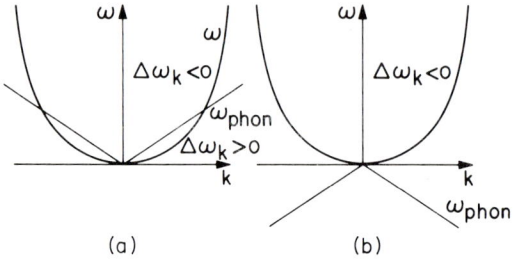

FIG. 19. (a) Schematic diagram of broadening and shift by phonon absorption near an M_0 critical point of the conduction band. (b) No broadening takes place for phonon emission since Eqs. (7.1) have no solution.

band minimum, the electrons have a finite lifetime in the presence of phonons: they can be excited to higher states by phonon absorption. This finite lifetime produces an energy uncertainty and thus a broadening of the singularities in the optical spectra. Phonon absorption is the only phonon process broadening electron states at the lowest conduction band minima and hole states at the highest valence band maxima. Because of the statistical factors involved, acoustical phonons play a dominant role in these processes except at high temperatures. Phonon absorption processes can be eliminated at low temperatures. This is the reason why very sharp optical spectra can be obtained near the lowest absorption edge of semiconductors and insulators at low temperatures. Unfortunately this conclusion does not hold for Van Hove edges above the lowest: phonon absorption and emission are now effective in broadening the spectra and very sharp singularities are not obtained even at the lowest temperatures. This fact has been a main stumbling block in the study of optical spectra of solids at energies above the lowest edge. However, since the Debye temperature θ is, in most solids, of the order of room temperature, some sharpening of the optical spectra results at low temperatures. The probability of phonon absorption processes, roughly proportional to:

$$f_B = \frac{1}{1 + e^{\theta/T}}, \tag{7.2a}$$

is decreased when the temperature is lowered and so is the probability of phonon emission, proportional to $1 + f_B$.

This qualitative consideration explains the results of Fig. 12. Considerable sharpening occurs at the E_1 and $E_1 + \Delta_1$ peaks of InSb, and of most other materials of its family,[46,117] when going from room temperature to 77°K. Almost no further sharpening occurs below 77°K since at these temperatures $f_B \ll 1$. Similar results are shown in Fig. 20 for ZnTe[117] and in Fig. 21 for HgTe.[125] The peaks in these

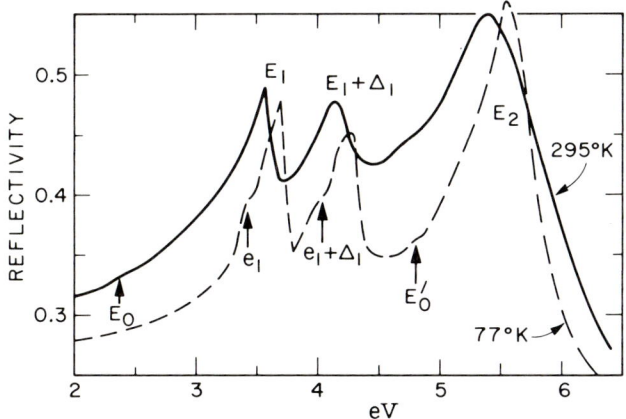

FIG. 20. Reflectivity spectrum of ZnTe at 77°K and at room temperature. [From M. Cardona and D. L. Greenaway, *Phys. Rev.* **131**, 98 (1963).]

figures become at low temperatures about 1.5–2 times sharper than at room temperature. Considerable amount of structure which is not resolved at room temperature in Fig. 21 (HgTe), obviously because of phonon broadening, is resolved in the low temperature spectrum. The e_1, $e_1 + \Delta_1$, and E_0' peaks of Fig. 20 are resolved only at low temperatures.

It is possible to obtain an analytic expression for the dielectric constant near a critical point which includes broadening in the phenomenological manner described by a finite η in Eq. (2.11). Assuming a constant optical matrix element and a parabolic density of states, we obtain for an M_0 three-dimensional critical point (the contour of Fig. 2 has been used for the integration)[126,127]:

$$\varepsilon - 1 \propto (\omega + i\eta)^{-2}[2\omega_g^{1/2} + (\omega_g - \omega - i\eta)^{1/2} - (\omega_g + \omega + i\eta)^{1/2}]. \quad (7.2b)$$

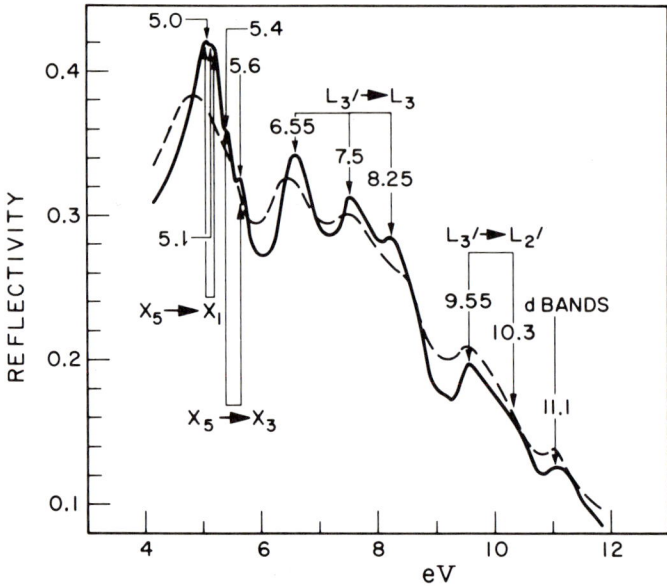

FIG. 21. Reflectivity spectrum of HgTe at 12°K (—) and at room temperature (- - -). [From W. J. Scouler and G. B. Wright, *Phys. Rev.* **133**, A736 (1964).]

The square roots with the smallest argument must be chosen for Eq. (7.2b). The broadening parameter η is an increasing function of T. It is trivial to generalize Eq. (7.2b) to other types of critical points. Associated with the broadening of electronic states discussed above, the electron phonon interaction produces energy shifts. These shifts can be interpreted as second-order processes involving virtual absorption and re-emission or emission and reabsorption of a phonon. The shift of state ω_k of the conduction band is[128]:

$$\Delta\omega_k(T) = \sum_{\substack{l \text{ empty} \\ \text{phon}}} \frac{|M_{k,l,\text{phon}}|^2}{\omega_k - \omega_l \pm \omega_{\text{phon}}}, \quad (7.3)$$

where $M_{k,l,\text{phon}}$ is the matrix element of the electron–phonon interaction, including phonon occupation numbers. The $+\omega_{\text{phon}}$ term corresponds to phonon absorption while $-\omega_{\text{phon}}$ corresponds to phonon emission. At an M_0 critical point the denominator of Eq. (7.3) is negative for phonon emission and can either be positive or negative, for phonon absorption. However, as shown in Fig. 19, the number of states for which this denominator is positive is, in general, much smaller than the number of states for which it is negative, provided lower valleys are either not present or not effective. Hence a temperature dependent lowering of the energy of the conduction band minimum results. A similar argument shows that a valence band maximum is raised with increasing temperature by the electron–phonon interaction. Hence, a decrease in the energy of M_0 edges with increasing temperature generally results. (The argument given above applies only to edges between valence band maxima and conduction band minima.) This increase is linear in temperature for temperatures close to or higher than θ. At low temperatures the rate of change of ω_g decreases with T.

To the effect discussed above, one must add the *implicit* effect of the thermal expansion which produces an increase in the average lattice constant with temperature. Typical temperature coefficients for many M_0 gaps are several times 10^{-4} eV × (°C)$^{-1}$ (typically $d\omega_g/dT \approx -5 \times 10^{-4}$ eV × (°C)$^{-1}$). Under special circumstances, and due possibly to a predominance of intermediate states with positive denominator for Eq. (7.3), increases of gaps with temperature due to electron–phonon shifts can occur for M_0 gaps. Such is the case of the lowest direct gap of the lead chalcogenides[129] $[d\omega_g/dT = +4 \times 10^{-4}$ eV × (°C)$^{-1}]$. Roughly one half of this effect is believed due to thermal expansion.[130]

The situation near a conduction band maximum or a valence band minimum would be expected to be the reverse of that described above; hence M_3 edges, which correspond to transitions between such minima and maxima, should exhibit, in general, a positive temperature coefficient. The fact that such coefficients are rarely observed (except for the lowest edge of the lead chalcogenides, which must be of the M_0-type), confirms our comments about the difficulty of observing M_3 edges made in Section 6b.

For M_1 or M_2 critical points one cannot *a priori* conclude whether the electron–phonon temperature coefficients should be positive or negative, especially if they do not correspond to transitions between band extrema, i.e., if Eq. (4.1) is fulfilled but for each band individually $\mathbf{V_k} \omega_{c,v} \neq 0$. The sign of the coefficient of the conduction (or valence) band states will depend, roughly speaking, on whether most states of the same band are at higher or lower energy. Hence it is not unreasonable to expect negative coefficients for an M_1 conduction band extremum and positive for a similar M_2 extremum (see Fig. 1). In particular the M_1 edges (E_1, $E_1 + \Delta_1$) of Fig. 12 (InSb) have a temperature coefficient[46] $d\omega_g/dT = -5 \times 10^{-4}$ eV (°C)$^{-1}$. The E_1, $E_1 + \Delta_1$ peaks of ZnTe (Fig. 16) have approximately the same coefficient (-6×10^{-4}).[117] As a general rule, equivalent transitions have similar temperature coefficients for all materials of the same family.

Broadening of optical singularities can also be produced by impurities. As we may infer from the well-known decrease of the carrier mobility with doping, the scattering time of electrons in a given valley is reduced by the presence of impurities. This can be translated into an uncertainty broadening of the energy levels and a broadening of the optical spectra. Indirect transitions (**k** not conserved) also become possible in the presence of impurities and contribute to the broadening of the optical Van Hove singularities. As an illustration of the effect of doping on the peaks in the optical spectra of solids, we show, in Fig. 22, the E_1, $E_1 + \Delta_1$, and E_2 peaks of the reflectivity spectrum of pure and heavily doped germanium ($N \sim 5 \times 10^{19}$ cm^{-3}, As doped) at 85°K.[131] The effect of this doping level on the optical spectrum is comparable to that of bringing the pure materials to room temperature. Similar broadening effects are observed for solid solutions of two crystals of the same structure.[132,133]

b. Excitons[134]

The broadening of the exciton lines can be treated in a manner similar to that used for the interband transitions. Due to the presence of phonons, indirect transitions to exciton states with $\mathbf{K} \neq 0$ are possible and a broadening of the exciton lines results. Because of the close spacing between the exciton ground state [$l = 1$, Eq. (6.10)] and the excited states ($l = 2, 3, \ldots$), indirect transitions and scattering to states within the same exciton band ($l = l'$, intraband transitions) and to different bands ($l \neq l'$, interband transitions) must be considered.[134] Acoustical phonon scattering is dominant in the intraband effects

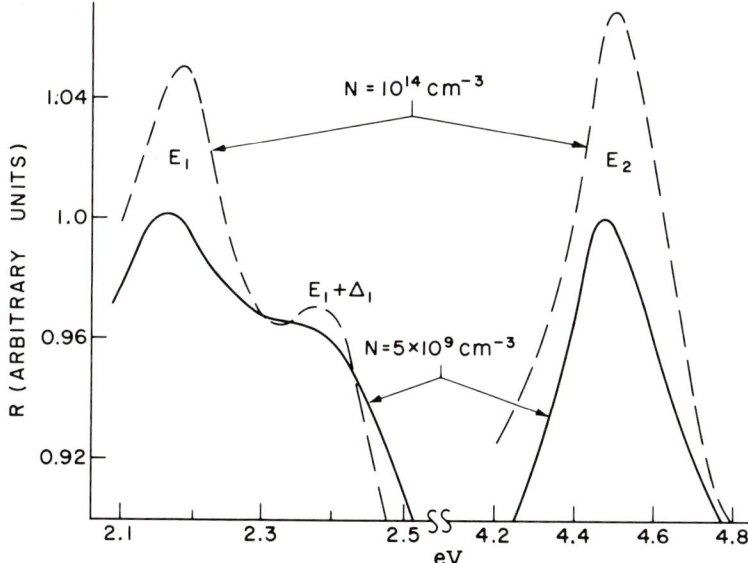

FIG. 22. Reflectivity spectrum of pure ($N \approx 10^{18}$ cm^{-3}) and heavily doped germanium ($N \approx 5 \times 10^{19}$ cm^{-3}, arsenic doped) at 85°K in the vicinity of the E_1, $E_1 + \Delta_1$, and E_2 peaks. [From M. Cardona and H. S. Sommers, Jr., Phys. Rev. **122**, 1382 (1962).]

while both, acoustical and optical phonons give comparable contributions to the interband ($l \neq l'$) effects. If the exciton–phonon coupling is weak, the δ-function associated with absorption to the discrete $\mathbf{K} = 0$, lth exciton level becomes, as a result of such coupling, an asymmetrically broadened Lorentzian line[134]:

$$\varepsilon_i \propto \frac{\Gamma_l/2 + 2A_l\{\omega - (\bar{\omega}_g + \Delta_l)\}}{\{\omega - (\bar{\omega}_g + \Delta_l)\}^2 + (\Gamma_l/2)^2}. \tag{7.4}$$

Equation (7.4) is the imaginary part of a function ε given by:

$$\varepsilon - 1 \propto \frac{1 - 2iA_l}{\bar{\omega}_g + \Delta_l - \omega - i\Gamma_l/2}. \tag{7.5}$$

Hence, the shape of the real part of ε is found from:

$$(\varepsilon_r - 1) \propto \frac{\bar{\omega}_g + \Delta_l - \omega - \Gamma_l A_l}{\{\omega - (\bar{\omega}_g + \Delta_l)\}^2 + (\Gamma_l/2)^2}, \tag{7.6}$$

where Δ_l is an energy shift due to the exciton–phonon coupling, Γ_l a broadening parameter, and A_l a broadening anisotropy parameter.

A_l arises only from interband scattering in the limit of extremely weak coupling.

An explicit expression for the broadened exciton absorption can also be obtained for the case of strong coupling.[134] This case will be of interest for materials with very narrow electron energy bands (and thus narrow exciton bands). The broadening of the exciton lines due to the phonon coupling may be, for these materials, larger than the width of the exciton bands. The imaginary part of the dielectric constant has, in this case, a Gaussian shape[134]:

$$\varepsilon_i \propto \exp\left[-\frac{(\omega - \bar{\omega}_g)^2}{2D_l^2}\right]. \tag{7.7}$$

Equation (7.7) peaks at $\omega = \bar{\omega}_g$ and has a half-value width equal to $2^{1/2}D_l \ln 2$. As an illustration of the broadening of exciton lines we show in Fig. 23 the temperature broadening and shift of two exciton

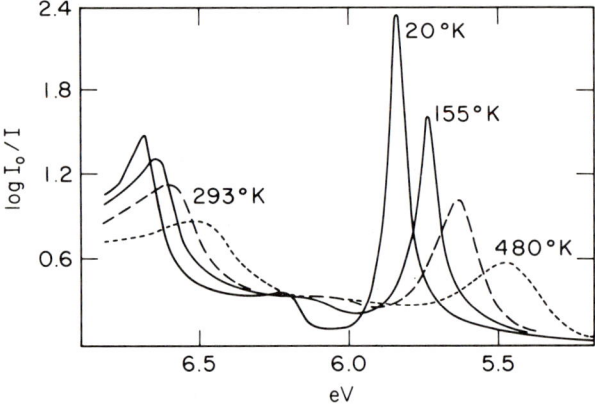

FIG. 23. Optical density of films of KI in the photon energy region near the lowest exciton peaks at several temperatures. This figure illustrates the broadening and shift of the exciton peaks due to phonons. [From W. Martienssen, *J. Phys. Chem. Solids* 2, 257 (1957).]

peaks of KI, as reported by Martienssen.[135] The width of the low energy exciton peak is much larger than kT, especially at 20°K (this fact, characteristic of the alkali halides, has already been mentioned in Chapter I). The increase in width between 20 and 480°K, however, is approximately kT. The shift of the exciton peak with temperature is much larger than the corresponding broadening. This fact, attributable in part to a large thermal expansion component of the shift (devoid of broadening), is common to many optical spectra of solids.

8. Experimental Techniques

a. Refractive Index below the Fundamental Edge of Semiconductors and Insulators

In the region of transparency, it is easy to determine the refractive index n_r of a material as a function of photon energy by a number of well-known methods. Best accuracy ($\approx 0.03\%$) is probably obtained by measuring the angle of minimum deviation of a prism made of the material under study.[136] This method requires the availability of crystals (single crystals for noncubic materials) large enough for the construction of a suitable prism. The data of Marple shown in Fig. 3 were obtained by this technique.

When only platelets or thin films of the materials under study are available, it is possible to determine the spectral dependence of n_r in the region of transparency by observing interference fringes by transmission or absorption in an accurately plane parallel film or platelet.[27] Results obtained using this method by Zemel et al.[27] for PbS, PbSe, and PbTe are shown in Fig. 24. The curves exhibit the typical square root singularity of ε_r (in this region $n_r^2 = \varepsilon_r$) near an M_0 critical point (Fig. 4) at the energy of the lowest direct gap.[129]

b. Transmission and Reflection of Platelets and Thin Films

Below the fundamental edge, the absorption coefficient (and ε_i) is nearly zero for undoped semiconductors and insulators and the optical behavior is determined solely by the refractive index n_r. When ε_i is not zero, two independent measurements are required to specify the optical constants n_r and n_i (or equivalently ε_r and ε_i). A convenient set of such two independent measurements is the transmissivity and the reflectivity of a plane parallel sample. Assuming that the sample thickness is much larger than the wavelength of the light (or that the sample faces are not very parallel so that multiple beam interference

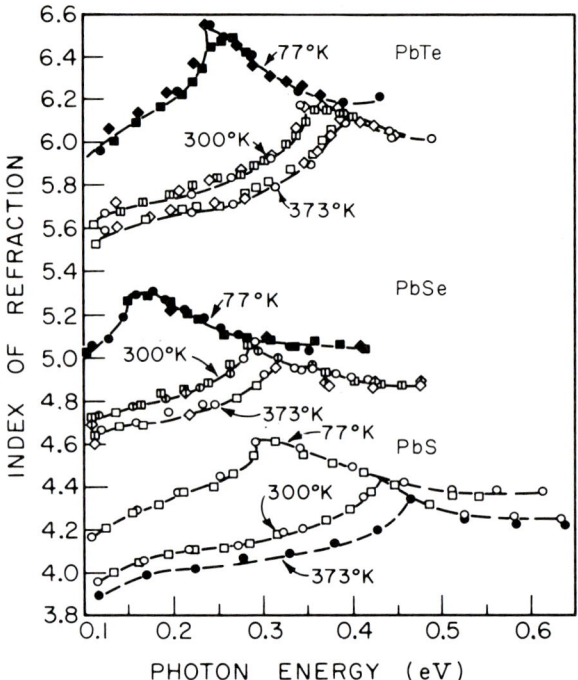

FIG. 24. Refractive index n_r of PbTe, PbSe, and PbS as a function of photon energy at several temperatures as measured by Zemel et al. [From J. N. Zemel, J. D. Jensen, and R. B. Schoolar, *Phys. Rev.* **140A**, 330 (1965).] The curves exhibit the characteristic singularity of $\varepsilon_r = n_r^2$ at an M_0 critical point.

effects can be neglected), the transmissivity $\mathscr{T}$ and the reflectivity $\mathscr{R}$ of the plate at normal incidence are given by[81]:

$$\mathscr{T} = \frac{(1-R)^2 e^{-\alpha d}}{1 - R^2 e^{-2\alpha d}}; \qquad \mathscr{R} = R[1 + \mathscr{T} e^{-\alpha d}], \qquad (8.1)$$

where R is the normal incidence reflection coefficient of the material with respect to the outside medium (air or vacuum in general) and d is the sample thickness. Equations (8.1) enable us to calculate R and α from the experimental values of $\mathscr{T}$ and $\mathscr{R}$. The optical constants n_r and n_i are obtained from α and R by solving:

$$R = \frac{(n_r - 1)^2 + n_i^2}{(n_r + 1)^2 + n_i^2}; \qquad \alpha = \frac{2\omega n_i}{c} \qquad (8.2)$$

[in Eq. (8.2) the outside medium is assumed to be vacuum]. In order to use this method we must have samples of a thickness such that $\alpha d \simeq 1$; if $\alpha d \ll 1$ the transmissivity $\mathcal{T}$ becomes independent of α and Eqs (8.1) only yield information about R. On the contrary, if $\alpha d \gg 1$, the absorption is very large and $\mathcal{T} = 0$: no information about α is obtained in this case either. The maximum value of αd for which the method is applicable is usually determined by the percentage of scattered light present in the monochromator output. When the intensity of scattered light transmitted by the material is higher than that of transmitted light at the frequency of the measurement, no significant data are obtained by transmission. Double-pass monochromators are used to decrease the amount of scattered light. Typical maximum measurable values of αd in the visible are 5 for single pass and 10 for double pass instruments.

The absorption coefficient α is of the order of 10^5 cm^{-1} for ω above the lowest direct edge, except in its immediate vicinity. Hence samples of a thickness of the order of 1000 Å must be prepared in order to study the region above the edge by the transmission–reflection method. The preparation of such thin samples is obviously beyond the capabilities of the conventional mechanical polishing and grinding methods The minimum sample thickness which can be prepared by conventional grinding and polishing methods lies in the neighborhood of 1 μ for most materials. It is, however, possible to prepare such samples by vacuum deposition. This method has been successfully used to observe critical point structure in a number of materials including germanium, III–V, (see Fig. 12), II–VI, and I–VII compounds.[46,137,138] Of particular interest is the study of epitaxial single crystal films such as that undertaken for the lead chalcogenides[27,139] and the alkali halides.[15]

Samples of germanium thin enough for studies of transmission well within the fundamental absorption region ($\alpha \leq 4 \times 10^5$ cm^{-1}) have been prepared from bulk material by mechanical polishing followed by chemical etching.[140,141] Using single crystals of low dislocation density as starting materials, platelets 2500 Å thick have been prepared.[141]

Very thin samples, suitable for transmission measurements, can be prepared by cleavage for certain materials with layer structures. Among these materials $Bi_8Te_7S_5$ is particularly easy to cleave[142]; very thin layers can be obtained by sticking adhesive tape to the cleavage surface of the material and peeling the layers off. The power of this method for studying optical properties in the fundamental absorption region has not yet been fully explored.

c. Reflection at Normal Incidence

As mentioned in Section 1 the two optical constants n_r and n_i are not completely independent. If one of them is known for *all* frequencies, the other is automatically determined by the dispersion relations (1.1) and (1.2). Therefore, the measurement of only one optical parameter at *all* frequencies should suffice for the determination of the optical constants n_r and n_i. In order to make this method practically applicable, we must reduce the infinite frequency range, required in principle, to an easily available range and use extrapolation methods to extend this range to $0 \leq \omega \leq \infty$. This can be easily accomplished since most of the interband contribution to the imaginary part of the dielectric constant [Eq. (4.3)] occurs at frequencies in the easily accessible visible and near ultraviolet range. Well below the smallest direct gap, the interband dielectric constant of an insulator is (neglecting infrared dispersion due to phonon absorption):

$$\varepsilon_r = \text{constant}; \qquad \varepsilon_i = 0. \tag{8.3}$$

Therefore, we do not have to perform measurements in the low frequency region where (8.3) is valid. Lattice absorption effects can be easily taken into account since they can usually be approximated by a single oscillator dispersion formula. They can be neglected in the evaluation of the electronic optical constants whenever the reststrahlen frequency is much smaller than that which corresponds to the lowest interband gap, because of the energy denominators which appear in the dispersion relations.

As we have already mentioned, the oscillator strengths in Eq. (2.13) become small for very large energies ω_{lk}. The main contribution to Eq. (2.13) occurs for ω_{lk} in the visible and near ultraviolet. It is therefore usually possible to approximate $\varepsilon(\omega)$ at frequencies above the near ultraviolet by:

$$\varepsilon = 1 - \frac{A}{\omega^2 - \bar{\omega}_{lk}^2}, \tag{8.4}$$

where A is a real constant and $\bar{\omega}_{lk}$ is an "average" interband transition frequency. For $\omega \gg \bar{\omega}_{lk}$, $\bar{\omega}_{lk}$ can be neglected in Eq. (8.4); the dispersion obtained is essentially the last term in the right-hand side

8. EXPERIMENTAL TECHNIQUES

of Eq. (2.8) with N replaced by an adjustable effective electron density.[143] An upper cutoff to the frequency region required for the determination of the optical constants by means of the dispersion relations can be set at a frequency $\omega \gg \bar{\omega}_{lk}$ for which Eq. (8.4) is valid.

Intraband or free carrier effects can be easily included in the above discussion since an analytic expression for them [Eq. (3.2) or the corresponding expression which includes scattering] is available.

The quantity usually measured for the determination of n_r and n_i by dispersion relation techniques is the normal incidence reflectivity R related to n_r and n_i by Eq. (8.2). R is the square of the magnitude of the complex reflectivity:

$$R^{1/2} e^{i\theta} = \frac{n_r - 1 + i n_i}{n_r + 1 + i n_i}, \tag{8.5}$$

where θ is the change in phase upon reflection, related to R by the dispersion relation[91,144]

$$\theta(\omega) = \frac{\omega}{\pi} P \int_0^\infty \log R(\omega') \frac{d\omega'}{\omega'^2 - \omega^2}. \tag{8.6}$$

Once $\theta(\omega)$ is evaluated by means of Eq. (8.6) from the experimental and the extrapolated values of $R(\omega)$, the optical constants n_r and n_i are obtained by solving Eq. (8.5).

In order to evaluate Eq. (8.6) by numerical methods, it is necessary to remove the singularity in the integrand for $\omega = \omega'$. This is readily done by adding to Eq. (8.6):

$$0 = \frac{\omega}{\pi} \log R(\omega) \, P \int_0^\infty \frac{d\omega'}{\omega'^2 - \omega^2},$$

and integrating the resulting equation:

$$\theta(\omega) = \frac{\omega}{\pi} P \int_0^\infty \frac{\log R(\omega') - \log R(\omega)}{\omega'^2 - \omega^2} d\omega' \tag{8.7}$$

with the aid of Simpson's rule. The integrand of Eq. (8.7) is nonsingular for $\omega \to \omega'$, since the experimental $R(\omega)$ never exhibits an infinite slope. $R(\omega) = $ constant is used for insulators and semiconductors as a low frequency extrapolation based on Eqs. (8.3). At high energies, the extrapolation:

$$R \sim \omega^{-4}, \tag{8.8}$$

based on Eq. (8.4), is commonly used. A discussion of the computational methods used are given by Cardona[91] and Roessler.[145]

The evaluation of the optical constants by the method described above requires reflecting surfaces of "good" crystalline quality. Such surfaces are normally not obtained by mechanical grinding and polishing, since this procedure leaves a heavily damaged surface layer of a depth comparable to the penetration depth of the light. This problem can be remedied by removing the damaged material with a chemical etchant after mechanical polishing. This is usually done at the expense of flatness and general cosmetic appearance of the sample surface. The production of a surface without polishing damage and with low enough diffuse reflection is usually a delicate problem which must be solved individually for each material. A description of chemical etchants for many semiconductors has been given by Gatos and Lavine.[146] Electropolishing also gives good quality specular surfaces and should be particularly suitable for metals.[147]

Good quality surfaces for reflection studies are obtained by cleavage for easily cleaving materials.[139] Surfaces of growth are also usually very reliable. They can sometimes be obtained by vacuum deposition, expitaxial growth, and by bulk growth from the vapor phase. Surfaces of growth from solution (in mercury) have been used for reflection studies of gray tin.[148]

The experimental determination of R requires a measurement at each frequency of the intensity of the incident light I_0 and that of the reflected light I in order to obtain $R = I/I_0$. This can be easily achieved by comparing the reflectivity of the sample to that of a standard front surface mirror, but the instability and inhomogeneity of such standard mirrors makes this method inaccurate, especially in the ultraviolet region. Similar criticism applies to double-beam measurement methods, which require several *identical* mirrors and possibly a neutral beam splitter. The method most commonly employed (see Fig. 25) involves the measurement of the reflected intensity followed by a removal of the sample from the beam and a measurement of the incident intensity by swinging the detector around an axis which passes through the original sample position. The mechanical motions required to perform this operation limit the speed and accuracy of such measurements. As we shall see in Chapter III, the absence of these mechanical operations constitutes one of the main advantages of the modulation methods.

Structure in ε_r and ε_i of the type associated with a Van Hove critical point produces usually similar structure in the reflectivity. It is, therefore, sometimes possible to identify critical points in the reflection spectrum without doing the Kramers–Kronig analysis. In order to obtain an idea of the shapes of such singularities in the reflection spectrum, let us consider a critical point and assume that its contribution to ε is relatively small (this can only be done for three-dimensional critical points since either ε_r and ε_i or both become

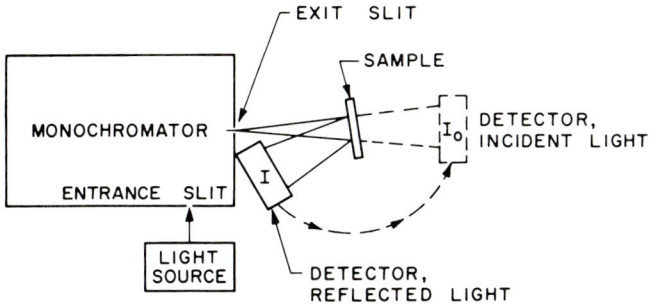

FIG. 25. Arrangement for the measurement of the reflection spectrum at normal incidence. The use of lenses or mirrors is usually avoided in the far ultraviolet region.

infinite at ω_g for one- and two-dimensional critical points; broadening can however be invoked to include these cases in our treatment). The contribution of the critical point to the reflectivity can be written in the vicinity of ω_g[149]:

$$\frac{\Delta R}{R} = \left(\frac{\partial \ln R}{\partial \varepsilon_r}\right)_{\omega_g} \Delta\varepsilon_r + \left(\frac{\partial \ln R}{\partial \varepsilon_i}\right)_{\omega_g} \Delta\varepsilon_i$$
$$= \beta_r(\omega_g)\, \Delta\varepsilon_r + \beta_i(\omega_g)\, \Delta\varepsilon_i, \qquad (8.9a)$$

where $\Delta\varepsilon_r$ and $\Delta\varepsilon_i$ are the contributions of the critical point to ε_r and ε_i respectively. β_r and β_i are easily calculated from Eqs. (1.3) and (8.2) if the optical constants at the frequency ω_g are known. We find[149]:

$$\beta_r = C_r[(\varepsilon_r - 1)A_+ + \varepsilon_i A_-]$$
$$\beta_i = C_i[(\varepsilon_r - 1)/A_+ - \varepsilon_i/A_-], \qquad (8.9b)$$

with

$$C_r = [(\varepsilon_r - 1)^2 + \varepsilon_i^2]^{-1}$$
$$C_i = 2\varepsilon_i[(\varepsilon_r - 1)^2 + \varepsilon_i^2]^{-1}(\varepsilon_r^2 + \varepsilon_i^2)^{-1}$$
$$A_\pm = \pm \frac{2^{1/2}[(\varepsilon_r^2 + \varepsilon_i^2)^{1/2} \pm \varepsilon_r]^{1/2}}{(\varepsilon_r^2 + \varepsilon_i^2)^{1/2}}.$$

Figure 26 shows β_r and β_i for germanium[149] and typifies the general trend of β_r and β_i for most semiconductors. Below and near the lowest direct gap, the structure in the reflection spectrum corresponds to structure in ε_r ($\beta_r > 0$, $\beta_i = 0$). Above this gap, β_i grows and becomes higher than β_r near the E_1 optical singularity. At higher frequencies, β_r becomes negative and larger in magnitude than β_i. At these frequencies the R singularity has again the shape of an ε_r singularity, but with opposite sign. For $\beta_r > 0$ and $\beta_i > 0$, the singularities

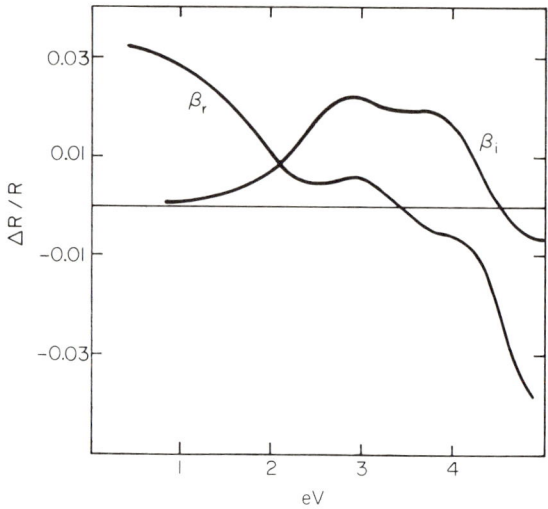

FIG. 26. Functions β_r and β_i which give the contribution of changes in the real and imaginary parts of the dielectric constant to the change in the reflectivity $\Delta R/R$, for germanium. [From B. O. Seraphin and N. Bottka, *Phys. Rev.* **145**, 628 (1966).]

in R have the same shape as those given for the ε_i singularities in Fig. 14 in the presence of a Koster–Slater electron hole interaction. Hence sharp exciton-like peaks can be obtained in the reflection spectrum which do not necessarily correspond to band-to-band peaks but to shoulders or edges in ε_r and ε_i.

d. Ellipsometry

If one reflects linearly polarized light on a solid at non normal incidence, the reflected light will, in general, be elliptically polarized except for the two normal modes of incidence: light polarized parallel or perpendicular to the plane of incidence. It is therefore possible to determine the two optical constants at a given wavelength by measuring the state of polarization of the reflected light, i.e., the orientation and axes ratio of the polarization ellipse. A number of schemes have been devised to perform this type of measurement; they are always restricted to the spectral region for which polarizing equipment and

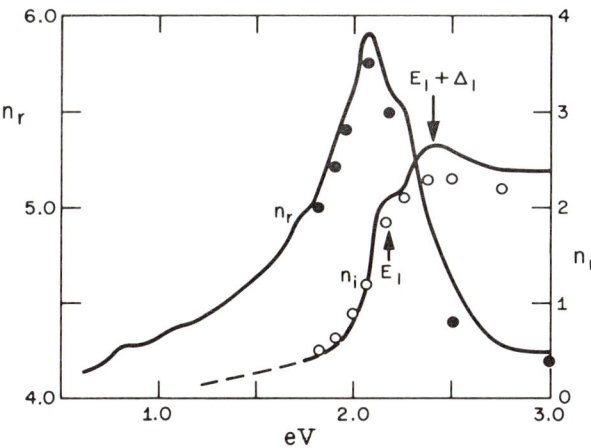

FIG. 27. Optical constants of germanium obtained by R. J. Archer, *Phys. Rev.* **110**, 354 (1958) (circles) with ellipsometric techniques and by R. F. Potter *Phys. Rev.* **145**, 628 (1966)] (solid lines) with the pseudo-Brewster angle method.

wave plates are available. In a method used by Archer,[24] incident elliptically polarized light of variable states of polarization, produced by means of a linear polarizer and a wave plate, is used. The state of polarization of the incident light is varied (by varying the azimuthal angle of the polarizer) until the ellipticity of the incident beam is cancelled by the reflection at oblique incidence on the solid under study. This condition is detected as an extinction with an analyzer. The optical constants are determined from the angle of the polarizer and the analyzer under extinction conditions. The optical constants n_r and n_i obtained for germanium by this method are represented by circles in Fig. 27, together with measurements by Potter to be discussed in the next section. The spin orbit splitting of the E_1 peak, resolved in the measurements of Potter, is not resolved in Archer's measurements, probably because of the wide spacing between the experimental points.

The ellipsometric technique has been recently used for studying the optical constants of the alkali metals. Figure 28 shows data obtained by Mayer and El Naby[39] for potassium. The peak in $\sigma = \omega\varepsilon_i/4\pi$ at 0.5 eV does not agree with the intraband theory of Section 3 nor can it be explained as due to interband transitions.[40] A number of theories have been put forward to explain this peak by invoking many-body effects.[41,150] Attempts to reproduce this peak in other optical measurements have, so far, failed.[42]

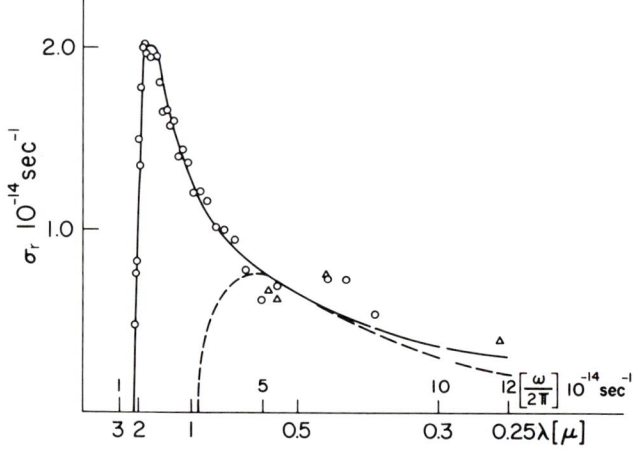

FIG. 28. Real part of the conductivity tensor of potassium as a function of photon energy with dashed line for σ_r calculated by Butcher, circles for σ_s (data of Mayer and El Naby, and triangles for σ_r data of Ives. From H. Mayer and M. H. El Naby, Z. Physik **174**, 289 (1963).

e. Reflectivity at Oblique Incidence

The optical constants can also be determined from the measurement of the reflectivity at oblique incidence for light polarized with the electric field perpendicular $(R_\perp)$ and parallel $(R_\parallel)$ to the plane of incidence.[23,151] In this type of measurement the angle of incidence of 45° is to be avoided[152] since for it $R_\parallel = R_\perp{}^2$ and thus the measurements of $R_\parallel$ and $R_\perp$ are not independent. It is also possible to extract the optical constants[152,153] from a measurement of the ratio $R_\parallel/R_\perp$ for two different angles of incidence.

An extremely accurate determination of the optical constants of germanium based on the determination of the pseudo-Brewster angle has been reported by Potter.[154] In the region of transparency, the reflectivity $R_\parallel$ for light polarized with the electric field in the plane of incidence goes to zero at the Brewster angle ϕ_B defined by tan $\phi_B = n_r$. If the material is absorbing, $R_\parallel$ still goes through a nonzero minimum $R_\parallel(\phi_B)$ at the pseudo-Brewster angle ϕ_B. The optical constants can be calculated from the measured values of $R_\parallel(\phi_B)$ and ϕ_B at any wavelength. The refractive indices n_r and n_i obtained for germanium by means of this technique are shown in Fig. 27 together with the results of the ellipsometric measurements of Archer.[24] The agreement between the results of such different experimental methods is certainly very satisfactory.

9. Assignment of Optical Structure to Interband Critical Points

a. Spin Orbit Splittings

The assignment of optical structure to direct interband critical points for specific values of **k** is a rather tortuous process which requires a great deal of guesswork.[34] The assignment must, very often, be changed as new experimental facts and better theoretical band structures become available. With few exceptions, most of the assignments made so far have to be considered as tentative.

The family of materials for which most reliable assignments exist and which has been most exhaustively studied is the germanium–zinc blende family. The existence of a number of critical points with spin orbit splittings has made assignments particularly easy and was actually responsible for the breakthrough in the field around 1960.[31]

The three-fold degeneracy of p-like states is preserved in cubic materials at $\mathbf{k} = 0$. Hence, predominantly p-like orbital triplets at $\mathbf{k} = 0$ show a spin orbit splitting into a quadruplet and a doublet. The magnitude of the spin orbit splitting bears a close relationship to the spin orbit splitting of the constituent atoms since most of the spin orbit interaction occurs very near the atomic core; the core potential and wave functions are not drastically affected by the presence of neighboring atoms. In a solid, however, the wave functions must be normalized to a unit cell while in the free atom they are normalized to the whole space. This increases the magnitude of the wave functions near the core in the solid as compared with the free atom. The matrix elements of the spin orbit interaction are correspondingly larger in the solid than in the free atoms.[155] The enhancement factor A, equal to the ratio of the spin orbit splitting of a p-like state at $\mathbf{k}=0$ to that of the corresponding atomic state, can be easily estimated for the outermost occupied p-like silicon ($3p$) and germanium ($4p$) valence states from the known splittings of the top valence band Δ_0 ($\Gamma_{25'}$, see Fig. 29) and either the atomic splitting derived from spectroscopic

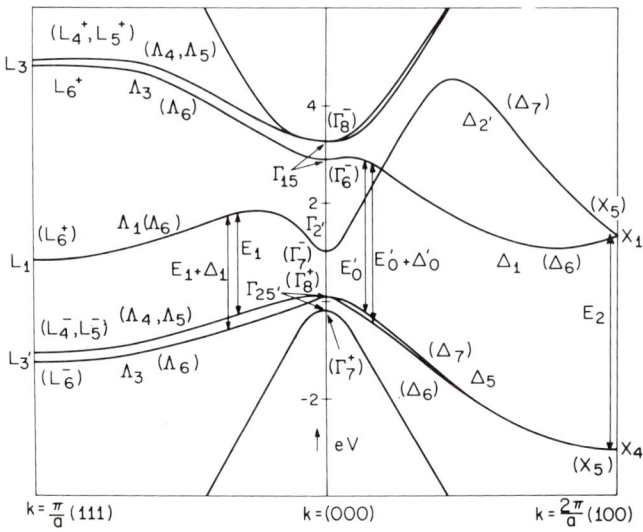

FIG. 29. Band structure of germanium including spin orbit interaction obtained by the $\mathbf{k} \cdot \mathbf{p}$ method [see M. Cardona and F. H. Pollak, *Phys. Rev.* **142**, 530 (1966)].

data[155,156] or that calculated by the Hartree–Fock method.[157] For both germanium and silicon, one obtains (see Table I) $A = 1.4$ from the calculated atomic splitting and also from the spectroscopic atomic splitting (the experimental value of Δ_0 for solid germanium is 0.29 eV).[155]

In order to estimate the spin orbit splitting of materials with two constituent atoms, one must introduce an extra degree of freedom to account for the fraction ξ of wave function of each constituent atom which makes up the wave function of the state under consideration near the core. The spin orbit splitting Δ_0 is:

$$\Delta_0 = A[\xi^{(1)}\Delta_0^{(1)} + \xi^{(2)}\Delta_0^{(2)}], \tag{9.1}$$

TABLE I. ATOMIC SPIN ORBIT SPLITTINGS (IN EV) OF p (AND SOME d) VALENCE LEVELS OF ELEMENTS WHICH FORM THE GROUP IV, GROUP IIIB–V, GROUP IIB–VI, AND GROUP I–VII COMPOUNDS AND THE RARE GASES[a]

Element	Spect.	Calc.	Element	Spect.	Calc.	Element	Spect.	Calc.	Element	Spect.	Calc.	Element	Spect.	Calc.	Element	Spect.	Calc.
IA			IB			III			V			VII					
Li	0.000					B	0.002	0.004	N	0.011	0.025	F	0.05	0.093			
Na	0.002					Al	0.016	0.018	P	0.046	0.062	Cl	0.11	0.15			
K	0.006		Cu	−0.15[b]	−0.19[b]	Ga	0.12	0.12	As	0.29	0.31	Br	0.46	0.55			
Rb	0.029		Ag	−0.33[b]	−0.35[b]	In	0.27	0.31	Sb	0.67	0.67	I	0.945	1.06			
Cs	0.067		Au	−0.91[b]	−1.0[b]	Tl	0.96	0.86	Bi	—[d]	1.68						
IIA			IIB			IV			VI			VIII					
Be	0.000	0.001				C	0.005	0.011	O	0.027	0.051	Ne	0.10[c]	0.12			
Mg	0.007	0.059				Si	0.028	0.035	S	0.069	0.096	Ar	0.18[c]	0.20			
Ca	0.020		Zn	0.071	0.054	Ge	0.21	0.21	Se	0.37	0.42	Kr	0.67[c]	0.69			
Sr	0.072		Cd	0.21	0.15	Sn	0.6	0.475	Te	0.89	0.86	Xe	1.31[c]	1.27			
Ba	0.15		Hg	0.76	0.46	Pb	—[d]	1.27									

[a] Obtained from spectroscopic data and from the calculations of Herman and Skillman.[157]
[b] d-level splittings.
[c] Obtained from spectroscopic data of singly ionized atom.
[d] Russel–Saunders coupling breaks down.

where $\Delta_0^{(1)}$ and $\Delta_0^{(2)}$ are the atomic spin orbit splittings (Table I) and $\xi^{(1,2)}$ indicate the relative fraction of each constituent atom which enters in Δ_0, ($\xi^{(1)} + \xi^{(2)} = 1$). We shall assume that the enhancement factor A is always the same for a given family of materials. Braunstein[155] suggested to use the values $\xi^{III} = 0.35$ and $\xi^{V} = 0.65$ for the Γ_{15} valence band state of the III–V compounds (see Fig. 9) while $\xi^{II} = 0.2$ and $\xi^{VI} = 0.8$ was suggested by the author[33] for the II–VI compounds. The spin orbit splitting of the alkali halides (IA–VII) is produced mainly by the anion ($\xi^{I} = 0$, $\xi^{VII} = 1$). Some anomalies, due to admixture of d-like wave function in the Γ_{15} valence band state, have been reported for the IB–VII[137] compounds and for ZnO.[158] In particular, this Γ_{15} multiplet is inverted for ZnO and for CuCl. Such inversion has been explained[159] as due to admixture of cation d-like wave function to the p-like contribution of the anion. The spin orbit matrix elements of atomic d-states have a sign opposite to those of p-states.

Table II lists the values of Δ_0 at the Γ_{15} (or $\Gamma_{25'}$ for germanium structures) valence band states calculated by the method described above from the spectroscopic values of Table I and the various experimental determinations for germanium–zincblende–type compounds and for the alkali halides. Many of these splittings were obtained by the modulation methods to be described later.

As indicated above, the wide range of the Δ_0 spin orbit splittings gives a good method of identifying whether optical structure involves a p-like state at $\mathbf{k} = 0$: if it shows a splitting close to that expected from Table II, the identification becomes plausible. A strong confirmation of this identification is obtained by applying uniaxial stress to the material: the spin orbit quadruplet splits into two doublets.[160]

In noncubic materials, the triple degeneracy of p-states is split, even at $\mathbf{k} = 0$, by the effect of the crystal field. Identification arguments based on spin orbit coupling cannot, in general, be made. However, if the crystal has at least one symmetry axis of order higher than two (3-, 4-, or 6-fold), the orbital crystal field splits p-states at $\mathbf{k} = 0$ into an orbital doublet and a singlet. The orbital doublet is split by the spin orbit interaction into two spin doublets and arguments similar to those mentioned above can still, in general, be applied to the identification of transitions. In this connection, the so-called two-thirds rule is of great utility. It states that the spin orbit splitting of an orbital doublet is approximately two-thirds of that calculated for a cubic material at $\mathbf{k} = 0$ (Table II), provided it is much

9. ASSIGNMENT OF OPTICAL STRUCTURE

smaller than the orbital crystal field splitting δ. The spin orbit Hamiltonian can be written, assuming only interactions within the p-state under consideration:

$$\Delta L \cdot S = \Delta[J^2 - S^2 - L^2], \qquad (9.2)$$

where Δ is a matrix element of the spin orbit operator and L, S, and J are operators which have the same transformation properties under the symmetry operations of the crystal as the orbital, spin, and total angular momentum respectively. For the cubic case at $\mathbf{k} = 0$ the wave functions which diagonalize the spin orbit Hamiltonian are:

$$\left(\frac{3}{2}, \frac{3}{2}\right) = \frac{1}{\sqrt{2}}(X + iY)\uparrow \qquad \left(\frac{3}{2}, -\frac{3}{2}\right) = \frac{1}{\sqrt{2}}(X - iY)\downarrow$$

$$\left(\frac{3}{2}, \frac{1}{2}\right) = \frac{1}{\sqrt{6}}(X + iY)\downarrow - \sqrt{\frac{2}{3}} Z\uparrow \qquad \left(\frac{3}{2}, -\frac{1}{2}\right) = \frac{1}{\sqrt{6}}(X - iY)\uparrow + \sqrt{\frac{2}{3}} Z\downarrow$$

$$\left(\frac{1}{2}, \frac{1}{2}\right) = \frac{1}{\sqrt{3}}(X + iY)\downarrow + \frac{1}{\sqrt{3}} Z\uparrow \qquad \left(\frac{1}{2}, -\frac{1}{2}\right) = \frac{1}{\sqrt{3}}(X - iY)\uparrow - \frac{1}{\sqrt{3}} Z\downarrow,$$

$$(9.3)$$

where X, Y, and Z transform as the coordinates x, y, and z under point group operations which leave *each* atom invariant. The spin orbit splitting is $\Delta_0 = 3\Delta$. If a large orbital crystal field of at least three-fold symmetry is present, the wave functions of the quadruplet are:

$$\frac{1}{\sqrt{2}}(X + iY)\uparrow \qquad \frac{1}{\sqrt{2}}(X - iY)\downarrow$$
$$\frac{1}{\sqrt{2}}(X - iY)\uparrow \qquad \frac{1}{\sqrt{2}}(X + iY)\downarrow \qquad (9.4)$$

We take as the z direction, the direction of the symmetry axis. If the spin orbit splitting is much smaller than the crystal field splitting δ, so that the spin orbit mixing of the quadruplet given above with the $Z\uparrow\downarrow$ doublet is negligible, the splitting of the quadruplet becomes $\Delta_0' = 2\Delta = \frac{2}{3}\Delta_0$. This can be shown by calculating the expectation value of (9.2) for the (9.4) states. This argument assumes, of course, that Δ is not changed by the orbital crystal field, which is rigorously true when no noncubic mixing of the p-states to other states takes place.

TABLE II. SPIN ORBIT SPLITTINGS AT $k = 0$ OF THE VALENCE BAND OF GROUP IV, IIIB–V, IIB–VI, AND IB–VII MATERIALS, THE ALKALI HALIDES, AND THE SOLIDIFIED RARE GASES†

Element	Transmission	Reflection	Reflectance modulation	Intravalence band transitions	Other experiment	Calc.
C				0.006[a]		0.007
Si				0.044[b]		0.039
Ge	0.29[c]		0.290[d]	0.29[e]		0.29
α-Sn						
AlN						0.018
AlP						0.049
AlAs						0.27
AlSb			~0.7[f]	0.75[e]		0.60
GaN						0.069
GaP	0.09[g]	0.1[h]	0.10[f]	0.127[i]		0.10
GaAs	0.35[j]		0.340[f]	0.33[e]		0.33
GaSb			0.80[f]	0.80[k]		0.66
InN						0.14
InP			~0.11[f]			0.17
InAs			0.38[l]	0.43		0.40
InSb			0.82[n]			0.72
ZnO		−0.0087[o]				+0.050
ZnS (cubic)		0.1[p]	0.068[q]			0.097
ZnSe	0.41[r]	0.43[s]				0.43
ZnTe		0.91[t]	0.93[f]	0.98[u]		1.01
CdO						0.089
CdS	0.065[v]		0.066[f]			0.14
CdSe		0.408[x]	0.404[f]			0.47
CdTe		0.9[t]	0.92[f]		0.9[z]	1.05
HgS						0.29
HgSe						0.63
HgTe						1.21
CuCl	−0.049[y]					+0.15
CuBr	0.197[y]					0.64
CuI	0.633[y]					1.3
AgCl	0.10[y]					0.15
AgBr	0.55[y]					0.64
AgI	0.837[y]					1.3
LiF						0.055
LiCl						0.12
LiBr	0.48[α]					0.51
LiI						
NaF						0.055
NaCl	0.13[α]	0.12[γ]				0.12
NaBr	0.50[α]					0.51
NaI	1.40[α]					1.04
KF	0.05[α]					0.055
KCl	0.11[α]	0.15[γ]				0.12
KBr	0.48[α]	0.47[γ]				0.51
KI	1.38[α]	1.40[γ]				1.04
RbF						0.055
RbCl	0.12[α]	0.15[γ]				0.12
RbBr	0.47[α]	0.46[γ]				0.51
RbI	1.20[α]	1.21[γ]				1.04
CsF						0.055
CsCl	1.13[β]					0.12
CsBr	0.56[α]					0.51
CsI	1.06[α]					1.04
Ne						0.10
Ar	0.2[δ]					0.18
Kr	0.65[δ]					0.67
Xe	1.17[δ]					1.31

The two-thirds rule given above also applies in cubic materials to points off $\mathbf{k} = 0$ along a direction of $\mathbf{k}$-space with at least three fold symmetry. The periodic part of an orbital wave function $u_{k,l}(\mathbf{r})$ for a general point in $\mathbf{k}$-space is obtained by solving the $\mathbf{k} \cdot \mathbf{p}$ Schrödinger equation with the orbital Hamiltonian[116,161]:

$$H = H_0 + \tfrac{1}{2}k^2 + \mathbf{k} \cdot \mathbf{p} + V(\mathbf{r}), \tag{9.5}$$

where H_0 is the Hamiltonian for $\mathbf{k} = 0$, $\mathbf{p}$ the linear momentum operator, and $V(\mathbf{r})$ the self-consistent crystal potential. The $\mathbf{k} \cdot \mathbf{p}$ term behaves as a crystal field with the symmetry of the $\mathbf{k}$-vector: n-fold if $\mathbf{k}$ lies along an n-fold symmetry axis. Except very near $\mathbf{k} = 0$, the $\mathbf{k} \cdot \mathbf{p}$ splittings are, in general, large compared with typical spin orbit splittings. The spin orbit contribution to the $\mathbf{k} \cdot \mathbf{p}$ Hamiltonian becomes[161] for a general $\mathbf{k}$:

$$H_{so}(\mathbf{k}) = \frac{1}{4c^2}[\nabla V \times \mathbf{p}] \cdot \boldsymbol{\sigma} + \frac{1}{4c^2}[\nabla V \times \mathbf{k}] \cdot \boldsymbol{\sigma}. \tag{9.6}$$

FOOTNOTES TO TABLE II

† The calculated splittings were obtained with the expression $\Delta_0 = A[\xi^{(1)}\Delta_0^{(1)} + \xi^{(2)}\Delta_0^{(2)}]$. The parameter A was taken equal to 1.4 for all germanium–zincblende–wurtzite–type materials, equal to 1.1 for the alkali halides and equal to 1.0 for the solidified rare gases. For the alkali halides, we took the contribution of the cation to the spin orbit splitting to be equal to zero, $\xi^{(1)} = 0$, $\xi^{(2)} = 1$. For the III–V compounds, $\xi^{III} = 0.35$, $\xi^{V} = 0.65$ and for the II–VI compounds $\xi^{II} = 0.2$, $\xi^{VI} = 0.8$. The spectroscopic values of Table I were used for Δ_0.

[a] C. J. Rauch, *Proc. Intern. Conf. Phys. Semiconductors*, Paris, 1964, p. 276. Dunod, Paris, 1965.
[b] S. Zwerdling, K. J. Button, B. Lax, and L. M. Roth, *Phys. Rev. Letters* **4**, 173 (1960).
[c] M. V. Hobden, *J. Phys. Chem. Solids* **23**, 821 (1962).
[d] B. O. Seraphin and R. B. Hess, *Phys. Rev. Letters* **14**, 138 (1965).
[e] R. Braunstein and E. O. Kane, *J. Phys. Chem. Solids* **23**, 1423 (1962).
[f] M. Cardona, K. L. Shaklee, and F. H. Pollak, *Phys. Rev.* **134**, 696 (1967).
[g] W. K. Subashiev and S. A. Abagyan, *Proc. Intern. Conf. Phys. Semiconductors*, Paris, 1964, p. 225. Dunod, Paris, 1965.
[h] R. Zallen and W. Paul, *Phys. Rev.* **134**, 1628 (1964).
[i] J. W. Hodby, *Proc. Phys. Soc. (London)* **82**, 324 (1963).
[j] M. Sturge, *Phys. Rev.* **127**, 768 (1962).
[k] B. B. Kosicki and W. Paul, *Bull. Am. Phys. Soc.* **11**, 52 (1966).
[l] C. R. Pidgeon, S. H. Groves, and J. Feinleib, *Solid State Commun.* **5**, 677 (1967).
[m] F. Matossi and F. Stern, *Phys. Rev.* **111**, 472 (1958).
[n] M. Cardona, F. H. Pollak, and K. L. Shaklee, *Phys. Letters*, **23**, 37 (1966).
[o] D. G. Thomas, *J. Phys. Chem. Solids* **86**, (1960).
[p] W. W. Piper, P. D. Johnson, and D. T. F. Marple, *J. Phys. Chem. Solids*, **8**, 457 (1959).
[q] R. Forman and M. Cardona, in "II–VI Semiconducting Compounds" (D. G. Thomas, ed.), p. 100. Benjamin, New York, 1967.
[r] M. Cardona and G. Harbeke, *J. Appl. Phys.* **34**, 813 (1963).
[s] M. Cardona, *J. Appl. Phys.* **32s**, 2151 (1961).
[t] M. Cardona and D. G. Greenaway, *Phys. Rev.* **131**, 98 (1963).
[u] N. Watanabe and S. Usui, *J. Appl. Phys. (Japan)* **4**, 467 (1965).
[v] D. G. Thomas and J. J. Hopfield, *Phys. Rev.* **116**, 573 (1959).
[x] J. O. Dimmock and R. J. Wheeler, *Phys. Rev.* **125**, 1805 (1962).
[y] M. Cardona, *Phys. Rev.* **129**, 69 (1963).
[z] D. T. F. Marple and H. Ehrenreich, *Phys. Rev. Letters* **8**, 87 (1962).
[α] K. Teegarden and G. Baldini, *Phys. Rev.* **155**, 896 (1967).
[β] J. E. Eby, K. J. Teegarden, and D. B. Dutton, *Phys. Rev.* **116**, 1099 (1959).
[γ] G. Baldini and B. Bosachi, *Phys. Rev.* **166**, 863 (1968).
[δ] G. Baldini, *Phys. Rev.* **128**, 1562 (1962).

The first term in the right-hand side of Eq. (9.6) is the spin orbit Hamiltonian for $\mathbf{k} = 0$. The $\mathbf{k}$-dependent second term is normally much smaller than the first since near the atomic core (where ∇V is largest) $k |u(\mathbf{r})| \ll p |u(\mathbf{r})|$ (u varies very rapidly near the core). Kane[161] has estimated that this $\mathbf{k}$-dependent term in Eq. (9.6) is one to two orders of magnitude smaller than the $\mathbf{k}$-independent term for InSb at the edge of the Brillouin zone. Therefore the $\mathbf{k}$-dependent term in Eq. (9.6) is usually neglected.

The two-thirds rule discussed above can be applied to the spin orbit splittings along the [111] direction for materials with germanium and zincblende structure. It has been shown[113] that the matrix element Δ for the valence band of these materials is approximately $\mathbf{k}$-independent for $\mathbf{k}$ along [111], in spite of the considerable mixing produced by the $\mathbf{k} \cdot \mathbf{p}$ term. For germanium-type lattices, the two-thirds rule also applies along the [100] direction but Δ is now strongly $\mathbf{k}$-dependent,[113] it must vanish by symmetry at the edge of the Brillouin zone. The $\mathbf{k} \cdot \mathbf{p}$ method has been used to calculate the effects of spin orbit coupling at a general point of $\mathbf{k}$-space in a parametrized fashion with only one adjustable parameter for each material of the germanium–zincblende family.[113] This parameter is adjusted so as to fit the most reliably known and experimentally identified spin orbit splitting of the material.

As mentioned earlier, materials of the germanium–zincblende family have spin orbit split M_1 critical points, usually referred to as E_1 and $E_1 + \Delta_1$, in the [111] direction. For most materials of the family, these critical points (see Figs. 9 and 29) seem to be inside the Brillouin zone, and hence there is another pair of M_0 critical points (e_1, $e_1 + \Delta_1$) at the edge of the zone. The E_1, $E_1 + \Delta_1$ critical points have been observed in the optical spectra of most materials of the family with the exception of silicon and diamond which, due to an inversion of the location of the Γ_{15} and $\Gamma_{2'}$ states, may not have E_1 critical points. The e_1, $e_1 + \Delta_1$ points are somewhat more elusive. They seem to have been observed by Potter[154] in germanium, slightly below the E_1, $E_1 + \Delta_1$ peaks, and in a number of III–V and II–VI[35] compounds, but their assignment is somewhat questionable since they are very weak. Spin orbit splitting should also appear in connection with the L critical point (see Fig. 29) $L_{3'} \to L_3$ (for germanium) or $L_3^v \to L_3^c$ (zincblende). Since both initial and final states have spin orbit splittings, the resulting structure (usually labeled E_1') should be a quadruplet. Actually, only a doublet may be resolved because the splitting of the conduction band state L_3 (or L_3^c) is roughly the same as that of the valence band state $L_{3'}$ (or L_3^v). Zincblende-type materials have also a spin orbit splitting of the X_5 valence band state (as mentioned earlier the corresponding splitting is zero for germanium). While this splitting may have been identified in some cases[35] (E_2 edge), the structure involved is weak and the experimental situation is not yet clear. The spin orbit splitting of the Γ_{15} conduction band state should also give optical structure. Such structure, which should appear in the $\Gamma_{25'} \to \Gamma_{15}$ transitions ($\Gamma_{15}^v \to \Gamma_{15}^c$), has not been unambiguously identified. Several orbital critical points, i.e., in the [100] direction, close to $\mathbf{k} = 0$, complicate the experimental picture and make such identification difficult. Table III lists the spin orbit splitting of the Γ_{15}^c, Γ_{15}^v, L_3^v and X_5 zinc blende states and those of the corresponding germanium states, as calculated by the $\mathbf{k} \cdot \mathbf{p}$ method.[103,113]

TABLE III. SPLITTINGS OF SEVERAL CONDUCTION AND VALENCE BAND STATES (IN eV) CALCULATED BY THE $\mathbf{k} \cdot \mathbf{p}$ METHOD[a,b] FOR GERMANIUM AND ZINCBLENDE-TYPE MATERIALS[c]

Material	Δ_0[d]	Δ_1[e]	Δ_{15}[f]	Δ_2[g]	$\Delta(L_3')$[h]	$\Delta(L_3)$[i]
Si	0.044	0.03	0.055	0	0.024	0.016
Ge	0.29	0.20	0.36	0	0.18	0.11
α-Sn	0.72	0.48	1.06	0	0.39	0.25
AlSb	0.75	0.41	0.24	0.34	0.34	0.09
GaP	0.10	0.073	0.22	0.019	0.067	0.073
GaAs	0.34	0.22	0.27	0.08	0.18	0.09
GaSb	0.80	0.46	0.39	0.32	0.38	0.12
InP	0.11	0.11	0.82	0.20	0.14	0.26
InAs	0.43	0.28	0.74	0.023	0.26	0.23
InSb	0.82	0.48	0.84	0.15	0.31	0.26
CdTe	0.92	0.51	0.63	0.30	0.44	0.22

[a] M. Cardona and F. H. Pollak, *Phys. Rev.* **142**, 530 (1966).
[b] F. H. Pollak, C. W. Higginbotham, and M. Cardona, *J. Phys. Soc. Japan Suppl.* **21**, 20 (1966); *Proc. Inter. Conf. Phys. Semiconductors*, Moscow 1968 (to be published).
[c] The numbers in boldface are experimental and have been used to fit adjustable parameters.
[d] Top valence band at $\mathbf{k} = 0$.
[e] E_1 and $E_1 + \Delta_1$ critical points at Λ.
[f] Lowest p-like conduction band at $\mathbf{k} = 0$.
[g] Top valence band at X.
[h] Top valence band at L.
[i] L_3 conduction band.

b. Band Structure and Dielectric Constant Calculations

In spite of the guidance of the spin orbit splitting, the complications mentioned above make identification of high symmetry critical points not always possible. Also, critical points may occur accidentally[162] at more general low-symmetry points; if they occur, the effect of the increased critical point multiplicity on the density of states makes the corresponding optical structure very strong. High-symmetry critical points can often be identified by studying reliable band structure plots along high symmetry directions of **k**-space. The identification of all possible general critical points is not an easy task since it requires a large number of such plots.

It is helpful to compute directly the dielectric constant ε_i (or the interband density of states, since optical matrix elements are often nearly constant) from the band structure by means of Eq. (2.13). In order to perform a calculation of ε_i, or of the density of states, an accurate band structure must be available. Calculations from first principles (such as the *O*rthogonalized *P*lane *W*aves or *A*ugmented *P*lane *W*aves methods) do not yield, in general, a good enough approximation to the experimental parameters (energy gaps) and hence parametrized calculations, with a few parameters adjusted to fit a few, well identified experimental gaps, are usually preferred. The pseudopotential method was the first one to be used for ε_i calculations[30,163] and it is particularly useful because of the small number of adjustable parameters involved: 3 Fourier coefficients of the pseudopotential are needed to give reasonable orbital energy bands for germanium type materials (without spin orbit coupling). For a zincblende-type semiconductor, the orbital bands are given by 3 additional parameters (Fourier components of the antisymmetric pseudopotential) plus the pseudopotential parameters of the isoelectronic group IV material. The 3 parameters of the group IV material can be adjusted so as to fit, for instance, the E_0 direct gap, the E_1 gap, and the E_1' gap, usually well identified. Once the pseudopotential parameters of the group IV materials are adjusted to fit 3 well identified gaps, the same gaps can be used to determine the antisymmetric pseudopotential parameters of the zincblende-type materials.

The full zone **k** · **p** method has also been successfully used[113,164] to calculate the energy bands throughout the entire Brillouin zone and the corresponding ε_i. While the number of parameters involved is larger than that for the pseudopotential method, spin orbit coupling

9. ASSIGNMENT OF OPTICAL STRUCTURE

can be easily introduced and the size of the Hamiltonian matrices to be diagonalized is somewhat smaller than for the pseudopotential method.

The technique of Fourier series expansion of the energy bands has been recently used for germanium and silicon.[165] The adjustable parameters are the first few expansion coefficients of the energy bands in three-dimensional Fourier series. Thirteen adjustable parameters are needed to represent reasonably well the important bands of germanium and silicon without spin orbit interaction.

Several methods have been used to compute ε_i once the energy bands are known. It is customary to compute the density of states and then multiply by an average matrix element of $\mathbf{p}$ since this matrix element does not vary very drastically for a given pair of bands. Also, the values of $|\langle|\mathbf{p}|\rangle|^2$ commonly used are essentially pseudopotential values and do not include core corrections,[43] which can amount to about 20% of $|\langle|p|\rangle|^2$. The multiplication of the contribution to the density of states of an element of constant energy surface $dS/|\mathbf{V}_k \omega_{kl}|$ [see Eq. (2.11)] by its appropriate matrix element $|\langle|\mathbf{p}|\rangle|^2$, however, does not pose any serious difficulties[163,165]; it does not, in general, introduce any serious changes in the calculated spectrum.[104,163]

The density of states can be obtained by diagonalizing the Hamiltonian at a large number of points in the Brillouin zone. We need only to consider the smallest volume from which the whole Brillouin zone can be generated by symmetry. These points can be either chosen to form a regular mesh[163] or they can be generated at random by a Monte Carlo method.[166] The density of states is obtained by counting the number of points with energies between ω and $\omega + \Delta\omega$ ($\Delta\omega$ is taken to be small but finite). Fluctuations due to a number of points moving through the boundaries of the $\omega - (\omega + \Delta\omega)$ shell are minimized by using Monte Carlo points, since the maximum number of points which can have the energy ω decreases when the points are determined at random.

The calculation of the density of states can be facilitated by the method of Gilat and Raubenheimer.[164,167] In this method, the Hamiltonian is diagonalized for all $\mathbf{k}$-points of a regular mesh (cubic for cubic materials) and the gradient of the energy, given by $\mathbf{V}_\mathbf{k} \omega = \langle\mathbf{p}\rangle$, is calculated at these points. The energy difference between bands is then approximated inside each cube (for cubic materials) by planes perpendicular to $\mathbf{V}_\mathbf{k}(\omega_c - \omega_v)$. All the planes inside each cube which correspond to evenly spaced values of $\omega_c - \omega_v$, e.g., each $(\omega_c - \omega_v) =$

0.01 eV, are constructed and the volume between a given set of contiguous planes is calculated. Summing all of the volumes between sets of contiguous planes which correspond to $\omega_c - \omega_v$ and $\omega_c - \omega_v + \Delta(\omega_c - \omega_v)$, we obtain the density of states at the energy $\omega_c - \omega_v$. The optical matrix element can be easily included in the ε_i calculation by multiplying it by the contribution of each cube to the density of states before adding all the contributions. The number of diagonalizations required can be reduced by fitting the energy bands to a quadratic expression instead of a plane.[113,166]

Figure 30 shows the results of ε_i calculations by Kane[162] and by

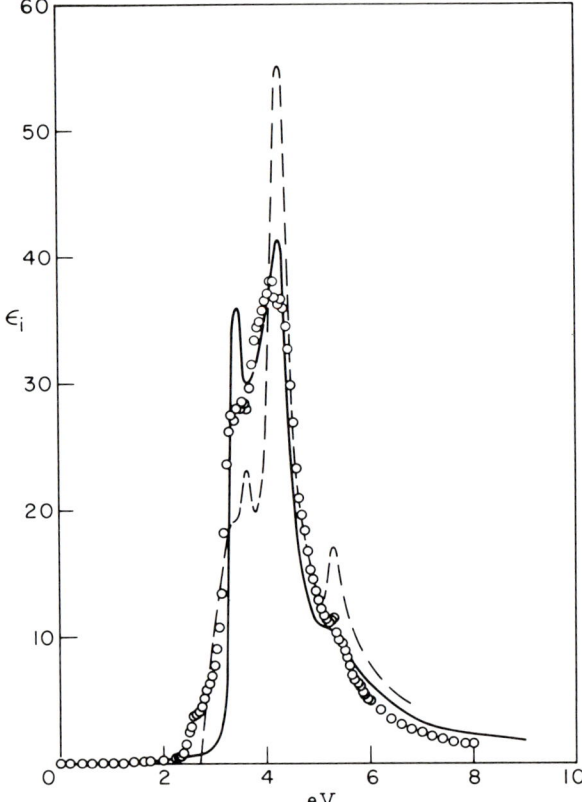

FIG. 30. Imaginary part of the dielectric constant of silicon calculated by G. Dresselhaus and M. S. Dresselhaus, *Phys. Rev.* **160**, 649 (1967) (circles) and by E. O. Kane, *Phys. Rev.* **146**, 558 (1966) (dashed line). The solid line represent experimental data of H. R. Philipp and H. Ehrenreich, *Phys. Rev.* **129**, 1550 (1963).

Dresselhaus and Dresselhaus[165] for silicon. Both calculations have used Monte Carlo generated points in **k**-space. It is interesting to point out that the main peak (E_2) at 4.3 eV does not correspond exclusively to critical point structure at high symmetry lines. In order to illustrate this fact, we have plotted in Fig. 31 the constant energy

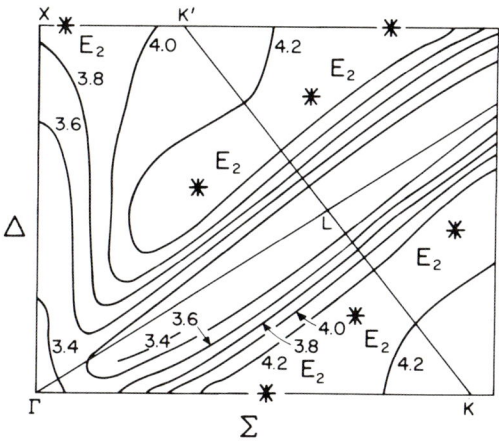

FIG. 31. Constant energy contours ($\omega_c - \omega_v =$ constant) of silicon in the (110) plane. [From E. O. Kane, *Phys. Rev.* **146**, 558 (1966).] Asterisks indicate critical points.

contours for the difference between a conduction band and a valence band in the (110) plane.[162] The asterisks labeled E_2 give the position of critical points; they all contribute to the E_2 optical structure but only one is in a high symmetry direction (Σ).

Figure 32 displays the real part of ε calculated by Dresselhaus and Dresselhaus[165] for germanium by applying the method sketched above to the interband part of Eq. (2.11). These authors found that the introduction of a finite broadening parameter ω_τ is necessary to fit the height of the experimental E_1 peak. The curve in Fig. 32 was calculated with an energy independent broadening parameter $\omega_\tau^{-1} = 2 \times 10^{-14}$ sec.

The curves in Figs. 30–32 were calculated for orbital bands without consideration of spin orbit interaction. As we have seen earlier, the spin orbit interaction is responsible for splittings in the optical structure of many cubic materials and is very helpful for its identification. The inclusion of its effect in the calculation of ε becomes imperative for atoms of large atomic number. The inclusion of spin orbit interaction, however, doubles the size of the Hamiltonian and hence greatly increases the amount of computational time required. Figure 33 shows ε_i for gray tin calculated by Higginbotham et al.[164] with inclusion of spin orbit effects. No experimental data on ε_i are

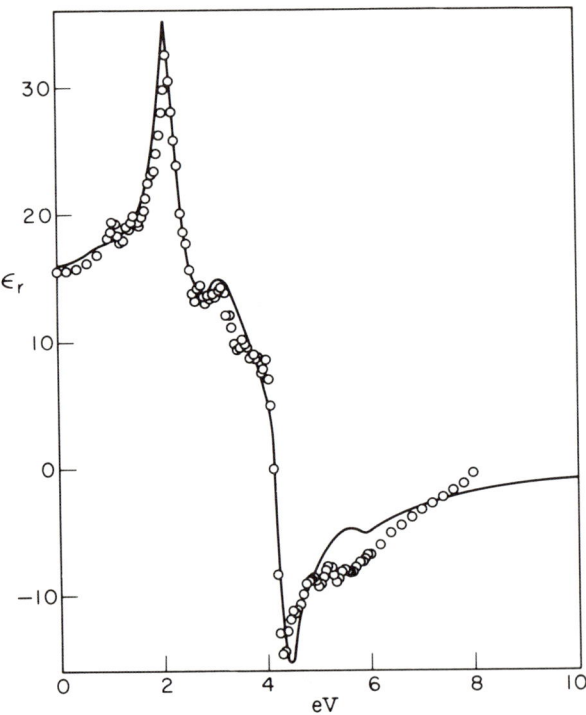

FIG. 32. Real part of the dielectric constant of germanium obtained by G. Dresselhaus and M. S. Dresselhaus *Phys. Rev.* **160**, 649 (1967) (dots) and experimental values (solid curve) obtained by H. R. Philipp and H. Ehrenreich, *Phys. Rev.* **129**, 1550 (1963).

available for gray tin and hence a comparison of Fig. 33 with experiment is not possible. The ε_i spectrum of gray tin, however, should be similar to that of InSb with all gaps shifted to lower energies. This is confirmed by comparing Fig. 33 to Fig. 1 of Philipp and Ehrenreich.[143]

As an example of another family of materials whose optical structure has been tentatively identified, in Fig. 34, we show the measured real and imaginary parts of the dielectric constant of PbS.[139] This material is cubic with rock salt structure; its properties and structure are similar to those of PbTe, PbSe, and SnTe. No direct calculation of ε_i is yet available for these materials but pseudopotential band structure calculations with inclusion of spin orbit effects have been performed[168] (see Fig. 35). On the basis of these calculations, and the obtained critical points along high-symmetry lines, the optical structure of Fig. 34 has been tentatively identified. We reproduce in

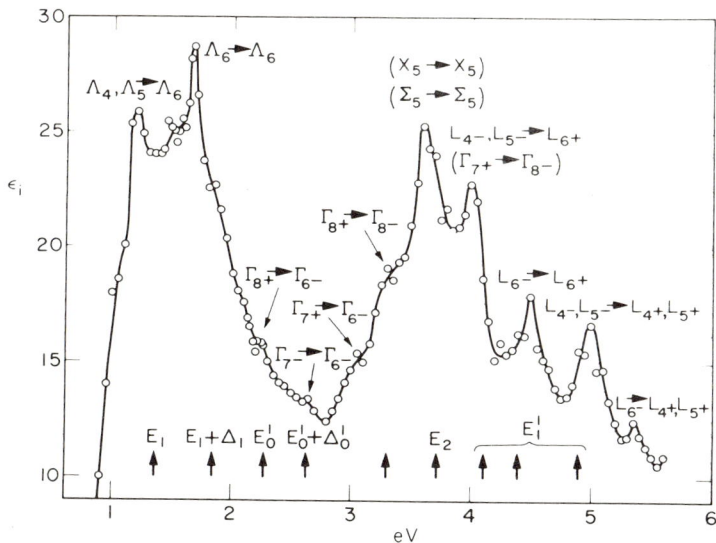

FIG. 33. Imaginary part of the dielectric constant of gray tin calculated with consideration of spin orbit interaction [see C. W. Higginbotham, F. H. Pollak, and M. Cardona, *Solid State Commun.* **5**, 513 (1967)].

Table IV the data of Table V of Lin and Kleinman.[168] with the identification of several optical gaps of the lead chalcogenides. Theoretical and calculated values of each gap are given. The gaps E_0, E_1, E_2, E_4, and E_5 were used for the adjustment of the five pseudopotential parameters in the calculation.

It is also possible to calculate, by the method described above, the various components of the dielectric constant tensor of anisotropic materials. Figure 36 shows the results of a calculation for antimony by Lin and Phillips[169] compared with experimental data.[139] Only the principal component of ε_i in the direction perpendicular to the optical (trigonal, c) axis is given; the principal component along the c-axis has also been calculated but it is difficult to measure, due to the existence of cleavage planes perpendicular to the c-axis. The pseudopotential procedure used by those authors to obtain the band structure of antimony is worth mentioning since it involves *no adjustable*

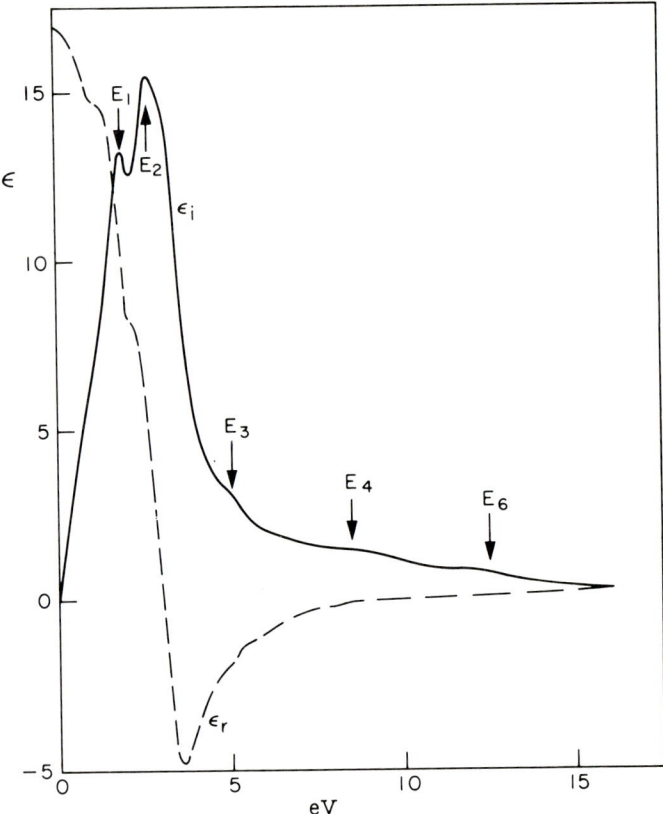

FIG. 34. Real and imaginary parts of ε for PbS, obtained from the Kramers–Kronig analysis of normal incidence reflection data [see M. Cardona and D. L. Greenaway, *Phys. Rev.* **133**, A1685 (1964)].

parameters. The required pseudopotential parameters of antimony are obtained by interpolation from the corresponding parameters of InSb (which were adjusted so as to fit the optical spectra of this material).[119,170] Since the germanium–zincblende materials cover a large fraction of the periodic table, and the pseudopotential parameters required to fit their spectra are fairly well known, we believe this method should prove to be very useful for calculating optical spectra of many materials without adjustable parameters.

Photoemission measurements[59,60,171,172] can also be helpful for the identification of optical structure: peaks in the spectral dependence of the photoelectric yield occur if the final state for the corresponding optical transitions lies above the vacuum level, while dips

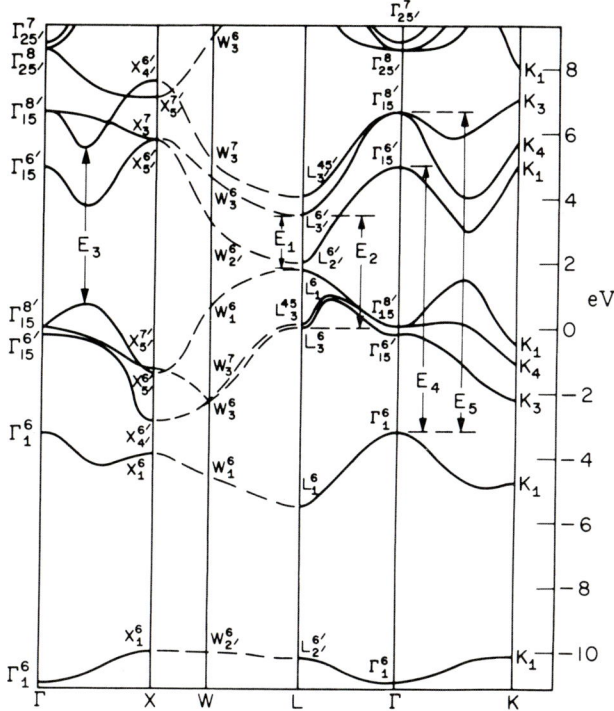

FIG. 35. Band structure of PbS. [From P. J. Lin and L. Kleinman, *Phys. Rev.* **142**, 478 (1966).] The solid lines were calculated at several points. The dashed lines are sketched between symmetry points.

occur if the final state lies below the vacuum level. Thus an idea of the position of the vacuum level with respect to the energy bands is obtained and certain possible assignments of the optical structure can be ruled out. Information can also be obtained by measuring the velocity distribution of the electrons emitted as a function of the incident photon energy. These experiments yield, in principle,[171] the energy of the final conduction states for a given incident photon energy. A choice of assignments can thus be made between two degenerate transition energies provided the final state energies are different. Unfortunately, the relationship between the phenomenon of

TABLE IV. ENERGY GAPS (IN eV) OF THE LEAD SALTS[a] COMPARED WITH THE POSITION OF SEVERAL SINGULARITIES IN THE REFLECTIVITY SPECTRA

		Calculated value			Experimental value		
		PbTe	PbSe	PbS	PbTe	PbSe	PbS
E_0	$L_1^6(2) - L_*^{6'b}$	0.19	0.16	0.28	0.19	0.16	0.285
E_1	$L_1^6(2) - L_{**}^{6'\,c}$	1.30	1.52	1.77	1.24	1.54	1.85
E_2	$L_3^6(1) - L_{**}^{6'\,c}$	2.52	3.34	3.49	2.45	3.12	3.66
E_3	$\Delta_1^6(3) - \Delta_5^6(2)$	3.59	4.61	5.29	3.50	4.50	5.30
E_4	$\Gamma_1^6(1) - \Gamma_1^6(2)$	6.30	7.33	8.15	6.30	7.10	8.11
E_5	$\Gamma_1^6(1) - \Gamma_{15}^8(2)$	7.87	8.96	9.85	7.79	9.10	9.80
E_6	$\Gamma_5^{7'}(1) - X_2^7(1)$	11.31	13.20	13.91	11.19	12.50	13.90

[a] Calculated by P. J. Lin and L. Kleinman, *Phys. Rev.* **142**, 478 (1966).
[b] $L_*^{6'}$ is $L_3^{6'}(1)$ for PbTe and $L_2^{6'}(2)$ for PbS, PbSe.
[c] $L_{**}^{6'}$ is $L_2^{6'}(2)$ for PbTe and $L_3^{6'}(1)$ for PbS, PbSe.

optical absorption and that of photoemission is not straightforward: transport effects and energy losses can occur between the time of excitation of the electron to the conduction band and that of emission to the vacuum.[173] It is not yet fully clear under what conditions such effects can be neglected.

Figure 37 shows the real and imaginary parts of the dielectric constant of copper as a function of photon energy.[38] At low energies (below 2 eV) the spectrum is dominated by free carrier intraband effects of the type described by Eq. (3.4). In order to identify interband effects, which occur at higher energies, it is convenient to subtract from the curves in Fig. 37 the intraband contributions. This can be done by fitting ε_i and ε_r in the low-energy region with Eq. (3.4) and subtracting Eq. (3.4) from ε_r and ε_i in Fig. 37.

Contributions to the interband sum of Eq. (2.10) come only from filled initial states k and empty final states l. For highly doped semiconductors and for insulators, the bands are either completely filled or completely empty and hence only critical point structure arises in the optical spectra. For metals, we have to cut the integral of Eq. (2.13) in the middle of a band as we go through the Fermi surface. Two typical cases can arise: (a) transition from a filled band to the neighborhood of the Fermi surface and (b) transitions from the neighborhood of the Fermi surface to an empty band. Sharp structure can

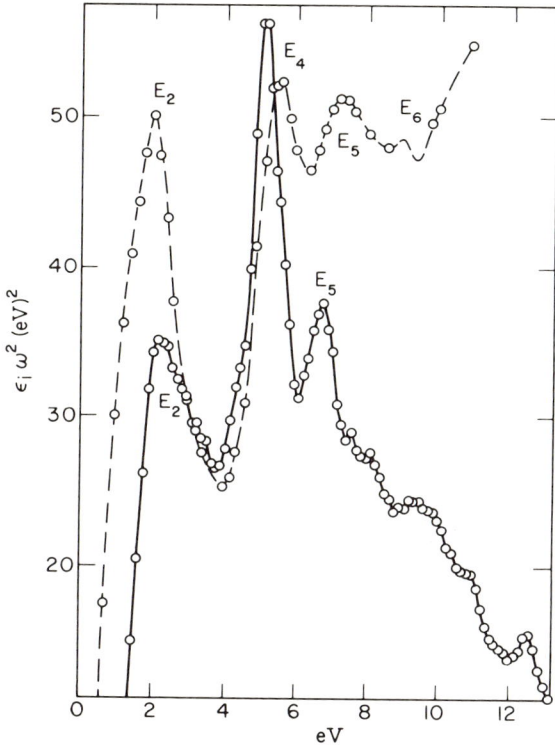

FIG. 36. Spectral dependence of ε_i for the ordinary ray in antimony. Dashed line represents experimental data. Solid line calculated by P. J. Lin and J. C. Philips, *Phys. Rev.* **147**, 469 (1966).

arise in case (a) when the filled band is nearly flat and in case (b) when the final empty band is nearly flat.[38,174] These cases and the corresponding ε_i spectra are sketched in Fig. 38. Because of the allowed–forbidden transition at the Fermi surface, steplike singularities arise which are similar to those in Fig. 1 for two-dimensional critical points.

A steplike singularity is obtained for transitions involving the Fermi

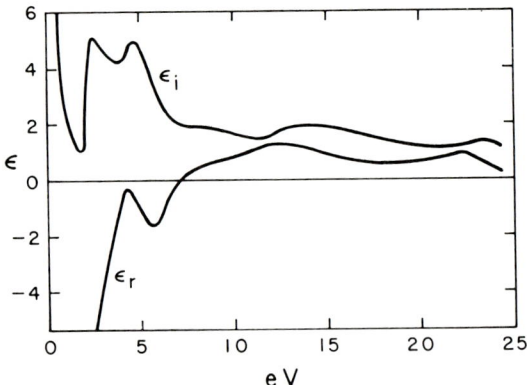

FIG. 37. Spectral dependence of ε_r and ε_i for copper obtained from the Kramers–Kronig analysis of the normal incidence reflectivity. [From B. R. Cooper, H. Ehrenreich, and H. R. Philipp, *Phys. Rev.* **138**, A494 (1965).] The data were taken at room temperature.

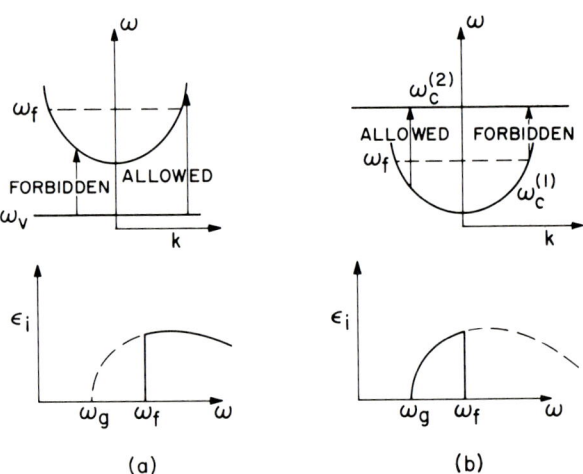

FIG. 38. (a) Transitions from a filled flat band to the neighborhood of the Fermi surface (b) and from such neighborhood to an empty flat band. The corresponding singularities in ε_i are also plotted.

9. ASSIGNMENT OF OPTICAL STRUCTURE

surface (Fig. 38) only if the Fermi surface coincides with the surface which is the locus in **k**-space of direct transitions, at a frequency ω. The Fermi surface is defined by $\omega_c(\mathbf{k}) = \omega_F$, while the locus of direct transitions at the frequency ω is:

$$\omega_c(\mathbf{k}) - \omega_v(\mathbf{k}) = \omega, \tag{9.6a}$$

or

$$\omega_c^{(2)}(\mathbf{k}) - \omega_c^{(1)}(\mathbf{k}) = \omega. \tag{9.6b}$$

Equation (9.6a) corresponds to transitions from a valence band to the neighborhood of the Fermi surface and Eq. (9.6b) to transitions from the Fermi surface to a higher conduction band. We assume, for the purpose of this discussion, that the surface described by (9.6a) is inside the Fermi surface for small ω and crosses it as ω increases. We have discussed earlier the case in which the surface of Eq. (9.6a) and the Fermi surface coincide at the crossing point: a steplike singularity is obtained for ε_i. This occurs when the ω_v band is nearly flat or when the surfaces $\omega_v(\mathbf{k})$ have the same shape as the Fermi surface. This is not the case in general: as ω increases, the surface of Eq. (9.6a) first touches the Fermi surface at several points equivalent by symmetry at an energy $\omega(\mathbf{k}_F)$ ($\mathbf{k}_F$ is the crystal momentum of a contact point). The volume outside the Fermi sphere corresponding to an energy $\omega(\mathbf{k}_F) + \Delta\omega$ is proportional to the total number of states N_a for which transitions are allowed:

$$N_a \propto (\Delta k)^2 \propto (\Delta\omega)^2. \tag{9.7}$$

The density of allowed states, and thus ε_i, is obtained by differentiating Eq. (9.7):

$$\varepsilon_i \propto (\Delta\omega) = [\omega - \omega(\mathbf{k}_F)] \tag{9.8}$$

Equation (9.8) shows that only a very weak singularity exists when the surface defined by Eq. (9.6a) touches the Fermi surface. The same argument applies to any other contact points and also to the transitions of Eq. (9.6b).

If the initial contact between Eq. (9.6a or b) and the Fermi surface occurs at a line, the singularity is of the form

$$\varepsilon_i \propto \Delta\omega^{1/2} = [\omega - \omega(\mathbf{k}_F)]^{1/2}; \tag{9.9}$$

correspondingly, a singularity similar to that at an allowed three-dimensional critical point is found.

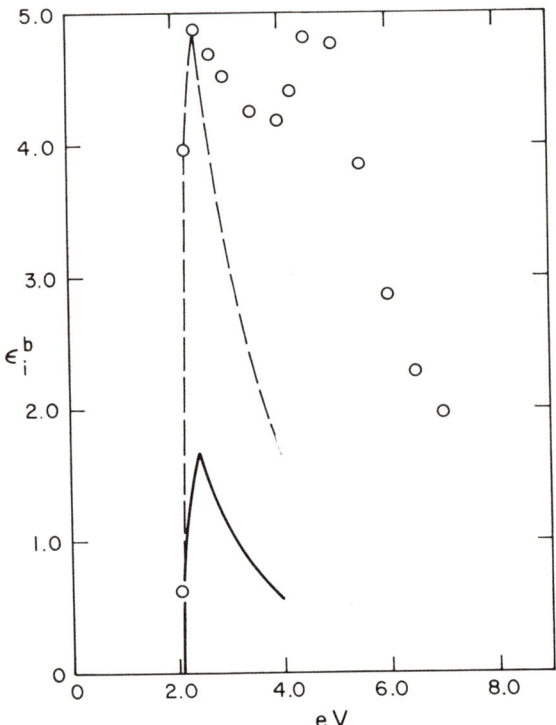

FIG. 39. Interband contribution to ε_i for copper, where the circles represent experimental points, the solid line represents absolute theory, and the dashed line is the theory scaled to the experimental amplitude. [From B. R. Cooper, H. Ehrenreich, and H. R. Philipp, *Phys. Rev.* **138**, A494 (1965).]

Figure 39 shows the interband contribution ε_i^b to ε_i obtained for copper from the data in Fig. 37 by the method described above. The peak near 2 eV has been assigned[174] to transitions of the variety described by Eq. (9.6a) between a nearly flat valence band (see Fig. 40) and a conduction band at the Fermi surface. The peak near 5 eV has been attributed[174] to an M_1 critical point at $X(X_5 \to X_{4'})$. Figure 39 also shows the results of calculations for the Fermi surface singularity based on the bands of Segall.[37] While the calculations reproduce well, the experimental line shape, an arbitrary scaling factor must be introduced in order to bring the experimental and calculated line intensities into agreement. The need for such a scaling fac-

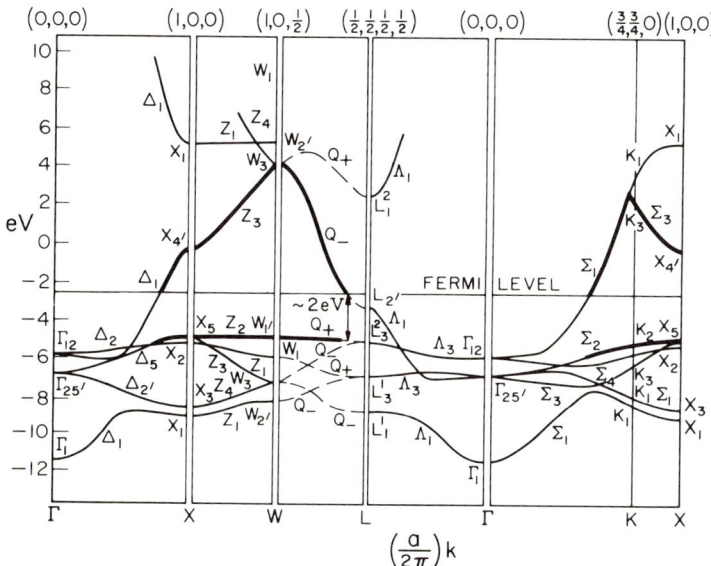

FIG. 40. Band structure of copper calculated by Segall. [From B. Segall, *Phys. Rev.* **125**, 109 (1962).] The thick lines are the bands which give rise to the optical structure of Fig. 39.

tor at the lowest interband absorption edge has been confirmed in recent calculations of ε_i from the band structure of copper by Mueller and Phillips.[175] Above this edge, however, the magnitude of the observed ε_i agrees well with that obtained from the band structure: many-body effects may be responsible for the enhancement of oscillator strength at the lowest interband edge.

III. Modulation Techniques

10. FUNDAMENTALS

The identification of a Van Hove singularity (see Section 4) in an optical absorption or reflection spectrum is not always an easy task. Singularities are usually superimposed on a broad, structureless background which is due to other transitions and they may often be lost in the noise of the background. This problem is particularly acute in the case of three-dimensional singularities since they do not produce infinite peaks in the optical structure but only changes in slope. Exciton effects may enhance the experimental structure but they may also reduce it, i.e., at an M_3 edge (see Section 6). Broadening will, in all cases, reduce the observed structure; as discussed in Section 7 this situation is only partially remedied by performing the measurements at low temperatures and by using very high quality crystals.

The optical structure associated with critical points can be greatly enhanced by means of modulation or derivative techniques. The foundations of these methods are as follows. As we have seen in Eq. (6.39), the dielectric constant near a three-dimensional critical point is:

$$\varepsilon = b(\omega - \omega_g)^{1/2} + \text{constant}. \tag{10.1}$$

The constant background in Eq. (10.1) may be much larger than the singular part and hence the observation of the singularity may be difficult since it is lost in the noise of the background. It would, therefore, be advantageous to measure, instead of ε (or ε_r and ε_i), the derivative of ε with respect to some as yet unspecified parameter ξ. The constant background would thus be eliminated and the singularity would become:

$$\frac{d\varepsilon}{d\xi} = \frac{b}{2}(\omega - \omega_g)^{-1/2}\frac{d(\omega - \omega_g)}{d\xi} + \frac{db}{d\xi}(\omega - \omega_g)^{1/2} \tag{10.2}$$

The first term of the derivative in Eq. (10.2) blows up at the critical gap ω_g and hence easy detectability of the singularity is expected. The term proportional to $(\omega - \omega_g)^{1/2}$ in Eq. (10.2) is usually negligible near ω_g. Two distinct possibilities for the derivative parameter ξ appear immediately: The frequency ω[67,176,177] and the energy gap ω_g. The derivative of ε (or any other optical constant) with respect to ω can be obtained by using, in the measurement of ε, a frequency modulated light beam. Let the dependence of the frequency of this light beam on time be:

$$\omega = \omega_0 + (\Delta\omega)\cos\Omega t. \qquad (10.3)$$

The dielectric constant ε is then given by:

$$\varepsilon = \varepsilon(\omega_0) + \Delta\varepsilon \cos\Omega t.$$

If $\Delta\varepsilon$ is measured, the derivative is immediately found with the equation:

$$d\varepsilon/d\omega = \Delta\varepsilon/\Delta\omega \qquad \text{for} \quad \Delta\omega \to 0. \qquad (10.4)$$

Hence, if $\Delta\omega$ is kept small and constant, $\Delta\varepsilon(\omega)$ is proportional to $d\varepsilon/d\omega$. The frequency modulation of the incident beam [Eq. (10.3)] can be accomplished by vibrating either the slit, or the grating, or a mirror in the monochromator[67,176] or another suitable arrangement.[177] We must keep in mind, however, that $\Delta\varepsilon$ is not measured directly; instead, the modulation in the transmissivity $\Delta\mathcal{T}$ (if the sample is transparent) or in the reflectivity ΔR are generally measured. Great accuracy and sensitivity in the measurement of ΔR and $\Delta\mathcal{T}$ is achieved by means of phase sensitive detection, i.e., with a lock-in amplifier; the modulation must be synchronous with its cause $(\Delta\omega)\cos\Omega t$. Great sensitivity is very desirable since an accurate measurement of a derivative requires a modulation amplitude as small as possible (see Chapter IV). As an illustration of the type of structure expected in a modulation spectrum we show in Fig. 41 the derivatives of ε_r and ε_i with respect to ω for the 4 kinds of three-dimensional critical points discussed in Section 4. The derivatives with respect to a gap modulating parameter ξ are obtained by multiplying the results of Fig. 41 by $-(d\omega_g/d\xi)$. Broadening and exciton effects have not been included but their qualitative influence on the line shapes of Fig. 41 can be easily inferred from the results of Sections 6 and 7. Whenever the critical point structure in ε amounts to only a small fraction of the total ε, the line shapes which appear in a reflectance modulation spectrum are a combination of the $d\varepsilon_r/d\omega$ and

$d\varepsilon_i/d\omega$ line shapes of Fig. 41 with coefficients β_r and β_i, according to Eq. (8.9). The coefficients β_r and β_i can be either positive or negative (see Fig. 26).

The frequency modulation experiments discussed above have a serious experimental difficulty. In a conventional reflection modulation experiment, the measured quantity is the reflected intensity $I = I_0 R$, where I_0 is the incident intensity. The measured derivative $dI/d\omega$ is:

$$dI/d\omega = I_0 \, dR/d\omega + R \, dI_0/d\omega. \tag{10.5}$$

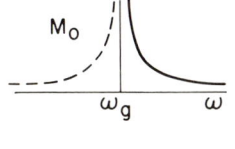
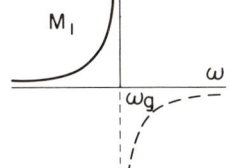
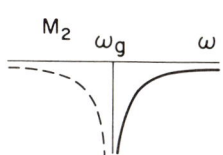

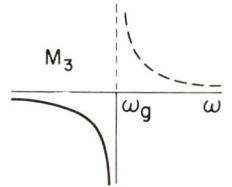

FIG. 41. Derivatives of the real and imaginary parts [$d\varepsilon_i/d\omega$ (solid curve) and $d\varepsilon_r/d\omega$ (dashed curve)] of the dielectric constant with respect to the frequency near a three-dimensional critical point.

Hence, we not only measure structure in $dR/d\omega$ but also the possible structure in the spectral distribution of the incident light. Since great sensitivity is desired, small structure in I_0 can yield spurious peaks in $dI/d\omega$ which are difficult to eliminate.

The difficulty mentioned above does not arise in the measurement of the derivative of the optical constants with respect to the energy gap ω_g. This measurement can be accomplished by applying to the sample a sinusoidally varying external parameter which modulates the energy gap (square wave modulation is sometimes used). Among the possible parameters ξ which change energy gaps we shall mention hydrostatic pressure and uniaxial stress.[62,63,178] The derivative measurement yields:

$$d\varepsilon/d\xi \propto (\omega - \omega_g)^{-1/2} \, d\omega_g/d\xi. \tag{10.6}$$

[the proportionality constant in Eq. (10.6) contains effective masses, and therefore can depend on ξ. The corresponding term in Eq. (10.6), however, has a $(\omega - \omega_g)^{1/2}$ dependence and therefore can be neglected.] The derivative can also be measured by means of phase sensitive methods as discussed above. Structure which is weak in the dc spectra can be strong in a derivative spectrum if $d\omega_g/d\xi$ is large, and vice versa, strong optical structure can be weak in the derivative spectrum if $d\omega_g/d\xi$ is small. Modulation by means of uniaxial stress changes the symmetry of the crystal and hence the derivative $d\omega_g/d\xi$ in Eq. (10.6) has tensorial character even for a cubic material. A study of the various tensor components (with polarized light) yields, in general, information about the symmetry of the critical points involved. We should recognize, however, that optical measurements in which the modulating agent acts on the sample (sometimes called internal modulation measurements) despite their experimental simplicity are not as simple in their theoretical interpretations as their wavelength modulation counterparts. The theory of a wavelength modulation spectrum involves only the theory of the optical constants. The theory of an internal modulation experiment involves not only the theory of the optical constants but also that of the effect of the perturbation on those optical constants.

We have limited the discussion above to allowed three-dimensional interband critical points. It is obvious that enhancement of structure by the derivative method also occurs for other types of critical points. In particular, one-dimensional singularities (interband transitions in a magnetic field) give [see Eqs. (4.6) and (4.13)] derivative singularities of the form[57,58]:

$$d\varepsilon/d\xi \propto (\omega - \omega_g)^{-3/2} d\omega_g/d\xi, \qquad (10.7)$$

considerably stronger than the structure in the ordinary spectra. Hence, as discussed in Chapter VIII, modulation methods are also very useful for the study of magnetooptical effects.[69,70] It is also worthwhile to consider the modulation spectra of indirect transitions: only a change in slope should be seen in these modulation spectra [Eq. (5.1)] if exciton effects are neglected. However, the inclusion of exciton interaction transforms the $(\omega - \bar{\omega}_g)^2$ shape of band-to-band indirect transitions into the shape of direct allowed transition (Fig. 41): sharp peaks should be seen at energies $\bar{\omega}_g = \omega_g - W_l \pm \omega_{\text{phon}}$, with the minus sign for phonon absorption and plus for phonon emission. The phonon absorption terms will usually be missing since exciton effects are often observed only at low temperatures.

It is also of interest to consider modulation effects of intraband transitions near the plasma frequency. (Similar effects are also produced near the plasma frequency of a group of valence electrons in filled bands.) Equation (3.2) gives the dielectric constant of a semiconductor or a metal well below the lowest interband gap, neglecting lifetime broadening. From Eqs. (3.2) and (8.5) we obtain:

$$\frac{dR}{d\omega_p} = \frac{4(1-n)}{n(n+1)^3} \frac{\omega_p}{\omega^2} \varepsilon_L. \tag{10.8}$$

with $n^2 = \varepsilon = \varepsilon_L(1 - \omega_p^2/\omega^2)$. $dR/d\omega_p$ can be easily measured by modulating the density of the material (hydrostatic pressure or temperature modulation). The behavior of R and $dR/d\omega_p$ is shown in Fig. 42 (solid curve). An infinite peak is seen for $dR/d\omega_p$ at the plasma

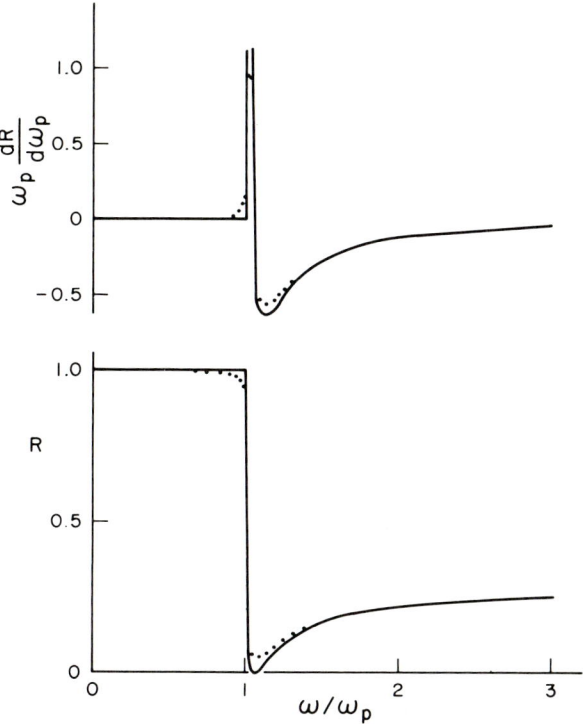

FIG. 42. Reflectivity R and $\omega_p \, dR/d\omega_p$ near the plasma frequency ω_p of a semiconductor or a metal. We have taken the nondispersive "lattice" dielectric constant $\varepsilon_0 = 10$. The solid lines correspond to infinite collision time. The dotted curves represent the effect of broadening.

frequency. The dashed curves indicate qualitatively the effect of broadening. Owing to the asymmetry of the $dR/d\omega_p$ line near ω_p, broadening shifts the maximum toward higher energies (similar effects are observed for square root singularities, see Chapter IV).

The application of a modulating stress transforms a three-dimensional crystal with a given translational lattice into another three-dimensional crystal with a perturbed translational lattice. If direct transitions were mainly responsible for the optical properties before the application of the stress, they remain so after its application and

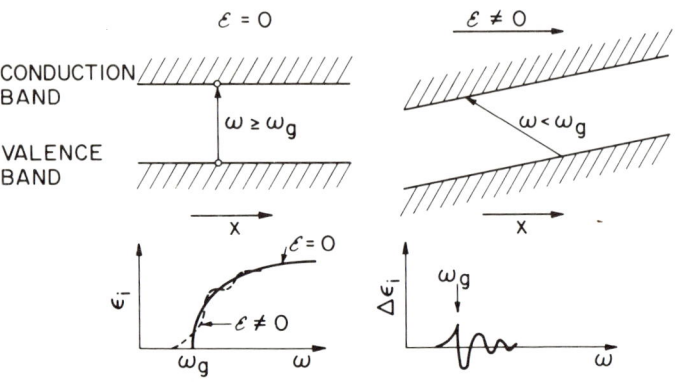

FIG. 43. Effect of an electric field $\mathscr{E}$ on interband transitions near an M_0 energy gap. The upper drawings represent the "local" energy gap as a function of the space coordinate along $\mathscr{E}$.

hence an equation of the form (10.1) remains valid with slightly perturbed values of b and ω_g. Equation (10.2) follows immediately. It is also possible to apply to the crystal a perturbation which destroys the translational invariance along, at least, some directions of real space. This occurs when an electric[61,71] or a magnetic field is applied: the electric field $\mathscr{E}$ destroys the translational symmetry of the crystal, at least along the direction of the applied field, since it adds to the Hamiltonian a potential energy of the form $-e\mathscr{E} \cdot \mathbf{r}$ (for a uniform field), which is not translationally invariant. The effect of adding a small uniform field is illustrated in Fig. 43; it can be treated by assuming that the perturbing field is much smaller than the crystalline fields and hence a local band structure exists at every point of real space. The term $-e\mathscr{E} \cdot \mathbf{r}$ shifts the origin of energy with $\mathbf{r}$ as indicated in Fig. 43. Transitions between the valence and the conduction band are now possible for energies smaller than ω_g, although their probability is small since the electron is brought from one point in space to another: The transition probability depends on a wavefunction overlap. The modification of ε_i by the electric field near an M_0 critical point is also sketched in Fig. 43. The theory of ε_i in the presence of an electric field will be discussed in detail in Chapter VII.

A similar situation obtains for a perturbing magnetic field: the translational symmetry operations are lost, except those along the

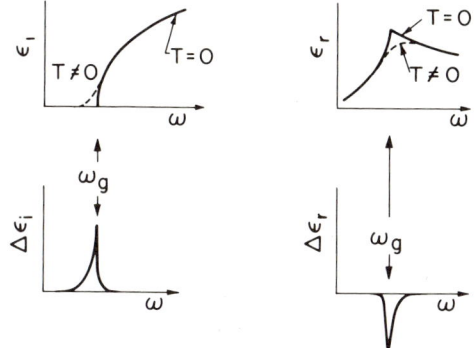

FIG. 44. Contribution of the phonon broadening to a temperature modulation spectrum near an M_0 critical point.

direction of the applied magnetic field. The gap does not disappear, but a three-dimensional critical point in ε is now transformed into a collection of one-dimensional singularities (Landau levels). The modulation spectrum obtained by application of a sinusoidally varying magnetic field is, therefore, not a derivative spectrum in the strict sense.

Temperature modulation[64,66,179] does not yield a strict derivative spectrum either. An increase in temperature can be decomposed into a dilation effect (equivalent to a hydrostatic stress) and a change in the phonon occupation numbers. The dilation effect is equivalent to a pure change in lattice constant and hence produces a strict derivative spectrum with respect to the gap. The change in the phonon occupation numbers also changes the amount of indirect transition allowed and, hence, produces a broadening as much as a shift in the energy gaps (see Section 7). The effect of the broadening is sketched qualitatively in Fig. 44 for ε_i near an M_0 edge. A more quantitative treatment of broadening modulation is given in Chapter V. Figure 44 suggests that while temperature broadening does not yield a strict gap derivative spectrum, it nevertheless achieves our objective of enhancing critical point structure and eliminating structureless background. As we shall see in Chapter V, the broadening produced by an increase in temperature is usually smaller than the corresponding gap shifts and therefore nearly true energy gap derivative spectra are often obtained by modulating the sample temperature.

Electric field modulation, while not a strict gap derivative method, also produces a great enhancement of the critical point structure, as shown in Fig. 43. This conclusion applies to almost any kind of nonderivative optical modulation experiment. A heuristic proof of this important conclusion, which applies to almost any kind of modulating perturbation, can be given as follows: consider the second-order perturbation produced by the modulating agent on a state via intermediate states in the same band. The perturbation due to intermediate states at higher energies will usually cancel that due to lower energy states if the density of states is smooth. Near a critical point an abrupt change in the density of states occurs and hence cancellation of second-order perturbation terms does not take place. The effect of the perturbation, is therefore expected to be large near critical points and small otherwise.

We have discussed above how to obtain the first derivative of an optical constant by means of phase sensitive detection. It is also possible to tune the lock-in amplifier to a harmonic of the modulating signal (second, third, etc.) using a frequency doubler to provide the reference signal for the lock-in amplifier. By this method the second or higher derivatives of the optical constants are obtained. Improved resolution with respect to the first derivative can sometimes be reached in higher derivative spectra. Whenever the response is an *even* function of the modulating signal, no signal is obtained at the fundamental frequency and at all odd harmonics. The measurement of the second harmonic, as described above, is indicated in such cases. An example is the temperature modulation produced by Joulean heat: the temperature is, for small amplitudes, a quadratic function of the modulating voltage for a purely ac modulation. A signal at the fundamental frequency can be obtained, however, by adding a dc bias to the ac modulating current.

11. General Techniques

A typical optical modulation setup has usually the following components:

(a) Light source,
(b) Monochromator,
(c) Detector,
(d) Lock-in amplifier(s) and modulation oscillator,
(e) Data reducing equipment and recorder,
(f) Modulating unit.

Components (a)–(e) are common to most modulation measurements while, obviously, (f) is a specific feature of each method. We shall discuss here components (a)–(e), the common features, and we shall leave those specific of each modulating method for subsequent sections.

As mentioned earlier, in order to obtain "true" derivative spectra, small modulations have to be employed. Actually when the gaps are modulated, only small modulations are usually obtained since the application of a large modulating agent (electric field,† uniaxial stress) results in the destruction of the sample.[180] It is therefore important to reduce the noise level, which is, in the best of the cases, photon shot noise proportional to the square root of the light intensity. Since the signal is proportional to the light intensity I the signal-to-noise ratio is proportional to $I^{1/2}$. Therefore, strong light sources are required for good signal-to-noise ratios. As we shall see later, source stability is not a very important consideration and hence high pressure xenon arcs are the most commonly used sources of radiation. Figure 45 reproduces the spectrum of a 2500 watt Hanovia xenon arc†† as measured with a Perkin–Elmer model 4000 spectrometer. This spectrum is relatively line free and smooth from 2500 Å (5 eV) to 8000 Å (1.5 eV). Its use becomes difficult beyond 8000 Å since in this region the spectrum is very rich in lines. These lamps are actually usable in the ultraviolet to 2000 Å. For the near infrared (line region of Fig. 45) intense tungsten filament lamps are

---

† Strong reflectance modulations (of the order of 50%) are sometimes obtained with electric fields on ferroelectric materials.[180]

†† Hanovia–Engelhardt Inc.

quite appropriate. In particular quartz envelope tungsten–iodine chemical transport lamps (of the type available for indoor photography) are commonly used. They cover the spectral range from 3500 to 25,000 Å. For longer wavelengths, conventional infrared sources (globar, Nernst glower) have to be used. Because of their low intensity, modulation measurements become difficult and not very sensitive in

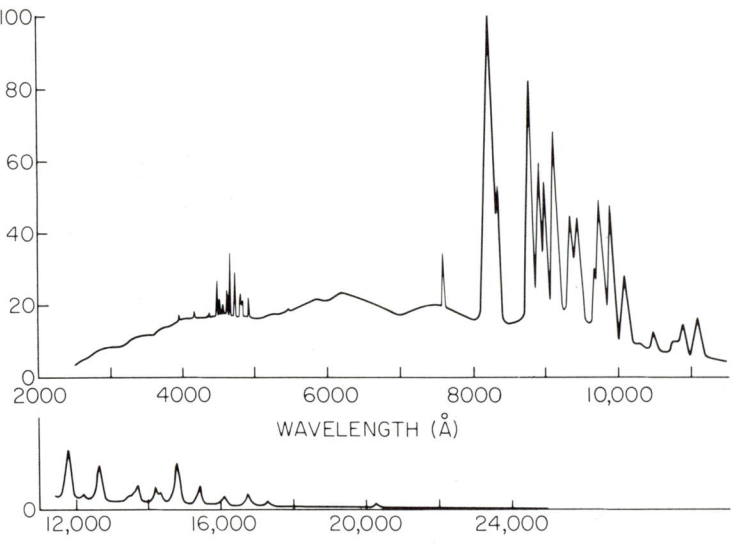

FIG. 45. Spectral distribution of a Hanovia 2500 xenon arc (measured by the Eppley Laboratories, Newport, R. I., with a Perkin–Elmer model 4000 spectrometer). Courtesy of Englehard–Hanovia, Inc.

this region. Similar problems arise in the ultraviolet for $\lambda < 2000$ Å. Hydrogen arcs, while relatively weak, give a continuous spectrum down to 1650 Å. The hydrogen line spectrum for $\lambda < 1650$ Å is strong but difficult to use. However, modulation measurements in this region have been performed by Scouler.[179] The possible use of the microwave continua of the rare gases[181] beyond 1650 Å should be explored. The solution of the light source problem in the vacuum ultraviolet may well be the use of synchrotron radiation.[182]

We shall not discuss in detail the question of monochromators for modulation work since it does not present any specific problems. The light intensity considerations made above indicate that they should have as high an aperture as possible.

Photomultiplier detectors have been used in the region from 11,000 Å to the ultraviolet. Cooling does not ordinarily improve signal-to-noise since we usually operate at light levels much higher than that equivalent to the dark current. PbS and PbSe cells have been used in the infrared beyond 11,000 Å. Cooled PbSe cells (77°K) can be used to a wavelength of 6 μ. Because of their relatively small room temperature dark resistance, cooling sometimes improves the signal-to-noise ratios obtained with PbS and PbSe detectors.

Let us, for the time being, assume we use a photomultiplier as a detector in a reflection modulation experiment. The photomultiplier output can be split into two channels: a dc channel, with a signal proportional to $I_0 R$, and an ac channel proportional to $I_0 \Delta R$ where ΔR is the amplitude of the modulation in the reflectance. These two channels can be separated: the dc signal can be amplified with a dc amplifier (such amplification may not be necessary) while the ac signal is sent to a lock-in amplifier. If we divide the dc output of the lock-in by that of the dc amplifier, we obtain $I_0 \Delta R / I_0 R = \Delta R / R$ and the incident light intensity is eliminated. As mentioned in Section 8c, the measurement of I_0 is one of the main sources of error in conventional optical experiments. Problems associated with the stability of the source are avoided, also, when the division mentioned above is done with an "on-line" computer element. If a photoconductive cell is used as a detector, the dc channel cannot be easily extracted since the cells have a rather low and unstable dark resistance. It is then possible to extract $I_0 R$ by chopping the light at a frequency different from that of the modulation (preferably higher) or any of its subharmonics. Two lock-in amplifiers must then be used to retrieve $I_0 \Delta R$ and $I_0 R$. Since normally $I_0 R$ is much higher than $I_0 \Delta R$, care must be taken to prevent the saturation of the lock-in which measures the modulated signal $I_0 \Delta R$. This can be accomplished with the use of filters of high rejection ratio.

Several methods have been used to perform the division of $I_0 \Delta R$ by $I_0 R$. It is not easy to construct a simple electronic analog circuit to perform this division,[183] especially when linearity over a broad dynamic range is required. A possible analog system[78] is obtained by feeding the signals from the amplifiers to two logarithmic converters and the outputs of the converters to a differential amplifier so as to obtain $\log \Delta R / R$.

An inexpensive and simple dividing system can be built[71] for photomultiplier detectors. The system consists of a servo which varies the high voltage applied to the photomultiplier so as to keep the dc output constant. If the dc output is kept equal to 1 V, the output from the lock-in which measures the modulation is equal to $\Delta R/R$. If a different reference voltage is used, the output is $\Delta R/R$ modified by the appropriate scaling factor. The high voltage is

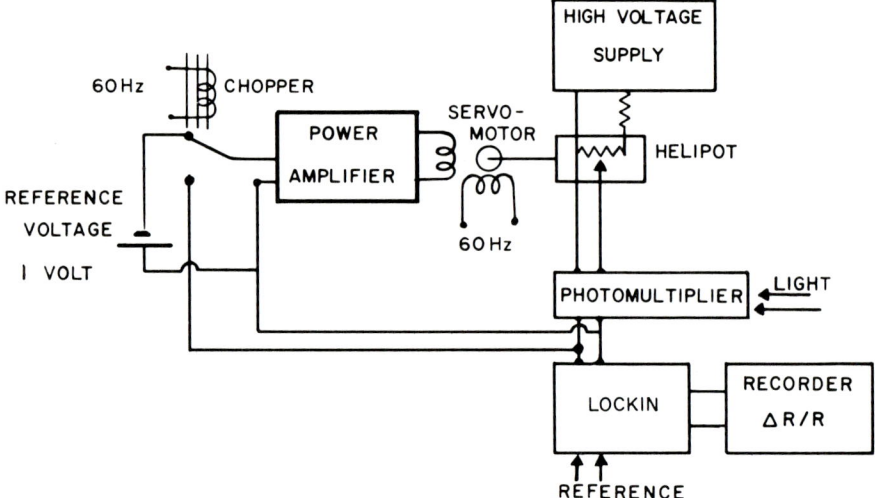

FIG. 46. Schematic diagram of the setup used to perform the $I_0 \Delta R/I_0 R$ division by varying the high-voltage applied to the photomultiplier.

varied by having the servo act on a potentiometer (helipot) connected as a voltage divider. The servomotor is activated by the difference between the reference voltage (1 V) and the dc output. A schematic diagram of the system is drawn in Fig. 46. The servo of an old x-t recorder can be used for this purpose. This servo system has the advantage of true linearity over an infinite dynamic range of light intensities. The high voltage can also be varied with an electronic rather than a mechanical servo. A diagram of such a system is shown in Fig. 47.[184]

The dividing methods described above cannot be applied to PbS and PbSe cells: a variation of the cell voltage produces large transients which make the servo unstable. It is, however, possible to link the servo to the spectrometer slits[62] (or to the power supply of the light source for tungsten lamps) so as to keep the cell output $I_0 R$ constant. An ingenious method has been used by Engeler et al.[65] for measurements with linearly polarized light. The polarized beam is intercepted by a second polarizer which is rotated by the servo motor. In this manner, the light intensity is controlled so as to balance the servo.

Another simple potentiometric method ($\Delta R/R$ exact over an infinite dynamic range) to effect the division is obtained (in principle) if

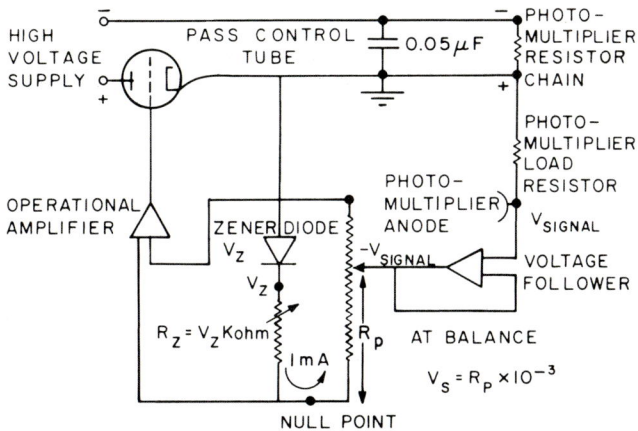

FIG. 47. Electronic "Servo" for performing the $I_0 \Delta R/I_0 R$ division. [From C. H. Anderson, Private communication.]

we replace the standard battery of an *x-t* strip chart recorder by the signal $I_0 R$ from the dc amplifier. Most *x-t* recorders, however, require a double ended (not grounded) standard battery. When this battery is replaced by the divisor $I_0 R$, one of the ends is usually put to ground. This makes the zero offset in the recorder inoperative[185] and, hence, only values of one sign of $\Delta R/R$ can be recorded. This can be corrected by means of the circuit of Fig. 48. The pen displacement is proportional to $A + (V_a/V_b)$ with the offset A a function of R_c, provided $R_a \gg R_b$. This condition is easily satisfied when V_a is a constant current source and V_b a constant voltage source. This method has been found particularly suitable for use with photoconductive cells.

Sinusoidal modulation is commonly used for modulation experiments. This type of modulation is quite suitable when true derivatives with respect to either ω or ω_g are measured (stress modulation, wavelength modulation). However, when electric or magnetic field

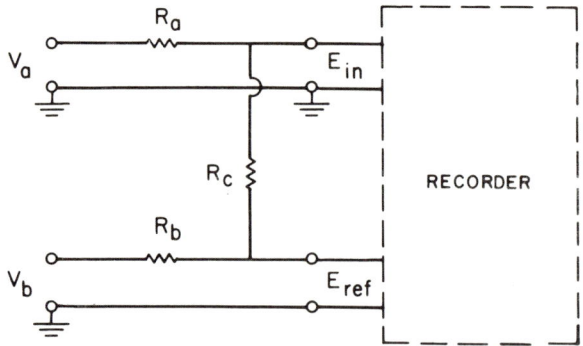

FIG. 48. Basic servo offset ratio recording system [from D. E. Aspnes, *Rev. Sci. Instr.* **38**, 1663 (1967)]. This circuit gives a pen displacement proportional to $A + (V_a/V_b)$ (A is the offset) provided $R_a \gg R_b$.

modulation is used, the modulation is a nonanalytic function of the modulating parameter and hence, even for small sinusoidal modulations, the output is an average of the outputs for each field in a cycle (dominated, of course, by the maximum). As a result, structure can be washed out and line shapes and intensities may be difficult to interpret. It is therefore advisable, in such cases, to operate with square wave modulation and thus measure the effect of a perturbation constant in time. The fact that the lock-in detects only the fundamental sinusoidal component of the square wave does not change this conclusion.

12. Modulation Techniques and Band Structure Dependence on Static Parameters

Optical measurements have been very useful for the study of the variation of the band structure with static external parameters. Optical measurements as a function of temperature,[33] hydrostatic pressure,[53] uniaxial stress,[160] concentration in mixed crystals,[186] and impurity concentration[187] have yielded the variation of energy gaps with the applied parameters. The coefficients of the variation of gaps with external parameters have very often been instrumental in the identification of transition.[53] At first glance it seems that the modulation methods may provide a very good way of determining these coefficients.[188] Nevertheless, while it is easy to determine accurately the magnitude of a static perturbation, it is often not easy to calibrate the magnitude of a sinusoidal perturbation. This is particularly true of temperature modulation experiments and of the stress modulation produced with an ultrasonic transducer. It is, however, possible to use a modulation method to enhance and sharpen optical structure at gaps and then to apply a static perturbation to measure gap shifts. Balslev[67] has used wavelength modulation and a static uniaxial stress to derive deformation potentials (strain coefficients of gaps) and also a combination of modulated stress (to sharpen structure) and static stress (to shift peaks).[189] Electric field modulation has been successfully used for studying composition dependence of gaps in alloy systems[190] and dependence of gaps on impurity concentrations.[71]

Temperature coefficients of gaps have been obtained by means of modulation techniques by a number of authors.[191] The determination of gap shifts with a static perturbation by means of modulation techniques will be discussed in Chapter VIII.

The application of a strong magnetic field to a solid produces drastic changes in its optical spectra. Oscillations in the optical constants due to Landau quantization appear near critical points.[57,58] The observation of these oscillations in the static optical spectra has been the object of many studies which yield very accurate values of energy gaps, effective masses, and g-factors.[192] The application of a large modulating magnetic field presents serious difficulties since photodetectors (photomultipliers in particular) are strongly affected by the modulating field. While this problem could be solved by carefully shielding the detector, no magnetic modulation data† have appeared in the literature.[193] Other modulation techniques used in connection with a static magnetic field enhance considerably the magnetooptical structure. Magnetoelectroreflectance,[70] magnetopiezoreflectance,[69] and magnetopiezotransmission[127,194] have been successfully used to determine masses, gaps, and g-factors for a number of semiconductors. These measurements will be discussed in Chapter VIII.

13. Kramers–Kronig Analysis

While it is possible to interpret a reflectance modulation spectrum with Eq. (8.9) if the optical constants of the material and the shapes of $\Delta\varepsilon_r$ and $\Delta\varepsilon_i$ are known,[149] it is usually more desirable to perform a

---

† It has been pointed out that magnetic field modulation is equivalent to modulating circularly polarized light from right-to-left circular polarization.[193]

Kramers–Kronig analysis[71,91,178] of $\Delta R/R$ and to extract from it $\Delta \varepsilon_r$ and $\Delta \varepsilon_i$. The former method involves a synthesis of the $\Delta R/R$ spectrum using the known values of β_r and β_i (determined, usually, by a Kramers–Kronig analysis of a normal incidence reflection spectrum) and the theoretical line shapes of $\Delta \varepsilon_r$ and $\Delta \varepsilon_i$. It therefore relies upon an *a priori* theoretical knowledge of these lines shapes.[149] The Kramers–Kronig analysis of ΔR yields the experimental line shapes of $\Delta \varepsilon_r$ and $\Delta \varepsilon_i$ and hence it gives more information for the theoretical analysis; both $\Delta \varepsilon_r$ and $\Delta \varepsilon_i$ must agree with theoretical shapes. Because of the qualitative nature of the agreement which one often finds, having two spectra to fit ($\Delta \varepsilon_r$ and $\Delta \varepsilon_i$) often helps and even proves decisive in sorting out the possible line shapes, i.e., types of critical points.

The Kramers–Kronig analysis of $\Delta R/R$ is performed by obtaining the modulation $\Delta \theta$ in the phase angle θ of Eq. (8.6). Since $\Delta \theta$ is small, we use the differential form of Eq. (8.7):

$$\Delta\theta(\omega) = \frac{\omega}{\pi} P \int_0^\infty \left[\left(\frac{\Delta R(\omega')}{R(\omega')} - \frac{\Delta R(\omega)}{R(\omega)} \right) d\omega' \bigg/ (\omega'^2 - \omega^2) \right]. \quad (13.1)$$

The integral of Eq. (13.1) is performed with a computer as discussed in Section 8. Extrapolation techniques can be used to extend $\Delta R/R$ beyond the experimental range. However, due to the sharpness of the $\Delta R/R$ lines, contributions of the extrapolation region to $\Delta \theta$ are nonsingular and often limited to the immediate neighborhood of a critical point. Hence the extrapolation is not necessary (except for metals at low frequencies, see Section 3). $\Delta \varepsilon_r$ and $\Delta \varepsilon_i$ are obtained with the expressions[71,178]:

$$\Delta\varepsilon_r = \frac{n_r}{2n_L}(n_r^2 - n_L^2 - 3n_i^2)\frac{\Delta R}{R} + \frac{n_i}{n_L}(3n_r^2 - n_L^2 - n_i^2)\Delta\theta,$$

$$\Delta\varepsilon_i = \frac{n_i}{2n_L}(3n_r^2 - n_L^2 - n_i^2)\frac{\Delta R}{R} + \frac{n_r}{n_L}(3n_i^2 + n_L^2 - n_r^2)\Delta\theta, \quad (13.2)$$

where n_r and n_i are the real and imaginary parts of the refractive index of the sample respectively and n_L is the real refractive index of the transparent medium surrounding the sample. In many modulation techniques $n_L = 1$ but in others (electrolyte method of electroreflectance) $n_L > 1$. In these cases Eq. (8.6) may not be valid but it can be shown that Eq. (13.2) is usually still applicable.[91]

IV. Wavelength Modulation

14. EXPERIMENTAL TECHNIQUES

The use of wavelength derivative spectra in molecular spectroscopy has been the object of considerable attention[195-197] since multiplet structure is more easily recognized in derivative spectra than in the corresponding conventional spectra. For instance, structure which appears only as a change of slope in the conventional optical spectra, appears as a step in the first derivative and as a peak in the second derivative. As an illustration, we show in Fig. 49 the absorption spectrum of the 4.25 to 4.28 μ CO_2 doublet and the first and second derivative spectra obtained by Collier and Singleton.[195] While the detection of the doublet in the nonderivative spectrum requires a rather trained eye, the doublet is resolved in the first derivative and even better in the second derivative.

Several of the systems proposed to obtain wavelength derivative spectra[195,196] consist essentially of an on-line digital or analog computer which performs a differentiation with respect to time while the wavelength is swept linearly in time. Hence they do not have the noise rejection features of the modulation schemes based on phase sensitive detection and they are, in principle, equivalent to performing a derivative after the nonderivative spectrum has been taken. We shall therefore not discuss these schemes in detail here.

In order to obtain a wavelength derivative spectrum by a modulation method one requires a wavelength modulated monochromatic beam. Several schemes for obtaining wavelength modulated monochromatic light have been reported.[67,198,199] Some of these methods involve the mechanical vibration of an entrance or exit slit[67] or, equivalently, the vibration of a mirror inside the monochromator.[198] The quantity directly obtained in any of these methods is the derivative of the reflected or transmitted light I with respect to the

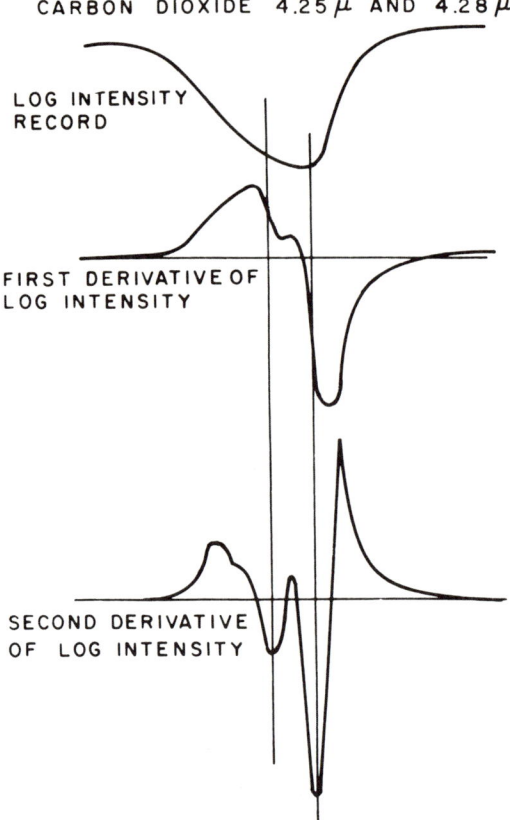

Fig. 49. First and second derivative spectra of a doublet which is just resolved together with the conventional absorption spectrum, obtained with a derivative spectrometer. [From A. Collier and C. Singleton, *J. Appl. Chem.* **6**, 495 (1956).]

horizontal displacement of the slit x. The wavelength derivative and the frequency derivative are

$$\frac{dI}{d\lambda} = \frac{dI}{dx} \cdot \frac{dx}{d\lambda}$$
$$\frac{dI}{d\omega} = -\frac{2\pi c}{\omega^2} \frac{dI}{dx} \frac{dx}{d\lambda}.$$
(14.1)

The dispersion $dx/d\lambda$ is usually a known function of the wavelength (for normal incidence, for instance, $d\lambda/dx \propto \cos \phi$ where ϕ is the angle of the diffracted beam with the normal to the grating) and therefore $dI/d\lambda$ and $dI/d\omega$ can be easily calculated from the measured dI/dx.

An ingenious system for producing a wavelength modulated monochromatic beam has been recently suggested.[177] It utilizes the lateral shift produced by a transparent plane parallel plate on a beam which traverses it at a nonzero angle of incidence. The modulation in the shift produced by an angular vibration of the plate around an axis perpendicular to the beam direction is used to obtain the wavelength modulation: the plate is placed inside the monochromator close to either the entrance or the exit slit. An alternative method[200] involves the rotation of the plate, placed at an angle of incidence α, around an axis which coincides with the direction of the light beam. A periodic displacement of the beam perpendicular to the slit is obtained. This displacement is a sinusoidal function of time with frequency equal to the frequency of rotation of the plate and amplitude Δx given by:

$$\Delta x = d \sin \alpha [1 - \cos \alpha / \sqrt{n^2 - \sin^2 \alpha}], \qquad (14.2)$$

where n is the refractive index (real) and d the thickness of the plate. In order to avoid corrections due to the wavelength dependence of n, it is convenient to use for the plate a material with little dispersion. Lithium fluoride is an excellent material for the visible and near ultraviolet and it remains usable in the vacuum ultraviolet to 1100 Å. The rotating plate must act on a parallel beam so as to avoid defocusing of the slit image and loss in resolution. Hence it could be placed in front of the monochromator grating or prism. It is, however, more convenient to place it in front of a slit where the beam is smaller. The corresponding loss in resolution can be kept small for small modulation amplitudes provided the aperture of the monochromator is not too large. The rotation of the plate produces beside the wavelength modulation, also a modulation in the intensity of the beam. This modulation is out of phase with the amplitude modulation and thus can, in principle, be eliminated.

As already mentioned in Chapter II, structure in I_0 (the dc response of the detector to the incident light) can be a serious problem when taking wavelength modulation spectra. This structure is usually due to structure in the spectral distribution of the source although it can also arise from the spectral response of the detector or from absorption in the optical system, i.e., CO_2 or water bands in the infrared. While this problem can be reduced by using a double beam system, and thus obtaining the derivative of I/I_0, small errors in the balance of the two beams (path differences, different mirror reflectivities, etc.) will still leave some residual spurious signal if I_0 has strong structure. Hence, the wavelength modulation method is most practical when the structure in I_0 is broader than the structure to be observed: it has been mainly used to study the sharp structure associated with exciton transitions near the lowest energy gap.[67,177]

We shall now discuss the errors introduced in the derivative measurements by the finite modulation amplitude and spectrometer band width. The discussion of the finite modulation amplitude is also applicable to other forms of modulation. Let us assume that the frequency of the monochromator output is given by Eq. (10.3) and consider the signal I in a reflectivity measurement (similar considerations apply to a transmission measurement[176]):

$$I(\omega) = R(\omega)I_0(\omega) = R[\omega_0 + (\Delta\omega)\cos\Omega t]I_0[\omega_0 + (\Delta\omega)\cos\Omega t]. \quad (14.3)$$

We shall assume, in order to simplify the treatment, that I_0 has no structure or else that we are measuring R directly with a double beam system. An expansion of R in power series of $(\Delta\omega)\cos\Omega t$ yields, to third order in $\Delta\omega$:

$$R = R(\omega_0) + \left(\frac{dR}{d\omega}\right)_{\omega_0}(\Delta\omega)\cos\Omega t + \frac{1}{2}\left(\frac{d^2R}{d\omega^2}\right)_{\omega_0}(\Delta\omega)^2\cos^2\Omega t$$

$$+ \frac{1}{6}\left(\frac{d^3R}{d\omega^3}\right)_{\omega_0}(\Delta\omega)^3\cos^3\Omega t + \cdots$$

$$= R(\omega_0) + \frac{1}{4}\left(\frac{d^2R}{d\omega^2}\right)_{\omega_0}(\Delta\omega)^2 + \left[\left(\frac{dR}{d\omega}\right)_{\omega_0} + \frac{1}{8}\left(\frac{d^3R}{d\omega^3}\right)_{\omega_0}(\Delta\omega)^2\right](\Delta\omega)\cos\Omega t$$

$$+ \frac{1}{4}\left(\frac{d^2R}{d\omega^2}\right)_{\omega_0}(\Delta\omega)\cos 2\Omega t + \frac{1}{24}\left(\frac{d^3R}{d\omega^3}\right)_{\omega_0}(\Delta\omega)^3\cos 3\Omega t + \cdots. \quad (14.4)$$

Equation (14.4) shows that the amplitude of the reflectance modulation at the fundamental frequency equals $(dR/d\omega)_{\omega_0}$ only for small modulation amplitudes. A correction proportional to the third derivative of R appears for higher amplitudes. The experimentally measured quantity is the ratio of the modulation at the fundamental frequency to the dc component of R:

$$\frac{\Delta R(\omega)}{\langle R \rangle} = \frac{R'}{R}\left[1 + \left(\frac{R'''}{8R'} - \frac{R''}{4R}\right)(\Delta\omega)^2\right]\Delta\omega, \quad (14.5)$$

where R', R'', R''' represent the derivatives of R evaluated at the frequency ω_0. In order to obtain a "true" derivative spectrum, $\Delta\omega$ must be kept small enough so as to make the correction terms in Eq. (14.5) negligible. This is not possible for a singular line, e.g., $(\omega - \omega_g)^{1/2}$ at the singularity, however, such theoretical singular lines never exist in nature: the singularity is removed by broadening and by the spectral band width of the radiation. Let us examine the effect of this spectral band width on a modulation spectrum. We assume that the radiation contains all frequencies between $\omega_0 - \delta\omega$ and $\omega_0 + \delta\omega$ with equal weight. The measured quantity is (for $\delta\omega$ very small):

$$\int_{\omega_0-\delta\omega}^{\omega_0+\delta\omega} R'(\omega)\,d\omega \bigg/ \int_{\omega_0-\delta\omega}^{\omega_0+\delta\omega} R(\omega)\,d\omega = \frac{R' + \tfrac{1}{6}R'''(\delta\omega)^2}{R + \tfrac{1}{6}R''(\delta\omega)^2}$$

$$= \left(\frac{R'}{R}\right)\left[1 + \frac{1}{6}\left(\frac{R'''}{R'} - \frac{R''}{R}\right)(\delta\omega)^2\right].$$

(14.6)

Hence we see that the finite spectral width $\delta\omega$ introduces corrections of order $(\delta\omega)^2$. These corrections are similar to those produced by the finite modulation amplitude [Eq. (14.5)]. It is therefore convenient, from the standpoint of signal-to-noise ratio and resolution, to keep the modulation amplitude and the spectral slit width approximately the same.

15. Line Shapes

We shall examine the expected line shapes of the wavelength derivative spectra of some typical types of structure which appear in the optical spectra of solids. Let us first consider the asymmetrically broadened excitons of Eqs. (7.4)–(7.6). The dielectric constant can be written as:

$$\varepsilon_r - 1 \propto \Gamma^{-1}(F_1 - 2\mathscr{A}F_2)$$
$$\varepsilon_i \propto \Gamma^{-1}(F_2 + 2\mathscr{A}F_1) \qquad (15.1)$$

with

$$F_1 = \frac{-W}{W^2 + 1}; \quad \text{and} \quad F_2 = \frac{1}{W^2 + 1},$$

where

$$W = \frac{\omega - (\bar{\omega}_g + \Delta_l)}{\Gamma/2}.$$

Equations (15.1) enable us to express ε_r and ε_i in terms of the Lorentzian functions F_1 and F_2 of the reduced frequency W. $\mathscr{A}$, the asymmetric broadening parameter, is usually small. The wavelength derivative spectra yield:

$$\begin{aligned} d\varepsilon_r/d\omega &\propto \Gamma^{-2}[F_1' - 2\mathscr{A}F_2'] \\ d\varepsilon_i/d\omega &\propto \Gamma^{-2}[F_2' + 2\mathscr{A}F_1']. \end{aligned} \quad (15.2)$$

The functions $F_1(W)$, $F_2(W)$, $F_1'(W)$ and $F_2'(W)$ are plotted in Fig. 50.

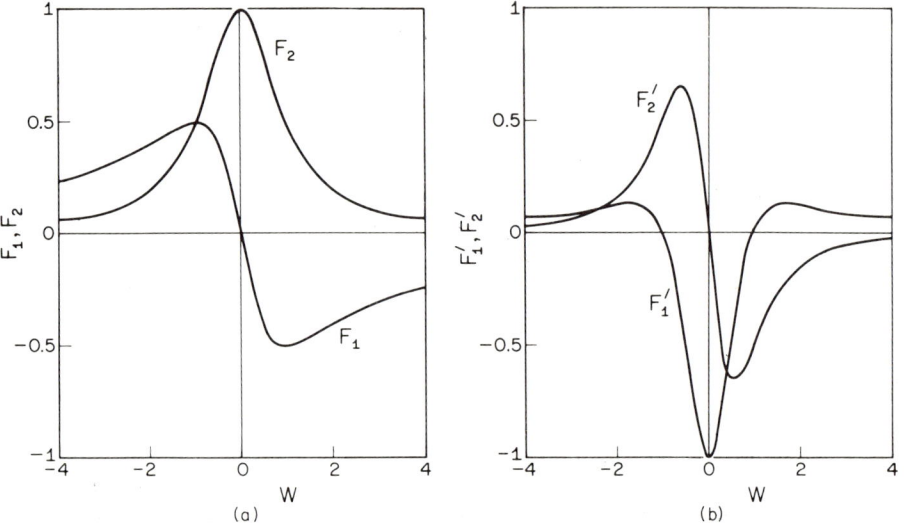

FIG. 50. (a) Lorentzian functions $F_1(W)$, $F_2(W)$ and (b) their derivatives $F_1'(W)$ and $F_2'(W)$.

The structure in the $\Delta R/R$ for a reflectance modulation spectrum usually corresponds to structure in ΔR since it is much sharper than the structure in R. Near an exciton line of the type described by Eq. (7.5) we have (assuming that the exciton represents only a small fraction of the total oscillator strength of the system):

$$\Delta R/R \propto \Gamma^{-2}[F_1'(\beta_r + 2\mathscr{A}\beta_i) + F_2'(\beta_i - 2\mathscr{A}\beta_r)], \quad (15.3)$$

with β_r and β_i defined in Eq. (8.9). In the vicinity of the smallest band edge $\beta_i \approx 0$ and the structure observed in a reflectance modulation spectrum correspond almost exclusively to structure in ε_r.

We shall now consider the derivative spectra of three-dimensional critical points of the square root type discussed in Section 4. These critical points appear at direct allowed edges and near indirect excitonic edges (Section 6d). We shall generalize the square root singularity to include broadening, as done in Eq. (7.2b). If the broadening parameter η is small ($\eta \ll \omega_g$), the main contribution to the derivative spectra comes from the $(\omega_g - \omega - i\eta)^{1/2}$ term in Eq. (7.2b), this contribution being singular for $\eta \to 0$ while the others are well behaved. In terms of the reduced frequency $W = (\omega - \omega_g)/\eta$ we obtain for an M_0 critical point[126, 201]

$$\begin{aligned}
d\varepsilon_r/d\omega &\propto \mathrm{Re}\left[\tfrac{1}{2}(\omega + i\eta)^{-2}(\omega_g - \omega - i\eta)^{-1/2}\right] \\
&= \tfrac{1}{2}\eta^{-1/2}[W^2 + 1]^{-1/2}[(W^2 + 1)^{1/2} - W]^{1/2} \\
&= \tfrac{1}{2}\eta^{-1/2}F(-W) \\
d\varepsilon_i/d\omega &\propto \mathrm{Im}\left[\tfrac{1}{2}(\omega + i\eta)^{-2}(\omega_g + \omega - i\eta)^{-1/2}\right] \\
&= \tfrac{1}{2}\eta^{-1/2}[W^2 + 1]^{-1/2}[(W^2 + 1)^{1/2} + W]^{1/2} \\
&= \tfrac{1}{2}\eta^{-1/2}F(W),
\end{aligned} \quad (15.4)$$

with $W = (\omega - \omega_g)/\eta$. (The frequency dependence of the factor $(\omega + i\eta)^{-2}$ can be neglected for $\eta \ll \omega_g$). The universal function $F(W)$ is shown in Fig. 51 (from Fig. 18.3 of Batz[201]).

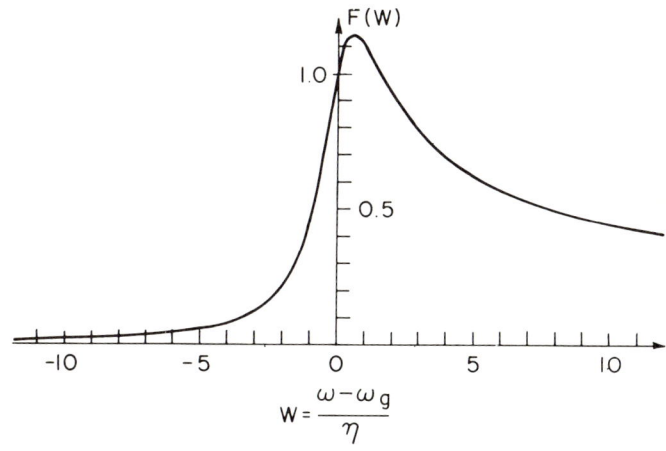

FIG. 51. Function $F(W) = [(W^2 + 1)^{1/2} + W]^{1/2}(W^2 + 1)^{-1/2}$ which determine the line shape of derivative spectra near direct allowed edges, from B. Batz, Ph.D. Thesis, Univ. Libre de Bruxelles, 1967.

The derivative spectra around any other kind of three-dimensional allowed critical points can also be expressed in terms of the function $F(W)$. The results are given in Table V. The derivatives with respect

TABLE V. DERIVATIVES $d\varepsilon_r/d\omega$ AND $d\varepsilon_i/d\omega$ FOR THE FOUR TYPES OF THREE-DIMENSIONAL CRITICAL POINTS (ALLOWED TRANSITIONS) IN TERMS OF THE FUNCTION $F(W)$ OF FIG. 51[a]

Type of critical point	$2\eta^{1/2}\, d\varepsilon_r/d\omega$	$2\eta^{1/2}\, d\varepsilon_i/d\omega$
M_0 (minimum)	$F(-W)$	$F(+W)$
M_1 (saddle point)	$-F(+W)$	$F(-W)$
M_2 (saddle point)	$-F(-W)$	$-F(+W)$
M_3 (maximum)	$F(+W)$	$-F(-W)$

[a] From B. Batz, Ph.D. Thesis. Univ. Libre de Bruxelles, 1967.

to ω_g are obtained by reversing the sign of those in Table V. The function $\eta^{-1/2}\, F[(\omega - \omega_g)/\eta]$ tends to $(\omega - \omega_g)^{-1/2}$ for $\eta \to 0$, as expected.

16. RESULTS

Data obtained by the wavelength modulation method for solids are not very abundant, probably due to the already mentioned fact that the method is only of practical interest for studying sharp structure. Figure 52 shows a transmission modulation spectrum obtained by Balslev[67] for germanium at 80°K. The main peak (at 0.761 eV) corresponds to indirect transitions with emission of a longitudinal acoustical phonon (L point of the B.Z.). The curve has been fitted to a square root singularity in ε_i [$d\varepsilon_i/d\omega \sim (\omega - \omega_g)^{-1/2}$ for $\omega > \omega_g$, $d\varepsilon_i/d\omega = 0$ for $\omega < \omega_g$]. This type of singularity is expected for absorption by indirect excitons (see Section 6d). The broadening, estimated at 0.7 meV by Balslev, is probably caused by instrumental resolution and modulation amplitude and, hence, a fit to the $F(W)$ function of Fig. 51 has not been attempted. The energy of the corresponding LA phonon has been determined to be 0.026 eV.[202] Minor structure, probably due to LO (0.030 eV) and TO phonons (0.034 eV), is also seen. The exciton binding energy has been estimated[67] to be about 5 meV.

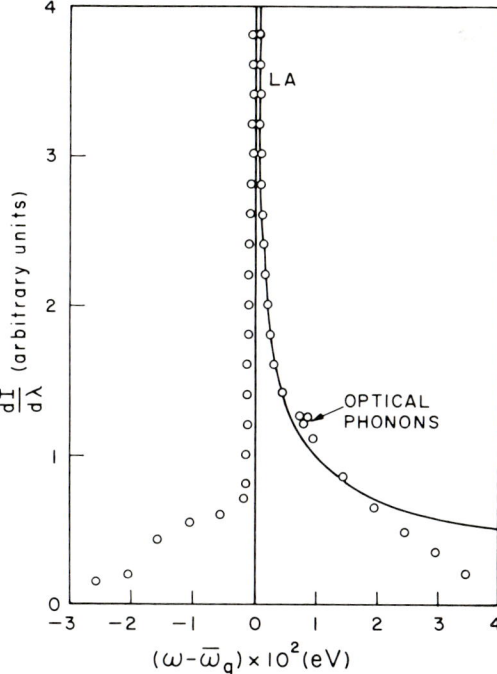

FIG. 52. Wavelength modulated transmission spectrum of germanium at 80°K (circles) near the indirect absorption edge [I. Baslev, *Phys. Rev.* **143**, 636 (1966).] The peak occurs at the energy $\bar{\omega}_g = \omega_g - \omega_{ex} + \omega_{phon} = 0.761$ eV. The solid curve is a fit to a $(\omega - \bar{\omega}_g)^{-1/2}$ function.

Figure 53 shows results for the indirect edge of silicon, also obtained by Balslev.[67] The peak is due to the formation of indirect excitons with emission of transverse optical phonons (0.058 eV).[202] The main peak has been fitted with the function $F(W)$ (Fig. 51) for a broadening parameter $\eta = 0.0012$ eV. This broadening parameter is identical with that found by Macfarlane *et al.*[22] although it differs from that mentioned by Balslev,[67] probably because of a difference in the definition of such a parameter. We shall discuss in Chapter VIII the splitting of the indirect exciton peaks of Figs. 52 and 53 under uniaxial stress.

Figure 54 shows the wavelength modulation spectra of wurtzite-type ZnS obtained by Drews[177] around the E_0 fundamental edge at room temperature for the two normal modes of polarization. The dotted portions of these spectra are caused by back reflection: the material is

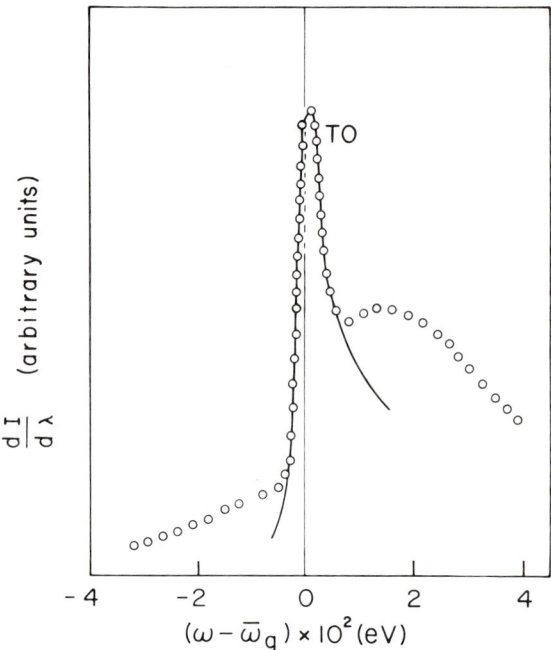

FIG. 53. Wavelength modulated transmission of silicon near the indirect edge at 80°K [I. Baslev, *Phys. Rev.* **143**, 636 (1966)]. The singularity is produced by indirect exciton absorption with TO phonon emission. The fit to the experimental points was made with the function $F[(\omega - \omega_g)/\eta]$, with the broadening parameter $\eta = 1.2$ meV.

transparent at the wavelength at which they occur. The contribution of the light reflected by the back of the crystal to a reflection modulation spectrum may be quite considerable in the region below the fundamental gap. It can be minimized by coarse grinding of the back surface or by the use of a wedge shaped sample, but seldom completely eliminated. This contribution usually contains the absorption modulation ($\Delta\varepsilon_i$) of the back reflected beam. Three peaks are otherwise visible in Fig. 54; they are labeled A, B, and C in the conventional notation. Their line shapes are somewhat difficult to identify owing to the partial overlap of three peaks but, since the reflectivity at the E_0 edge is mainly determined (at room temperature) by the *real* part of the dielectric constant ε_r, the spectrum is expected to have the shape of $F_1'(W)$ shown in Fig. 50. Hence, to a good approximation, the energy of the corresponding exciton peaks should be that shown

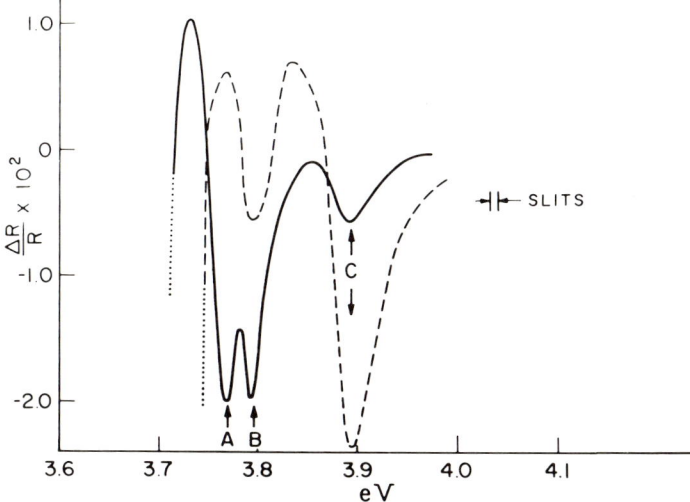

FIG. 54. Wavelength modulation spectrum of wurtzite-type ZnS at room temperature (297°K) (R. E. Drews, *Bull. Am. Phys. Soc.*, **12**, 384 (1967) and Private Communication). The dotted signal is due to back reflection (film thickness $\sim 30\,\mu$), the solid curve is $E \perp C$, and the dashed curve is $E \parallel C$.

by the arrows, $\omega_A = 3.770$ eV, $\omega_B = 3.797$ eV, and $\omega_C = 3.895$ eV. The $\omega_B - \omega_A$ splitting, mostly due to the hexagonal orbital crystal field, is $\omega_{BA} = 27$ meV, while the ω_{CB} splitting, mostly due to spin orbit interaction, is 98 meV. These room temperature splittings are in agreement with less accurate values previously reported[203] and also with measurements at low temperatures[204] ($\omega_{CB} = 97$ meV at 77°K).

V. Temperature Modulation

17. General Considerations

a. Direct Transitions

A temperature change usually produces two effects on an absorption edge (see Section 7): a shift of the gap ω_g and a change of the broadening parameter η. The shift in energy gap produces essentially the same type of singularity (except possibly for a sign reversal) as wavelength modulation since near a critical point we have:

$$\left(\frac{\partial \varepsilon}{\partial T}\right)_{\eta=\text{constant}} = \frac{\partial \varepsilon}{\partial \omega_g}\frac{\partial \omega_g}{\partial T} = -\frac{\partial \varepsilon}{\partial \omega}\frac{\partial \omega_g}{\partial T}. \quad (17.1)$$

The line shapes of $\Delta\varepsilon_r$ and $\Delta\varepsilon_i$ are those of Table V multiplied by $-(\partial\omega_g/\partial T)$. The quantity $(\partial\omega_g/\partial T)$ is usually negative, although it can be positive in some cases, e.g., for the lowest direct gap of the lead chalcogenides.[130]

The contribution of the temperature modulation of the broadening parameter η to a thermal modulation spectrum can be found from Eq. (7.2b). We obtain for an M_0 three-dimensional critical point (for $\eta \ll \omega_g$):

$$d\varepsilon_r/d\eta \sim -\tfrac{1}{2}\eta^{-1/2}[(W^2+1)^{1/2} + W]^{1/2}[1+W^2]^{-1/2} = -\tfrac{1}{2}\eta^{-1/2}F(W). \quad (17.2)$$

The broadening modulation in ε_r and ε_i for other types of critical points is given in Table VI in terms of the universal function $F(W)$ which also determines wavelength and energy gap modulation. This function was given in Fig. 51.

The temperature modulation caused by the shift of ω_g (equivalent to hydrostatic pressure modulation) is usually larger than the modulation caused by broadening since temperature coefficients of gaps are

TABLE VI. Derivatives $d\varepsilon_r/d\eta$ and $d\varepsilon_i/d\eta$ for the Four Types of Three-Dimensional Critical Points (Allowed Transitions) Expressed in Terms of the Universal Function $F(W)$ of Fig. 51

Type of critical point	$2\eta^{1/2}\, d\varepsilon_r/d\eta$	$2\eta^{1/2}\, d\varepsilon_i/d\eta$
M_0 (minimum)	$-F(W)$	$F(-W)$
M_1 (saddle point)	$-F(-W)$	$-F(W)$
M_2 (saddle point)	$F(W)$	$-F(-W)$
M_3 (maximum)	$F(-W)$	$F(W)$

around 4×10^{-4} eV/°C while the derivative of the broadening parameter with respect to T is usually close to Boltzmann's constant ($\approx 10^{-4}$ eV/°C). Hence, temperature modulation spectra of critical point transitions are usually similar to lattice constant, e.g., hydrostatic pressure, modulation spectra. This is not necessarily the case for transitions to or from the Fermi level of metals (such as those discussed in Section 9) and heavily doped semiconductors[205]: the modulation of the broadening of the Fermi distribution around the Fermi level may contribute significantly to the optical modulation.

b. Excitons

Some of the considerations given above also apply to the thermal modulation of exciton spectra. Since excitons are associated with band edges, the modulation caused by a shift in ω_g is usually several times larger than that owing to broadening, although this last contribution is not completely negligible. The expressions which correspond to Eq. (15.2) for broadening modulation are:

$$d\varepsilon_r/d\Gamma \propto \Gamma^{-2}[F_2' + 2\mathscr{A}F_1'];$$
$$d\varepsilon_i/d\Gamma \propto \Gamma^{-2}[F_1' - 2\mathscr{A}F_2'].$$
(17.3)

As already mentioned, the asymmetry parameter $\mathscr{A}$ is usually small.

c. Plasma Resonance

As discussed in Section 10, sharp structure in the optical modulation spectra can result near a plasma resonance. This structure can be due either to a modulation of the plasma frequency ω_p or to a modulation of the broadening parameter ω_τ [Eq. (3.4)]. A temperature modulation of ω_p can be produced either by thermal expansion (ω_p^2 is proportional to the electron density) or by a temperature dependence of the effective mass m^*. The effective mass may be strongly temperature dependent in small band gap semiconductors[206] such as InSb. In such cases the mass modulation effect may overwhelm the carrier concentration effect. No mass modulation effects should take place when the plasma resonance is that caused by completely filled bands since the mass is then the free electron mass (see Section 2).

In order to give a qualitative idea of the expected line shapes we shall treat the case of a broadening parameter small compared to the plasma frequency. The dielectric constant can then be written, to first-order in η/W [see Eq. (3.4)]:

$$\varepsilon_r = \varepsilon_L[1 - (1/W^2)]; \qquad \varepsilon_i = (\varepsilon_L/W^3)\,\eta, \qquad (17.4)$$

where $W = \omega/\omega_p$ is the reduced frequency, $\eta = \omega_\tau/\omega_p$ the reduced broadening parameter, and ε_L the nondispersive dielectric constant of the lattice. In order to obtain the reflectance modulation spectrum, we must evaluate the derivatives of the reflectivity with respect to ω_p, i.e., W, and η. If we write the change in reflectivity as a linear combination of $\Delta\varepsilon_r$ and $\Delta\varepsilon_i$ with coefficients β_r and β_i, the structure in ΔR comes from structure in β_r and β_i [contrary to the cases treated above under (a) and (b) in which β_r and β_i are well behaved at the singular frequency]. The derivative of the reflectivity with respect to either the frequency ω or the plasma frequency ω_p can be obtained from the function:

$$dR/dW = A[2n_r(n_r^2 - 3n_i^2 - 1) + 3(\eta/W)n_i(n_i^2 - 3n_r^2 + 1)], \qquad (17.5a)$$

with

$$A = \frac{3\varepsilon_L}{W^3[(n_r + 1)^2 + n_i^2]^2(n_r^2 + n_i^2)}.$$

The derivative of R with respect to the broadening parameter is obtained from:

$$dR/d\eta = An_i[3n_r^2 - n_i^2 - 1]. \qquad (17.5b)$$

We have plotted in Fig. 55 and 56 the functions $(1/R)\,dR/dW$ and $(1/R)\,dR/d\eta$, for $\varepsilon_L = 10$ and $\eta = 0.1$ and 0.3 respectively. We see that the peak in $(1/R)\,dR/dW$ occurs very close to the plasma frequency

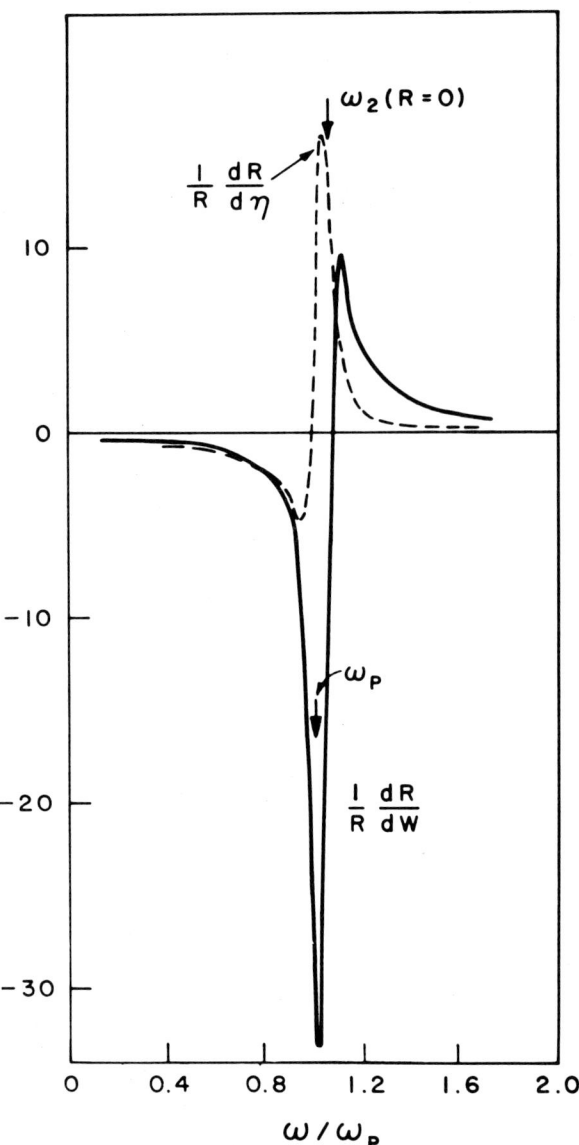

Fig. 55. Reflectance modulation spectra of a plasma resonance for a broadening parameter $\eta = 0.1$ [see Eqs. (17.5)] $(1/R)\, dR/d\eta$ corresponds to the broadening modulation and $(1/R)\, dR/dW$ to plasma frequency modulation.

($W = 1$) while the peak in $(1/R)\,dR/d\eta$ occurs very close to the minimum in the reflectivity: the loss modulation modulates very strongly the value of the reflectivity at the minimum which would be zero for $\eta = 0$. This conclusion remains quite general provided the loss parameter η

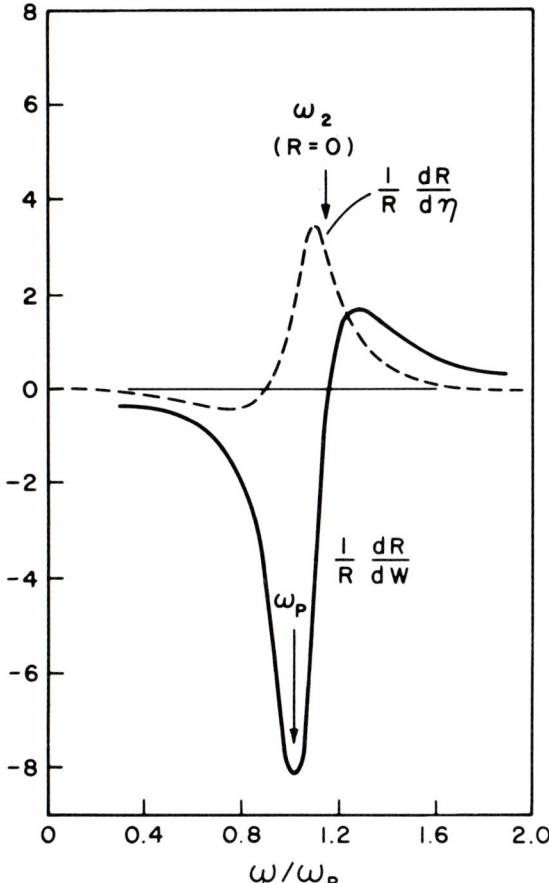

FIG. 56. Reflectance modulation spectra of a plasma resonance for $\eta = 0.3$ [see Eqs. (17.5)] $(1/R)\,dR/d\eta$ corresponds to broadening modulation and $(1/R)\,dR/dW$ to plasma frequency modulation.

does not become too large: 2 peaks, usually of opposite sign, should appear in the thermoreflectance spectrum, one close to the plasma frequency and the other at the frequency at which the reflectivity has a minimum ($\varepsilon_r = 1$). Thus from the positions of these peaks ω_1 and ω_2, we can determine $\omega_p \simeq \omega_1$ and $\varepsilon_L \simeq [1 - (\omega_p{}^2/\omega_2{}^2)]^{-1}$, where $\omega_2 > \omega_1$. The loss parameter can be conveniently determined from the value of the reflectivity at the minimum:

$$R(\omega_2) \simeq [\varepsilon_L \omega_p{}^3 \eta / 4\omega_2{}^3]^2. \tag{17.6}$$

d. Indirect Transitions

Since the strength of indirect transitions depends on the phonon occupation numbers, it may seem at first glance that temperature modulation would be a good way to observe them. If we assume that the gap is independent of temperature, we obtain from Eq. (5.1) for the contribution to the derivative $d\varepsilon_i/dT$ due to the change in the phonon occupation number[66]:

$$\frac{d\varepsilon_i}{dT} = \varepsilon_i [1 + f_B] \frac{\omega_{\text{phon}}}{kT^2}, \tag{17.7}$$

for indirect transitions with phonon absorption ($\omega \geq \omega_g - \omega_{\text{phon}}$). For phonon emission processes ($\omega \geq \omega_g + \omega_{\text{phon}}$) we find:

$$\frac{d\varepsilon_i}{dT} = \varepsilon_i f_B \frac{\omega_{\text{phon}}}{kT^2}. \tag{17.8}$$

For a combination of phonon absorption and phonon emission ($\omega \geq \omega_g + \omega_{\text{phon}}$) we obtain:

$$\frac{d\varepsilon_i}{dT} = \varepsilon_i [1 + f_B] \left[1 + \frac{(\omega - \omega_g - \omega_{\text{phon}})^2}{2[(\omega - \omega_g)^2 + \omega_{\text{phon}}^2] f_B} \right]^{-1} \frac{\omega_{\text{phon}}}{kT^2}. \tag{17.9}$$

Equations (17.7)–(17.9) indicate that the main advantage of modulation spectroscopy is lost in this type of modulation for indirect transitions: the spectrum is proportional to ε_i and not to its derivative with respect to ω. Hence, no enhancement of gap structure results. Equations (17.7) and (17.8) indicate, however, that the method may be useful to enhance phonon absorption structure at low temperatures (difficult to observe in conventional spectra) with respect to phonon emission structure. Equation (17.9) shows that $d(\log \varepsilon_i)/dT$ is constant for $\omega \leq \omega_g + \omega_{\text{phon}}$ and deviates from linearity (becomes smaller) above this energy.

Beside the structure caused by change in phonon occupation numbers, there is structure in $d\varepsilon_i/dT$ caused by temperature modulation of the indirect gap ω_g. For band-to-band transitions, this structure is weak since it corresponds to the derivative of $(\omega - \omega_g \pm \omega_{phon})^2$ (there is only a change in slope at $\omega_g \pm \omega_{phon}$). Exciton effects, however, transform this weak singularity into a strong $(\omega - \omega_g)^{-1/2}$ singularity at low temperatures as shown in Section 6d.

18. Experimental Techniques

Thermal modulation is usually achieved by passing a modulating current along a thin rectangular sample which is glued to a heat sink. In order to simplify our analysis we shall assume that the temperature of the sample is uniform (this is true for a thin sample of good thermal conductivity) and that the increase in the average temperature of the sample $\langle T \rangle$ above that of the heat sink T_0 is much larger than the amplitude ΔT of the temperature modulation. Let us call C the heat capacity of the sample and Q the heat leak per unit time and unit temperature difference $T-T_0$ between sample and sink. The temperature of the sample is governed by:

$$C\, dT/dt = I^2 R - Q(T - T_0), \qquad (18.1)$$

where I is the current and R is the resistance of the sample. Let us use square pulses of duty cycle 1 for our modulation current: $I = I_0$ for $0 < t < \tau/2$, $I = 0$ for $\tau/2 < t < \tau$. Since we have assumed $T - T_0$ much larger than the modulation amplitude, we can replace, in the thermal conductivity term of Eq. (18.1), T by the average temperature $\langle T \rangle$. We then obtain:

$$\langle T \rangle - T_0 = I^2 R/2Q,$$
$$T - \langle T \rangle = (I^2 R/C)(t - \tfrac{1}{4}\tau) \quad \text{for } 0 < t < \tfrac{1}{2}\tau, \qquad (18.2)$$
$$T - \langle T \rangle = (I^2 R/C)(\tfrac{3}{4}\tau - t) \quad \text{for } \tfrac{1}{2}\tau < t < \tau.$$

In order to calculate the response at the fundamental frequency $\Omega = 2\pi/\tau$, as measured with a lock-in amplifier, we calculate the amplitude ΔT of the first Fourier component of $T - \langle T \rangle$ in Eq. (18.2). We find:

$$\Delta T(\Omega) = 2I^2 R\tau/\pi^2 C. \qquad (18.3)$$

The "modulation" efficiency γ can be defined as the ratio of the modulation amplitude ΔT to the rise in the average temperature:

$$\gamma = \Delta T/(\langle T \rangle - T_0) = 4Q\tau/\pi^2 C = 0.405 Q\tau/C. \qquad (18.4)$$

We see that this efficiency decreases as the frequency of the modulation increases. For an efficient thermal modulation we must keep the heat capacity C small and the heat leak Q large. The heat leak is, however, limited in our calculation by our assumption of a uniform temperature for the sample. Hence, the maximum permissible value of Q in Eq. (18.4) is K (the thermal conductivity) divided by the thickness of the sample (per cm^2 of sample area). Typical values of these parameters for a $1 \times 1 \times 0.01$ cm sample are: $K = 0.01$ W cm^{-1} (°C)$^{-1}$, $C = 10^{-1}$ J (°C)$^{-1}$, $Q = 1$ W (°C)$^{-1}$, and $\gamma = 4\tau$ (τ in seconds).

It is also interesting to consider sine wave instead of square wave modulation. In order to obtain thermal modulation at the fundamental frequency, we must add a dc bias. Optimum γ is obtained (as in the case of square pulses) for a bias equal to the root mean square value of the modulating current:

$$I = I_0[2^{-1/2} + \cos \Omega t].$$

The efficiency γ is in this case:

$$\gamma = 2^{1/2}Q\tau/2\pi C = 0.22Q\tau/C. \tag{18.5}$$

By comparing Eq. (18.4) with (18.5) we realize the advantage of using square pulses for thermal modulation: the efficiency γ is almost a factor of 2 higher than for sine waves. It is also interesting to compare the efficiencies for modulation at the fundamental frequency Ω with those for second harmonic modulation. It can be easily shown that maximum second harmonic efficiency is obtained for a purely sinusoidal modulating signal without dc bias. The maximum efficiency is:

$$\gamma(2\Omega) = 0.08Q\tau/C. \tag{18.6}$$

We have just seen the convenience of using as thin a sample as possible in order to obtain a large γ. However, a minimum thickness is imposed by the type of optical spectrum whose modulation is being measured. If one measures reflectance modulation, the sample should be thick enough to exhibit bulk properties and hence $d > \lambda_0/(n_r^2 + n_i^2)^{1/2}$ where λ_0 is the wavelength of the light in vacuum. In transmission measurements the sample should be thick enough to absorb a significant fraction of the light, i.e., $d \geq (1/\alpha)$ (α = absorption coefficient). Optimum thicknesses for thermoreflectance measurements lie between 0.1 and 1 μ. Such samples can be easily obtained[179,207] by vacuum deposition. When ground and polished bulk samples are required, it is usually necessary to use thicker samples (10–100 μ) at the expense of the modulation efficiency.

19. Results

a. Semiconductors

The thermoreflectance spectrum of germanium in the photon energy region below 5 eV has been investigated by Batz over a wide temperature range.[64,126,201] Figure 57 shows a recorder trace of a section of the spectrum around the E_1 and $E_1 + \Delta_1$ reflectivity peaks at 90°K. Two strong s-shaped spectral lines appear in correspondence to the

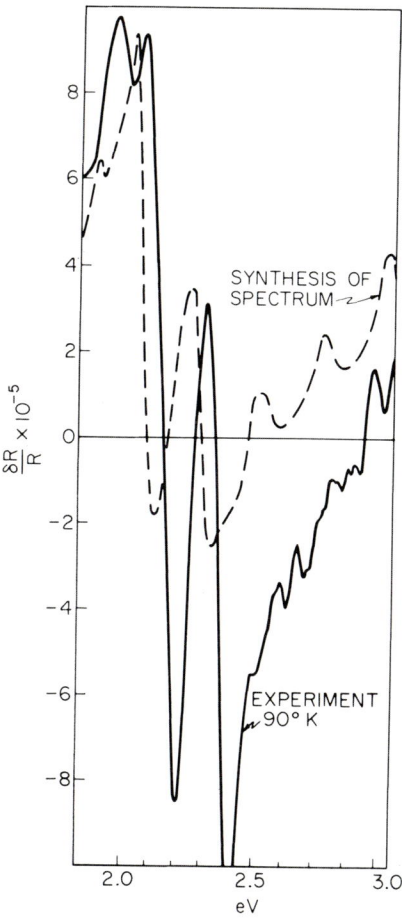

FIG. 57. Thermoreflectance spectrum of germanium (recorder trace) obtained by B. Batz, Ph.D. Thesis. Univ. Libre de Bruxelles, 1967, at 90°K and synthesis of the spectrum with 2 major M_1 critical points and a number of tentative M_0 critical points.

E_1, $E_1 + \Delta_1$ peaks seen in the reflectivity spectrum (Fig. 58). These peaks are believed due to transitions at two M_1 critical points in the [111] (Λ) direction of **k**-space, split by spin orbit interaction (see band structure of germanium in Fig. 29). The dotted curve in Fig. 57 shows a tentative synthesis of the observed spectrum assuming no broadening modulation (only gap modulation). This synthesis was

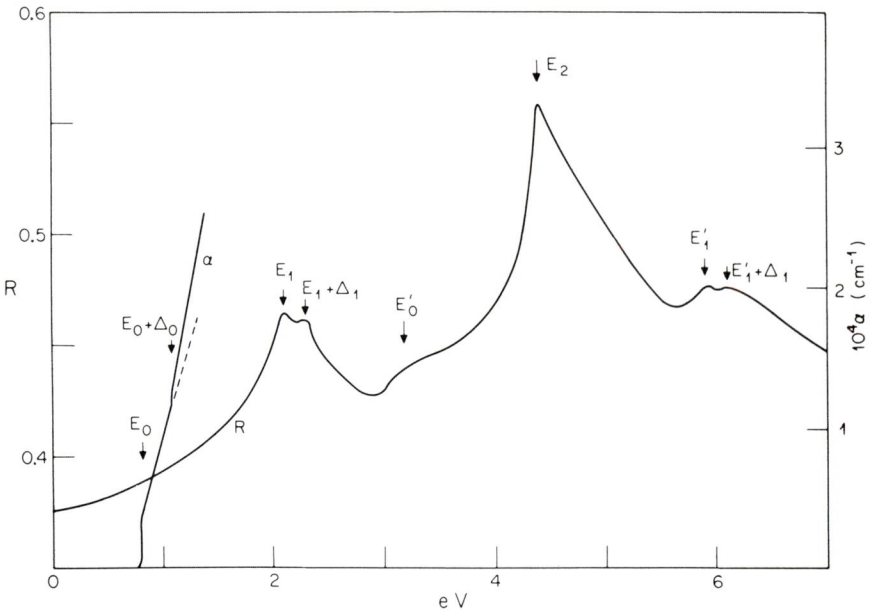

FIG. 58. Normal incidence reflectivity spectrum of germanium at room temperature and transmission spectrum near the fundamental direct edge (E_0, $E_0 + \Delta_0$). [From M. Cardona, K. L. Shaklee, and F. H. Pollak, *Phys. Rev.* **154**, 696 (1967).]

produced by Batz using the curve in Fig. 51, the relationships in Table V, and the β_r and β_i parameters of Eq. (8.9) appropriate to germanium. The close analogy between the experimental and the calculated M_1 structures confirms the M_1 nature of the corresponding critical points. The M_0 structure at 1.97 eV is quite distinct and it may seem reasonable to assign it to the M_0 critical point which occurs in Fig. 29 at the L point ($L_3' \rightarrow L_1$), slightly below the E_1 critical point. This M_0 peak (sometimes labeled e_1) should have a mate $e_1 + \Delta_1$, split by spin orbit interaction, which is not seen in Fig. 57 but has been reported by several authors.[154, 208] The positions of the

e_1, $e_1 + \Delta_1$ peaks reported by the various authors, however, are not in agreement with each other (see Table VII) and we feel that further

TABLE VII. POSITION OF THE E_1, $E_1 + \Delta_1$, e_1, AND $e_1 + \Delta_1$ PEAKS OBSERVED IN GERMANIUM BY SEVERAL METHODS AND AUTHORS NEAR LIQUID NITROGEN TEMPERATURE[a]

Author	E_1	$E_1 + \Delta_1$	e_1	$e_1 + \Delta_1$
Batz[b]	2.18	2.36	1.97	—
Ghosh[c]	2.22	2.42	2.15	2.35
Potter[d]	2.22	2.41	1.84	2.04

[a] The temperature of Batz's measurements may be somewhat higher due to the root mean square value of the modulating current. Agreement between the various measurements is good for the $E_1, E_1 + \Delta_1$ but not for the $e_1, e_1 + \Delta_1$ peaks.
[b] See B. Batz, *Solid State Commun.* **4**, 241 (1965).
[c] See A. K. Ghosh, *Solid State Commun.* **4**, 565 (1966). An energy of 0.1 eV has been added to the measured room temperature energies so as to make comparision with the low-temperature data of other authors possible.
[d] See R. F. Potter, *Phys. Rev.* **150**, 562 (1966).

work is required in order to confirm their assignment. Also questionable is the structure, assigned to three M_0 singularities (after a suggestion by Herman et al.[209]) which appears between 2.5 and 3 eV since it is almost within the noise level.

Batz[201] also reported the thermoreflectance spectrum of germanium around the E_2 peak (4.4 eV) and the lowest direct edge (E_0, $E_0 + \Delta_0$). He found at $320°K$, $\omega_g(E_0) = 0.76$ eV and $\omega_g(E_0 + \Delta_0) = 1.06$ eV.

Figure 59 shows a tracing of the thermoabsorption spectrum of germanium[201] around the direct (E_0) edge at room temperature. The structure at 1 eV could be due to *indirect* transitions between the top valence band (Fig. 29) and the Δ_1 conduction band minimum (in the [100] direction). If this assignment is confirmed thermoabsorption would prove useful to study indirect transitions superimposed on the direct ones.

Figure 60 shows the thermoabsorption spectrum of silicon obtained at room temperature by Berglund[66] around the lowest indirect edge. We have plotted both $d\alpha/dT$ and $d\log\alpha/dT$, the uncertainty in $d\log\alpha/dT$ becomes large at low energies (small α). We see in Fig. 60

that $d \log \alpha/dT$ is indeed constant below 1.165 eV at which point phonon-emission-aided transitions begin.

Figure 59 also shows the thermoabsorption spectrum of germanium at room temperature obtained by Batz.[201] In this spectrum $d (\log \alpha)/dT$ also shows a nearly flat region but now 4 indirect exciton peaks are apparent. The indirect gap determined from these measurements is 0.660 eV, in good agreement with that obtained by a careful

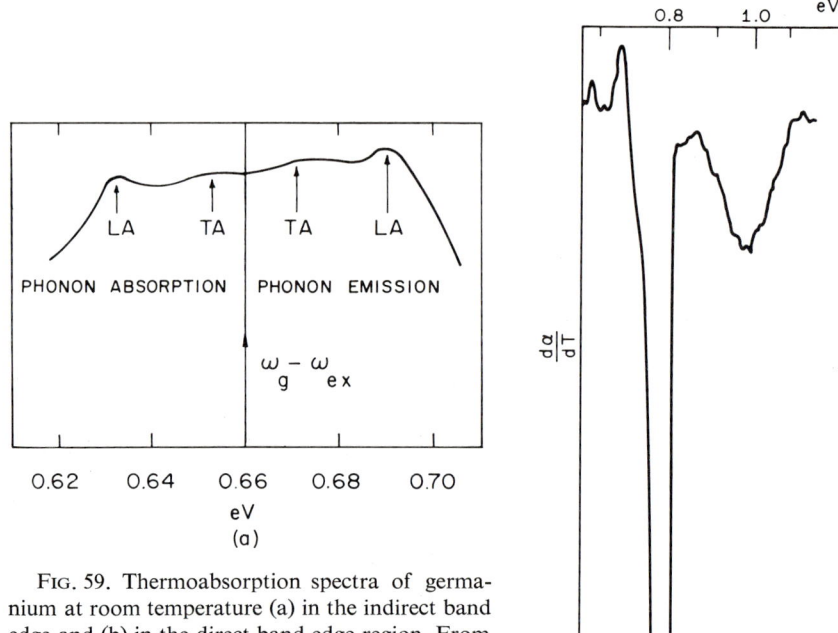

FIG. 59. Thermoabsorption spectra of germanium at room temperature (a) in the indirect band edge and (b) in the direct band edge region. From B. Batz, Ph.D. Thesis, Univ. Libre de Bruxelles, 1967.

analysis of the conventional absorption spectrum.[22] The corresponding phonon energies are 0.029 eV (LA) and 0.01 eV (TA), in reasonable agreement with those reported in Section 16.

Arthur et al.[205] have performed a rather unique thermal modulation experiment of the intravalence bands transitions of p-type germanium. These transitions take place between filled and empty top valence band states ($\Gamma_{2'}$, Fig. 29) in p-type germanium. The Fermi level for those experiments was above the top of the valence band (the doping level was less than 10^{16} cm^{-3} and the temperature around

19. RESULTS

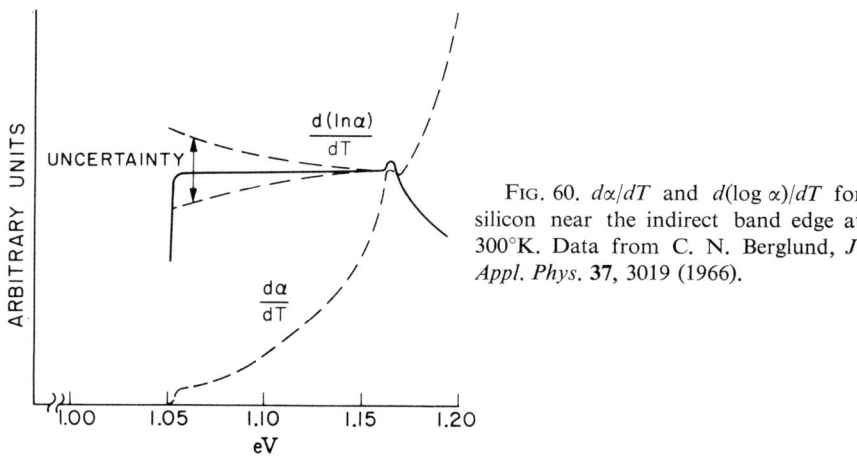

Fig. 60. $d\alpha/dT$ and $d(\log \alpha)/dT$ for silicon near the indirect band edge at 300°K. Data from C. N. Berglund, *J. Appl. Phys.* **37**, 3019 (1966).

77°K). When the temperature is raised, the spread in the Fermi distribution increases and thus strong modulation in the absorption can be obtained if the temperature is modulated. The experimental results obtained by Arthur et al.[205] are shown in Fig. 61. The measurement

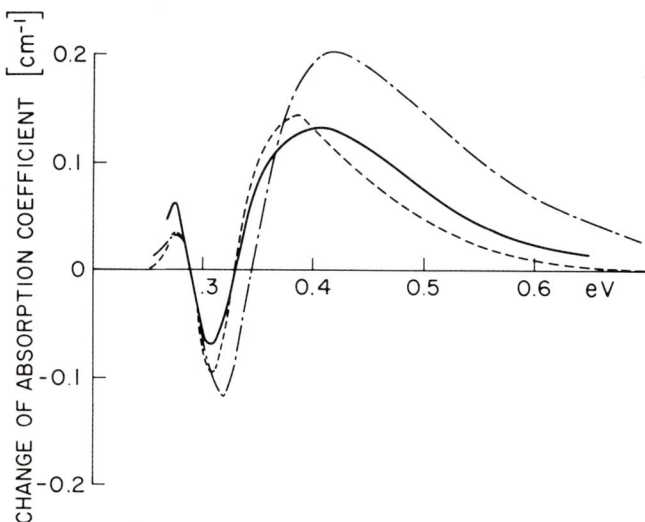

Fig. 61. Theoretical and experimental changes in absorption induced by a 15.5°K increase in the lattice temperature for *p*-type germanium at 77°K. The free carrier concentration is 4.4×10^{15} cm^{-3}. The figure shows the changes for the experiment (—) the theory of Fawcett (- - -), and the theory of Kane (— - —). Data from J. B. Arthur, A. C. Baynham, W. Fawcett, and E. G. S. Paige, *Phys. Rev.* **152**, 740 (1966).

technique was somewhat different from the conventional modulation techniques described here: owing to the small absorption coefficients under consideration, thick samples and hence small modulation frequencies (1 cps) must be used and some of the advantages of phase sensitive detection are lost. Arthur *et al.* used chopped radiation detected with a lock-in amplifier and balanced the signal at a given wavelength with a reference voltage. They measured directly, on a scope, the change induced on this signal by a heat pulse (2 sec). We believe some improvement in sensitivity and ease of measurement would ensue if the standard modulation techniques were used. Figure 61 also shows calculations of the expected change in absorption performed by Arthur *et al.* with the expression [see Eq. (2.11)]:

$$\alpha_{ij} = \frac{4\pi^2 N}{3cn_r \omega \sum_{i,k} e^{-\omega_i(k)/T}}$$

$$\times \sum_{k} |\langle i, \mathbf{k}|\mathbf{p}|j, \mathbf{k}\rangle|^2 \times [e^{-\omega_i(k)/T} - e^{-\omega_j(k)/T}] \times \delta(\omega - \omega_{ij}), \quad (19.1)$$

which represents the contribution to α of transitions between the i-th and the j-th bands. Equation (19.1) was evaluated by using the band structures of Kane[161] and Fawcett[210] (N is the hole concentration). Kane's band structure is valid only very near the top of the valence band ($\Gamma_{25'}$) since it treats the $\mathbf{k} \cdot \mathbf{p}$ interaction of $\Gamma_{25'}$ with the other bands by perturbation theory. Fawcett includes the $\Gamma_{2'}$ and Γ_{15} conduction bands (Fig. 29) exactly in the Hamiltonian and treats other bands ($\Gamma_{12'}$) by perturbation theory. The agreement between calculations and experiment is better with Fawcett's bands, as expected. The negative peak in Fig. 61 occurs at an energy (0.31 eV) only slightly higher than the spin orbit splitting Δ_0 ($\Delta_0 = 0.295$ eV). We believe similar measurements should yield values of Δ_0 for a number of materials for which intra-valence-band transitions have not been reported (InP, Si).

b. *Metals*

Figure 62 shows the thermoreflectance spectrum of gold near liquid nitrogen temperature obtained by Scouler and Wright[125] together with the ordinary reflectivity[125] and the piezoreflectance spectrum obtained by Garfinkel *et al.*[178] The structure in the modulation spectra is much sharper than that in the reflectivity, as expected. A remarkable similarity exists between both types of modulation spectra, as expected from the discussion in Section 17. However, the relative strength of the structure at 2.4 to 2.5 eV is a lot higher in the thermoreflectance spectrum. This suggests the possibility of transitions to or from the Fermi surface (Fig. 38) causing this structure: the modulation would be due, in large part, to the smearing of the Fermi distribution as the temperature is raised. The sign of the corresponding

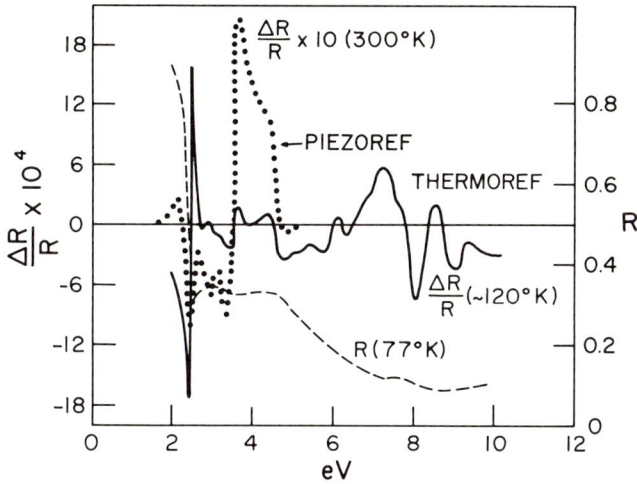

FIG. 62. Thermoreflectance and reflectivity spectra of gold near liquid nitrogen temperature [from W. J. Scouler, *Phys. Rev. Letters* **18**, 445 (1967)], together with the room temperature piezoreflectance spectrum [M. Garfinkel, J. J. Tiemann, and W. E. Engeler, *Phys. Rev.* **148**, 698 (1966)].

structure in $\Delta\varepsilon_i$ (see Fig. 63), suggests transitions to the Fermi surface (the Fermi smearing should produce an increase in ε_i at low energies followed by a decrease at higher energies). A rough estimate of the strength of the signal causd by this phenomenon, based on the reflectivity data of Fig. 62, yields, for the $\simeq 2°C$ modulation amplitude, a modulation in the reflectivity $\Delta R/R \simeq 10^{-3}$, in good qualitative agreement with experiment. This type of transitions has been proposed previously[174] to explain the structure under consideration (see Fig. 40).

Additional structure is seen in Fig. 62 (and 63) at about 2.9, 3.2, 3.6, 4.4, 5.0, 5.4, 6.1, 8.1, and 9 eV. A definite assignment of this structure is certainly premature, especially in view of the lack of

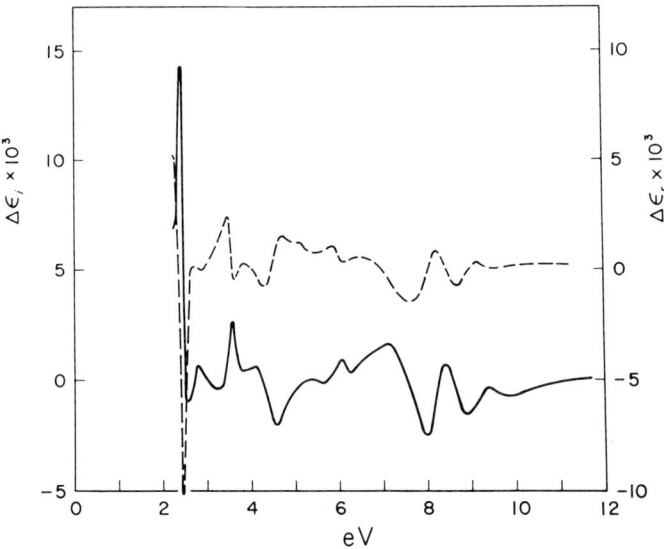

FIG. 63. Modulation in ε_r and ε_i which corresponds to the thermoreflectance data for gold at room temperature of Fig. 62. These results were obtained by Kramers–Kronig analysis from the data of Fig. 62. The dashed curve represents $\Delta \varepsilon_r$ and the solid curve $\Delta \varepsilon_i$.

energy band calculations for gold. We have, however, compiled in Table VIII a very tentative list of assignments based on recent band calculations for copper (Fig. 40, also Ref. 175) and the expected analogy of copper and gold.[174]

Hanus et al.[207] have recently reported thermoreflectance measurements for nickel which are shown in Fig. 64. Most remarkable in these data is the extremely sharp structure which occurs at low photon energies. In order to interpret the observed thermoreflectance spectrum, these authors have proposed a new band structure of nickel (based on the majority spin band minority spin band model[211]) near the Fermi surface. This band structure is shown schematically in Fig. 65. The three peaks at 0.25, 0.4, and 1.3 eV involve transitions to the Fermi surface. The 0.25 and 1.3 eV line shapes can be understood as due to an increase in the Fermi energy with respect to the initial states for the spin-up band. The shape of the 0.4 eV, somewhat similar to a derivative of a Lorentzian (Fig. 50) cannot be completely understood as caused by interband transitions.

19. RESULTS

TABLE VIII. SOME OF THE PEAKS OBSERVED ON THE THERMO-
REFLECTANCE SPECTRUM OF GOLD (FIG. 62) AND TENTATIVE
BAND STRUCTURE ASSIGNMENTS[a]

Peak (eV)	Tentative assignment
2.4	near $L_3^{(2h)} \to E_F$ (near $L_{2'}$)
2.9	near $\Delta_5 \to E_F$ (near Δ_1)
3.2	near $L_3^{(2l)} \to E_F$ (near $L_{2'}$)
3.6	$X_5 \to X_{4'}$
4.4	near $L_3^{(1)} \to E_F$ (near $L_{2'}$)
8.1	X_1 or $X_3 \to X_{4'}$
9.0	$W_1 \to W_{2'}$ or $Z_2 \to Z_1$

[a] The superscript after L-states indicates the order in which several states with the same symmetry occur, starting with the lowest. The superscripts h and l are used for transitions "near" a degenerate state of a high-symmetry point to indicate whether the split state is that of higher or lower energy.

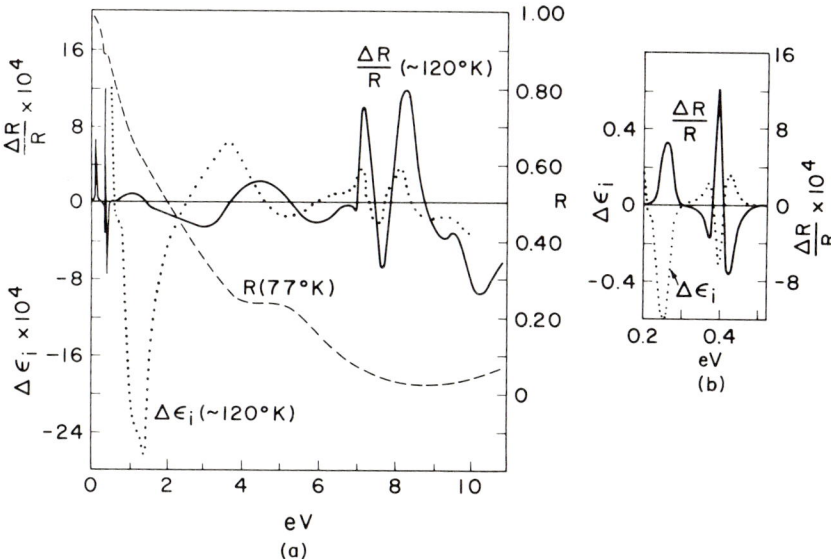

FIG. 64. (a) Reflectivity and thermoreflectance spectrum of nickel near liquid nitrogen temperature [see J. Hanus, J. Feinleib, and W. J. Scouler. *Phys. Rev. Letters* **19**, 16 (1967)]. (b) ε_i in the 0.2–0.5 eV region obtained by the Kramers–Kronig analysis of the thermo-reflectance data.

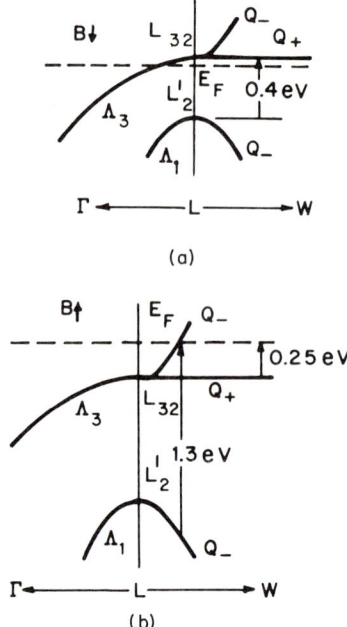

FIG. 65. Band structures of nickel for (a) minority spin band (B↓) and (b) the majority spin band (B↑) proposed by J. Hanus, J. Feinleib, and W. J. Scouler, *Phys. Rev. Letters* **19**, 16 (1967), on the bases of thermoreflectance measurements. The transitions responsible for the 0.25, 0.4, and 1.3 eV thermoreflectance peaks are indicated.

Matatagui and Cardona[212] have recently performed thermoreflectance measurements on evaporated films of alkali metals (K, Rb, and Cs) in the photon energy region between 1 and 6 eV. A curious result of these experiments is the appearance of very strong interference fringes at photon energies above the plasma frequency for very thin samples (see Fig. 66). These fringes, are very difficult to see in the conventional transmission and reflection spectra.†

For thick samples the interference fringes disappear (Fig. 66) and a strong peak, with a shape similar to that discussed in Section 17c appears at a frequency near that expected for the plasma frequency. The peak observed for potassium at higher frequencies is possibly

† These fringes may be related to *p*-wave plasma resonances predicted by Melnyk and Harrison [*Phys, Rev. Letters* **21**, 85 (1968)].

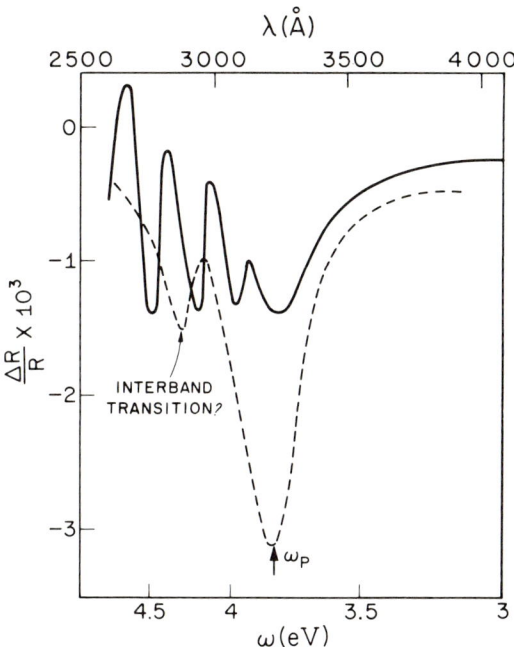

FIG. 66. Thermoreflectance spectrum of 2 potassium films. The thinner film (dashed curve) exhibits strong interference fringes at photon energies above the plasma frequency. The thicker film shows a peak at the plasma frequency and another peak due possibly to interband transitions. From E. Matatagui and M. Cardona, *Solid State Commun.* **6**, 313 (1968).

due to interband transitions. No evidence for Mayer–El Naby collective resonances[39] has been obtained.

The peak in thermoreflectance associated with the minimum in reflectivity (Section 17c) has only been observed for cesium: in other alkali metals it should fall beyond the range of those measurements ($\omega_2 > 6$ eV) since ω_p is high and ε_L is small. The values of ω_p determined by this method and those determined by Mayer and Hietel[213] are given in Table IX together with the estimated ε_L and free electron plasma frequencies (for $m^* \simeq m$ and $\varepsilon_L \simeq 1$). The plasma frequencies obtained from electron energy loss experiments[214] are also given in Table IX.

TABLE IX. CALCULATED PLASMA FREQUENCIES AND "LATTICE" DIELECTRIC CONSTANTS ε_L OF THE ALKALI METALS COMPARED WITH THOSE OBSERVED BY SEVERAL METHODS, INCLUDING THERMOREFLECTANCE

Metal	Calculated[a]		Estimated[b]	Calculated[c]	Thermoreflectance	Energy loss[d]	Polarimetry[e]	Thermoreflectance	Polarimetry[f]
	ω_p		ε_L	ω_p	ω_p	ω_p	ω_p	ε_L	ε_L
Li	8.1		1.03	8.0					1.10
Na	6.0		1.11	5.7		5.87	5.49		1.28
K	4.4		1.27	3.9	3.82	3.87	4.25		1.50
Rb	4.0		1.38	3.4	3.40		3.09		1.37
Cs	3.6		1.55	2.9	2.85		2.86	2.5	

[a] Calculated for $m^* = 1$ and $\varepsilon_L = 1$, from D. Pines, in "Elementary Excitations in Solids," p. 209. Benjamin, New York, 1963, and references therein.

[b] Estimated from the core polarizabilities given by J. H. Van Vleck, "The Theory of Electric and Magnetic Susceptibilities," p. 225. Oxford Univ. Press, London and New York, 1932.

[c] Calculated for ε_L of [b] and $m^* = 1$, from D. Pines, in "Elementary Excitations in Solids." p.209. Benjamin, New York, 1963, and references therein.

[d] J. L. Robins and P. E. Best, Proc. Phys. Soc. (London) **79**, 110 (1962).

[e] From H. Mayer and B. Hietel, in "Optical Properties and Electronic Structure of Metals and Alloys" (F. Abeles, ed.). Wiley, New York, 1966.

[f] M. H. Cohen, Phil. Mag. **3**, 762 (1958).

VI. Stress Modulation

20. General Considerations

a. Phenomenological Analysis

Stress modulation experiments can, in principle, be performed with hydrostatic or uniaxial stress. Hydrostatic stress modulation preserves the symmetry of the crystal and hence the corresponding spectra are similar to those obtained by means of wavelength modulation or, if broadening modulation is negligible, to temperature modulation. Differences in the stress coefficients of the various gaps may, however, make certain transitions more or less easily observable in hydrostatic pressure modulation than in wavelength modulation spectra.

No hydrostatic pressure modulation spectra have yet been reported. Hence, in our discussion we shall emphasize uniaxial pressure modulation. All materials measured so far by this technique are cubic. We shall therefore limit our discussion to cubic materials.

Let e_{ij} be the components of the strain tensor e produced by the application of the modulating stress. The change $\Delta\varepsilon$ induced by the strain on the dielectric constant tensor can be written, under the assumption of a small modulating stress, as:

$$\Delta\varepsilon = \Delta\varepsilon_r + i\,\Delta\varepsilon_i = \mathsf{W} \cdot \mathsf{e}. \tag{20.1}$$

The components W_{ijkl} of the electrooptic tensor W are in general complex and, hence, two *real* fourth rank tensors are required to describe completely the electrooptic effect. In a cubic material of the O_h, O_d, and T_d point groups (we shall treat only materials with one of these point groups) only 3 independent components are needed to specify each one of these tensors. They are, in the standard notation[215a] †:

$$W_{1111} = W_{11} = W_{11}^{(r)} + iW_{11}^{(i)}; \quad W_{1122} = W_{12} = W_{12}^{(r)} + iW_{12}^{(i)}$$
$$W_{1212} = W_{44} = W_{44}^{(r)} + iW_{44}^{(i)}. \quad (20.2)$$

The modulation is usually accomplished by the application of a uniaxial stress with a nonzero diagonal component along a given direction. All other components of the stress tensor are zero. This stress, and the corresponding strain tensors, can be decomposed into the sum of 2 parts, a pure hydrostatic part [$e_{hyd} = (1/3)$ tr e for the strain tensor] and a pure shear part [$e_{shear} = e - (1/3)$ tr e]. The effects of the hydrostatic component of the strain can be reduced to those discussed in Chapter IV. In this chapter, we shall consider in detail the effects of the pure shear strain.

A cubic material becomes birefringent upon application of a shear stress. Thus the reflectivity at normal incidence depends on $\hat{n}$, the unit vector along the direction of the electric field of light. To first order in e, we can write:

$$(\Delta R/R)_{\hat{n}} = \hat{n} \cdot \Delta \mathscr{R} \cdot \hat{n}, \quad (20.3)$$

with

$$\Delta \mathscr{R} = \beta_r \, \Delta \varepsilon_r + \beta_i \, \Delta \varepsilon_i.$$

The coefficients β_r and β_i are obtained with Eq. (8.9) from the optical constants of the unperturbed material. Hence, the behavior of $(\Delta R/R)_{\hat{n}}$ is characterized by the tensor $\Delta \mathscr{R}$, which is related to the strain tensor e through the equation:

$$\Delta \mathscr{R} = \mathbf{Q} \cdot \mathbf{e}, \quad (20.4)$$

with

$$\mathbf{Q} = \beta_r \, \mathbf{W}^{(r)} + \beta_i \, \mathbf{W}^{(i)}.$$

The piezoreflectance tensor Q is characterized by 3 independent real components:

$$Q_{11} = Q_{1111}; \quad Q_{12} = Q_{1122}; \quad Q_{44} = Q_{1212}.$$

We emphasize again that this conclusion is only valid for small modulation amplitudes and for cubic materials.

The effect of a hydrostatic strain $e = e_{11} = e_{22} = e_{33}$ is governed for cubic materials by $Q_{11} + 2Q_{12}$ and by $W_{11} + 2W_{12}$:

$$\Delta R/R = (Q_{11} + 2Q_{12})e; \quad \Delta \varepsilon = (W_{11} + 2W_{12})e, \quad (20.5)$$

† We use W_{44} as defined in Gobeli and Kane[63] and Engeler et al.[65] The W_{44} of Garfinkel et al.[178] differs from this by a factor of 2.

where Q is real and W is complex. 3 sets of measurements are required to determine Q_{11}, Q_{12}, and Q_{44}. Measurements are usually performed for a stress along [100] and [111] (or along [100] and [110]), with **E** perpendicular and parallel to the stress direction. These measurements suffice to determine all independent components of Q (3 components), with one extra condition which can be used to check self-consistency. The components of $W^{(r)}$ and $W^{(i)}$ are then determined[178] with the dispersion relations [Eqs. (13.1) and (13.2) for each independent tensor component].

b. Direct Transitions

We shall first consider optical structure caused by an allowed critical point at $\mathbf{k} = 0$. We shall treat the important case of transitions between a p-like valence band state of $\Gamma_{25'}$ (or Γ_{15}) symmetry (see Figs. 9 and 29) and an s-like conduction band of $\Gamma_{2'}$ (or Γ_1) symmetry. The valence band is usually split by the spin orbit interaction in the manner described in Section 9. The stress Hamiltonian is assumed to couple only states of the $\Gamma_{25'}$ (or Γ_{15}) band with each other and the $\Gamma_{2'}$ (or Γ_1) state with itself: energy gaps other than those produced by the spin orbit coupling are large. The stress Hamiltonian representing the energy difference between the conduction and the valence band states can be written as[215b]:

$$a \operatorname{tr} e - 3b[(L_x^2 - \tfrac{1}{3}L^2)e_{xx} + \mathrm{cp}] - 2d3^{1/2}[\{L_x L_y\}e_{xy} + \mathrm{cp}], \quad (20.6)$$

where cp stands for circular permutation, L is the orbital angular momentum operator and a, b, and d are three parameters (deformation potentials). The deformation potential a represents the change in the gap with hydrostatic strain while the deformation potentials b and d are responsible for splittings of the $\Gamma_{25'}$ (or Γ_{15}) bands under uniaxial compression along [100] and [111], respectively. In the small modulation experiments the splittings produced by the stress are much smaller than the spin orbit splittings; hence we shall neglect here the coupling of the Γ_8 top valence bands to the Γ_6 lowest spin orbit split component of $\Gamma_{25'}$ (or Γ_{15}). We shall consider this interaction and its effects in Chapter VIII. We first treat the case of a modulating stress with a nonzero diagonal component X along [001]. The wave functions which diagonalize the Hamiltonian are those of Eq. (9.3) if coupling between $(\tfrac{3}{2}, \tfrac{1}{2})$ and $(\tfrac{1}{2}, \tfrac{1}{2})$ is neglected.

The energy gaps for the stressed crystals are (in the conventional notation):

$$E_0 + \begin{cases} \delta\omega_H + \frac{1}{2}\delta\omega_{001} & \text{for the } (\frac{3}{2}, \pm\frac{3}{2}) \text{ valence bands} \\ \delta\omega_H - \frac{1}{2}\delta\omega_{001} & \text{for the } (\frac{3}{2}, \pm\frac{1}{2}) \text{ valence bands} \\ \delta\omega_H + \Delta_0 & \text{for the } (\frac{1}{2}, \pm\frac{1}{2}) \text{ valence bands,} \end{cases} \quad (20.7)$$

with $\delta\omega_{001} = 2b(S_{11} - S_{12})X$ and $\delta\omega_H = a \text{ tr e} = a(S_{11} + 2S_{12})X$. The constants S_{ij} are the elastic compliance constants of the crystal. In order to calculate the expected modulation spectrum we must evaluate the matrix elements and density of states masses for the transitions between the various split bands and the conduction band. The optical density of states masses of the $(\frac{3}{2}, \frac{3}{2})$ and $(\frac{3}{2}, \frac{1}{2})$ bands after splitting are not the same. An estimate based on the F, G, and M $\mathbf{k} \cdot \mathbf{p}$ parameters[116] indicates, however, that for most germanium-like materials, these masses are not very different. Since the optical reduced mass is determined mostly by the small conduction band mass [Eq. (6.11)] we shall assume that all 3 valence bands have the same mass [if the spin orbit splitting is large, the mass of the $(\frac{1}{2}, \frac{1}{2})$ band can become larger but this fact is easily taken into account]. The contribution of the various bands to the change in $\Delta\varepsilon_r$ is:

$$E_0 \begin{cases} (\frac{3}{2},\frac{3}{2}) \begin{cases} \Delta\varepsilon_{33}^{(r)} = \Delta\varepsilon_{\parallel}^{(r)} = 0 \\ \Delta\varepsilon_{11}^{(r)} = \Delta\varepsilon_{22}^{(r)} = \Delta\varepsilon_{\perp}^{(r)} \\ \quad \propto -\text{Re}(\omega_g - \omega)^{-1/2} \times \frac{1}{2}[\delta\omega_H + \frac{1}{2}\delta\omega_{001}] \end{cases} \\ (\frac{3}{2},\frac{1}{2}) \begin{cases} \Delta\varepsilon_{\parallel}^{(r)} \propto -\text{Re}(\omega_g - \omega)^{-1/2} \times \frac{2}{3}[\delta\omega_H - \frac{1}{2}\delta\omega_{001}] \\ \Delta\varepsilon_{\perp}^{(r)} \propto -\text{Re}(\omega_g - \omega)^{-1/2} \times \frac{1}{6}[\delta\omega_H - \frac{1}{2}\delta\omega_{001}] \end{cases} \end{cases} \quad (20.8)$$

$E_0 + \Delta_0; (\frac{1}{2}, \frac{1}{2}); \Delta\varepsilon_{\parallel}^{(r)} = \Delta\varepsilon_{\perp}^{(r)} \propto -\text{Re}(\omega_g + \Delta_0 - \omega)^{-1/2}\frac{1}{3}\delta\omega_H.$

Equations (20.8) show that the $E_0 + \Delta_0$ gap gives an isotropic singularity in the stress modulation spectra, while the E_0 gap gives anisotropic spectra (the material becomes uniaxial under stress). Adding the contributions of the $(\frac{3}{2}, \frac{3}{2})$ and $(\frac{3}{2}, \frac{1}{2})$ transitions we obtain:

$$\begin{aligned} \Delta\varepsilon_{\parallel}^{(r)} &\propto -\text{Re}(\omega_g - \omega)^{-1/2} \times \frac{2}{3}[\delta\omega_H - \frac{1}{2}\delta\omega_{001}] \\ \Delta\varepsilon_{\perp}^{(r)} &\propto -\text{Re}(\omega_g - \omega)^{-1/2} \times \frac{2}{3}[\delta\omega_H + \frac{1}{4}\delta\omega_{001}]. \end{aligned} \quad (20.9)$$

The effect of broadenings is easily taken into account by replacing $(\omega_g - \omega)^{-1/2}$ in Eqs. (20.9) by $\eta^{-1/2}F[(\omega_g - \omega)/\eta]$, as discussed in Section 15.

The E_0 edge under consideration is often the lowest direct edge and, therefore, $\beta_i \approx 0$: in this case the reflectance modulation reproduces the structure in $\Delta\varepsilon^{(r)}$. For light polarized with **E** at an angle ϕ with the [001] stress axis, we find:

$$\Delta R_\phi/R = \beta_r[\Delta\varepsilon_\parallel^{(r)} \cos^2\phi + \Delta\varepsilon_\perp^{(r)} \sin^2\phi]. \tag{20.10}$$

The ratio of the strengths of the peaks for **E** perpendicular and parallel to the stress gives the useful relationship:

$$\left(\frac{\Delta R_\parallel/R}{\Delta R_\perp/R}\right)_{X\parallel[100]} = \frac{\delta\omega_H - \tfrac{1}{2}\delta\omega_{001}}{\delta\omega_H + \tfrac{1}{4}\delta\omega_{001}}$$

$$= \frac{a(S_{11} + 2S_{12}) - b(S_{11} - S_{12})}{a(S_{11} + 2S_{12}) + \tfrac{1}{2}b(S_{11} - S_{12})}. \tag{20.11}$$

Since a is usually known from hydrostatic pressure experiments,[53] Eq. (20.11) enables us to determine the shear deformation potential b without knowing the amplitude of the applied stress.

For a [111] stress the wave functions are also those of Eq. (9.3) with the z-axis along the [111] direction. The expressions in Eqs. (20.9)–(20.11) remain valid for the polarization parallel and perpendicular to [111] provided one replaces $\delta\omega_{001}$ by $\delta\omega_{111} = 3^{-1/2} dS_{44}X$. A relationship similar to Eq. (20.11) can be used to obtain d from piezoreflectance data if the hydrostatic deformation potential a is known. The components of the **W** tensor, and likewise those of **Q**, are obtained (in arbitrary units) by multiplying the functions of Table V by the scaling factors:

$$W_{11} \propto -a + b; \quad W_{12} \propto -a - \tfrac{1}{2}b; \quad W_{44} \propto (3^{1/2}/4)d. \tag{20.12}$$

Complete isotropy of the piezoreflectance lines discussed here is obtained whenever $\delta\omega_{111} = \delta\omega_{001}$. This condition is approximately satisfied in many cases.

We shall now discuss transitions at critical points away from $\mathbf{k} = 0$. We assume that there is no degeneracy (other than spin) at a given point of **k**-space. In general the shear strain lifts some of the degeneracy between the various equivalent points of **k**-space. Let $\hat{\mathbf{m}} = \mathbf{k}/|\mathbf{k}|$ be a unit vector along the direction of the critical point. The strain induced change in the energy gap is[216]:

$$\Delta\omega_{\hat{\mathbf{m}}} = \mathscr{E}_1 \operatorname{tr} \mathbf{e} + \mathscr{E}_2[\hat{\mathbf{m}} \cdot \mathbf{e} \cdot \hat{\mathbf{m}} - \tfrac{1}{3}\operatorname{tr} \mathbf{e}], \tag{20.13}$$

where $\mathscr{E}_1$ is a hydrostatic and $\mathscr{E}_2$ a pure shear deformation potential.

The hydrostatic term produces an isotropic effect in piezoreflectance while the shear term produces anisotropy. We shall now calculate the tensors $W_{ij}^{(r)}$ and $W_{ij}^{(i)}$ expected for the various types of critical points along the high-symmetry directions {100}, {111}, and {110}. The shape of the lines is always that given in Table V. The differences between the various independent components of $W^{(r)}$ and $W^{(i)}$ lies only in the coefficients by which the results of Table V must be multiplied. Let us first consider equivalent critical points for $\hat{m}$ = [001] and all equivalent directions. A [111] stress does not split these critical points, hence $W_{44} = 0$. Optical transitions along [001] are polarized with the electric field **E** either along [001] or perpendicular to [001]. We obtain for transitions polarized with **E** along the direction $\hat{m}$ of the critical point:

$$W_{11} - W_{12} \propto -\mathscr{E}_2; \quad W_{11} + 2W_{12} \propto -3\mathscr{E}_1; \quad W_{44} = 0 \quad (20.14)$$

and for transitions polarized with **E** perpendicular to $\hat{m}$:

$$W_{11} - W_{12} \propto \mathscr{E}_2; \quad _{11} + 2W_{12} \propto -6\mathscr{E}_1; \quad W_{44} = 0. \quad (20.15)$$

A similar treatment yields for {111} ellipsoids with transitions polarized with **E** along $\hat{m}$:

$$W_{11} - W_{12} = 0; \quad W_{11} + 2W_{12} \propto -4\mathscr{E}_1; \quad W_{44} \propto -(4/9)\mathscr{E}_2 \quad (20.16)$$

and for transitions polarized with **E** perpendicular to $\hat{m}$:

$$W_{11} - W_{12} = 0; \quad W_{11} + 2W_{12} \propto -8\mathscr{E}_1; \quad W_{44} \propto (4/9)\mathscr{E}_2. \quad (20.17)$$

For {110} critical points we can have 3 different kinds of polarization: **E** parallel to either [110], [1$\bar{1}$0], or [001]. The corresponding piezooptical tensors are given by:

(i) Polarization **E** along [110] for [110] critical points

$$W_{11} - W_{12} \propto -\tfrac{1}{2}\mathscr{E}_2; \quad W_{11} + 2W_{12} \propto -6\mathscr{E}_1; \quad W_{44} \propto -(1/2)\mathscr{E}_2. \quad (20.18)$$

(ii) Polarization **E** along [1$\bar{1}$0] for [110] critical points

$$W_{11} - W_{12} \propto \tfrac{1}{2}\mathscr{E}_2; \quad W_{11} + 2W_{12} \propto -6\mathscr{E}_1; \quad W_{44} \propto (1/2)\mathscr{E}_2, \quad (20.19)$$

(iii) Polarization **E** along [001] for [110] critical points

$$W_{11} - W_{12} \propto \mathscr{E}_2; \quad W_{11} + 2W_{12} \propto -6\mathscr{E}_1; \quad W_{44} = 0.$$

(20.20)

The results given above are summarized in Table X. It is clear that a

TABLE X. Factors by Which the Results of Table V Must Be Multiplied in Order to Obtain the Differential Piezooptical Tensors

Critical point	Polarization	$W_{11} + 2W_{12}$	$W_{11} - W_{12}$	W_{44}
$\mathbf{k} = 0$	$\Gamma_8 \to \Gamma_6$	$-3a$	$(3/2)b$	$3^{1/2}d/4$
[100]	[100]	$-3\mathscr{E}_1$	$-\mathscr{E}_2$	0
	perp. to [100]	$-6\mathscr{E}_1$	$\mathscr{E}_2$	0
[111]	[111]	$-4\mathscr{E}_1$	0	$-(4/9)\mathscr{E}_2$
	perp. to [111]	$-8\mathscr{E}_1$	0	$(4/9)\mathscr{E}_2$
[110]	[110]	$-6\mathscr{E}_1$	$-(1/2)\mathscr{E}_2$	$-(1/2)\mathscr{E}_2$
	[1$\bar{1}$0]	$-6\mathscr{E}_1$	$(1/2)\mathscr{E}_2$	$(1/2)\mathscr{E}_2$
	[001]	$-6\mathscr{E}_1$	$\mathscr{E}_2$	0

measurement of the ratio of $W_{11} - W_{12}$ to W_{44} (or equivalently that of $Q_{11} - Q_{12}$ to Q_{44}) by piezoreflectance techniques can yield information about the symmetry of the transitions involved. For instance $W_{11} - W_{12} = 0$ immediately implies {111} transitions while $W_{44} = 0$ implies either {100} transitions or {110} transitions polarized along {001}.† An approximate equality of $W_{11} - W_{12}$ and W_{44} usually means $\mathbf{k} = 0$ ($\Gamma_8 \to \Gamma_6$) transitions while for [110] transitions with either [110] or [1$\bar{1}$0] polarization one finds $W_{11} - W_{12} = W_{44}$. Hence it may be difficult to distinguish this case from that of $\mathbf{k} = 0$ transitions discussed above by means of optical stress modulation techniques.

c. Direct Excitons

Within the Wannier model it is possible to describe the effect of stress on an exciton level as a combination of the effect of stress on

† Thus the statement found sometimes in the literature (see Gerhardt *et al.*[188]) that $W_{44} = 0$ implies [100] transitions is not correct.

the band edges plus the effect on the binding energy for the excitons. If this binding energy is small its dependence on stress is usually small and the behavior of the exciton under stress is governed by the behavior of the corresponding band edges. If the exciton line shapes are Lorentzian, the corresponding modulation spectra have the shape of Fig. 50 multiplied by the factors of Table X.

If the exciton binding energy is large, we must consider the effect of stress on the symmetry of the full exciton state.[217] Let us treat as an example the case of the exciton levels associated with a $\Gamma_8^- \to \Gamma_6^+$ band-to-band transition in alkali halides. The levels are formed from a Γ_6^+ electron wave function, a Γ_8^- hole wave function, and the envelope function (Section 6b). Optical transitions will be most prominent for s-like (Γ_1) envelope functions; the symmetry of these exciton states is:

$$\Gamma_1 \times \Gamma_8^- \times \Gamma_6^+ = \Gamma_{15} + \Gamma_{25} + \Gamma_{12}. \qquad (20.21)$$

The Γ_{15}, Γ_{25}, and Γ_{12} states in general will be split from each other. Only transitions to Γ_{15} are electric dipole allowed. If the stress splittings are smaller than the $\Gamma_{15} - (\Gamma_{25}, \Gamma_{12})$ splitting, and if this splitting is well resolved, we can consider the effect of stress on the Γ_{15} exciton state alone and neglect the stress coupling to the Γ_{25} and Γ_{12} states. Using the equivalent of Eq. (20.6) for the exciton we find for a Γ_{15} state[217]:

$$W_{11} + 2W_{12} \propto -3a_{\text{ex}}; \qquad W_{11} - W_{12} \propto -3b_{\text{ex}}; \qquad W_{44} \propto -d_{\text{ex}}\, 3^{1/2}/2, \qquad (20.22)$$

where a_{ex}, b_{ex}, and d_{ex} are the deformation potentials of the Γ_{15} exciton state. These parameters are not necessarily related to those of the corresponding valence band edge since we have assumed the exciton binding energy large and, hence, the stress could possibly have an effect on the envelope function of the exciton. Also, because of the large exciton binding energy, the exciton state may involve contributions from more than one band edge: in the alkali halides, severe mixing of the Γ_7^- and Γ_8^- valence band edges may occur.[218] If we neglect these effects, we find $a_{\text{ex}} = a$, $b_{\text{ex}} = -\frac{1}{2}b$ and $d_{\text{ex}} = -\frac{1}{2}d$, where $\frac{1}{2}b$ and $\frac{1}{2}d$ are the band edge, parameters and the coefficients of Eqs. (20.22) revert to those found for interband transitions.

d. Indirect Transitions

Since the energy dependence of ε_i for band-to-band allowed indirect transitions is $\varepsilon_i \sim (\omega - \omega_g \pm \omega_{\text{phon}})^2$, the structure expected for these transitions in optical modulation spectra is rather weak. Fortunately exciton effects transform this singularity near $\omega = \omega_g \pm \omega_{\text{phon}}$ into a familiar square root singularity and hence sharp peaks appear in the corresponding modulation spectra for $\omega = \omega_g \pm \omega_{\text{phon}}$ as already discussed in Section 16.† The line shapes of these modulation spectra will be those already given for direct allowed transitions. In the usual case (germanium and silicon) of transitions from the top degenerate valence band at $\mathbf{k} = 0$ to conduction band-valleys off $\mathbf{k} = 0$ both valence and conduction band splittings and shifts contribute to the stress modulation spectra. In order to calculate the relative intensities of these lines we must examine the matrix element for the various possible transitions.

The phonon spectrum of germanium for $\mathbf{k}$ along [100] and [111] is shown in Fig. 67. The corresponding spectrum of silicon is similar to it (same ordering of states and symmetries). Figure 29 shows that indirect transitions can occur for germanium below E_0 from $\Gamma_{25'}$ (Γ_8^+) to L_1. These transitions take place through an intermediate

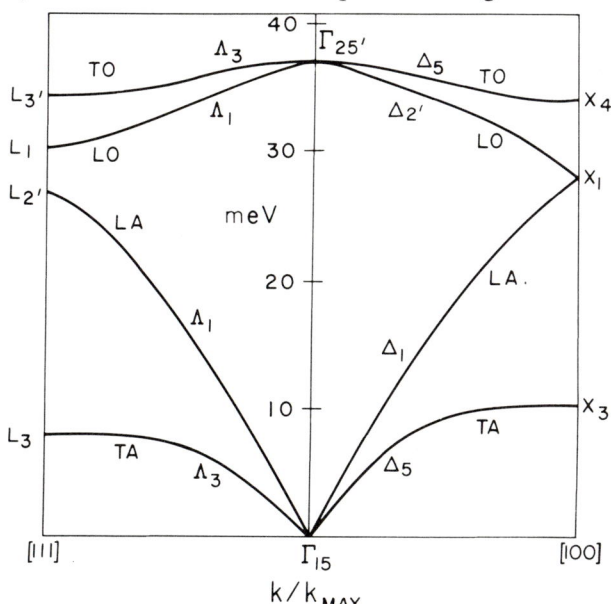

FIG. 67. Phonon spectrum of germanium [B. N. Brockhouse and P. K. Iyengar, *Phys. Rev.* **111**, 747 (1958)] for $\mathbf{k}$ along [111] and [100].

† The exciton binding energy is usually quite small. For the purpose of this discussion, we assume it is negligible.

state either at $\Gamma(\Gamma_i)$ or at $L(L_i)$. In the case of a Γ_i intermediate state the photon takes the electron from $\Gamma_8{}^+$ to Γ_i and the phonon takes it from Γ_i to the final state. For an L_i intermediate state the phonon takes the electron from $\Gamma_8{}^+$ to L_i and the photon from L_i to L_1. Two intermediate states appear possible at $\mathbf{k}=0$ within the range of Fig. 29: $\Gamma_{2'}$ and Γ_{15} (we neglect all other Γ_i states because of the large energy denominators involved). The characters of the representation of the possible phonons which produce the $\Gamma_i \rightarrow L_1$ transitions are found from the characters of the Γ_i and L_1 representations as the product[219]:

$$\chi^*_{\Gamma_i}(R_L) \cdot \chi_{L_1}(R_L), \qquad (20.23)$$

where R_L are the elements of the group of the **k**-vector at L. By using (20.23) and standard character tables[220] we obtain:

$$\begin{aligned} L_1 \times \Gamma_{2'} &= L_{2'}(\text{LA}) \\ L_1 \times \Gamma_{15} &= L_{1'} + L_{3'}(\text{TO}) \end{aligned} \qquad (20.24)$$

Within the range of Fig. 29, we find L_3 and $L_{3'}$ as the only possible intermediate states at L. The possible phonons which could produce these transitions are:

$$\begin{aligned} L_3 \times \Gamma_{25'} &= 2L_3(\text{TA}) + L_1(\text{LO}) + L_2 \\ L_{3'} \times \Gamma_{25'} &= 2L_{3'}(\text{TO}) + L_{1'} + L_{2'}(\text{LA}). \end{aligned} \qquad (20.25)$$

A consideration of the energy denominators involved in the indirect transitions $[\varepsilon_i \sim (\omega_f - \omega_i)^{-2}$, see Eq. (5.1)] suggests that the LA phonon transitions of Eq. (20.24) must be dominant while the transitions with L_3 as intermediate state should be negligible. Equations (20.24) and (20.25) show that allowed transitions with the aid of a transverse optical (TO) phonon should play an important role. Transitions with the aid of LO and TA phonons should be forbidden at L but they may become allowed if we bring the electrons to points in the neighborhood of L; these transitions are first-order forbidden and the energy dependence of the corresponding modulation spectrum is, for exciton transitions, $(\omega - \omega_g \pm \omega_{\text{phon}})^{1/2}$. In this case, the derivative spectrum has the shape of a threshold rather than a peak and hence we have an easy method for distinguishing between allowed and forbidden transitions.

The transitions which use the $\Gamma_{2'}$ state as the intermediate state are particularly simple. The selection rules and matrix elements for the photon part of the transition are those derived in Section 20b for the $\Gamma_{25'} \to \Gamma_{2'}$ direct transitions. The phonon part is isotropic: the matrix element is the same for transition to any of the L_1 equivalent valleys. For germanium we shall only treat in detail these dominant transitions because of their simplicity. For a [100] stress no splitting of the conduction band valleys occurs and therefore the W_{11} and the W_{12} components of the strain optical tensor are [Eqs. (20.12)]:

$$W_{11} \propto -\mathscr{E}_1 + a_v + b; \quad W_{12} \propto -\mathscr{E}_1 + a_v - \tfrac{1}{2}b, \quad (20.26)$$

where $\mathscr{E}_1$ and a_v are the hydrostatic coefficients of the $\Gamma_{25'}$ valence band and the L_1 conduction band, respectively, and b is the shear coefficient of the valence band. For a [111] compression we have, beside the splitting of the valence band, a splitting of the L_1 conduction bands according to Eq. (20.13). However, since the phonon transitions from $\Gamma_{2'}$ to all L_1 valleys are equally allowed, we can take for the L valleys their *average energy* (their splitting does not contribute to the strain optical tensor) and therefore $W_{44} \propto 3^{1/2}d/4$ [Eq. (20.12)].

In silicon, indirect transitions take place from $\Gamma_{25'}$ to Δ_1 conduction band minima ([100] direction). The $\Gamma_{2'}$ and the Γ_{15} possible intermediate states are now almost degenerate, although density of states considerations lead us to believe that the transitions with Γ_{15} as an intermediate state are dominant. They are allowed for LA, TO and TA phonons. A Δ_5 intermediate state is also possible and should lead to allowed transitions for all phonon branches.[219] However, indirect transitions involving either TA or TO phonons seem to be dominant.[202] Let us consider the transitions aided by TO phonons, since they are the only indirect transitions in silicon whose modulation spectrum is strong.[67] For a [111] stress, the conduction bands are not split, therefore the only contribution to W_{44} comes from the splitting of the valence band. However, W_{44} is going to be different from that given in Eq. (20.12) since now the photon transitions take place between states of symmetry other than $\Gamma_{25'} \to \Gamma_{2'}$. The matrix elements for these indirect transitions have been calculated by Erlbach.[221] We find from his results for the strain optical tensor which corresponds to transitions via the Γ_{15} intermediate state:

$$W_{11} \propto -\mathscr{E}_1 + a_v - \tfrac{1}{2}b - \tfrac{1}{6}\mathscr{E}_2; \quad W_{12} \propto -\mathscr{E}_1 + a_v + \tfrac{1}{4}b + \tfrac{1}{12}\mathscr{E}_2;$$

$$W_{44} \propto -3^{1/2}d/8. \quad (20.27)$$

and for transitions via the Δ_5 intermediate state:

$$W_{11} \propto -\mathscr{E}_1 + a_v - \tfrac{1}{2}b + \tfrac{1}{3}\mathscr{E}_2 \, ; \qquad W_{12} \propto -\mathscr{E}_1 + a_v + \tfrac{1}{4}b - \tfrac{1}{6}\mathscr{E}_2 \, ;$$
$$W_{44} = 0. \tag{20.28}$$

Equations (20.27) and (20.28) indicate that we could, from piezo-transmission measurements, determine the relative strengths of indirect transition via the various possible intermediate states (such as $\Gamma_{2'}$, Γ_{15}, and Δ_5).

21. Experimental Techniques: Piezoabsorption and Piezoreflectance

Several methods have been used to apply a modulated stress to a sample for optical measurements. The original piezoreflectance measurements made use of an electromechanical transducer built with a piezoelectric material such as quartz[63] or lead zirconate–titanate.[62] If measurements are performed on evaporated films, the films can be evaporated directly on a polished surface of the transducer.[62] Thin single crystals prepared from bulk by grinding and polishing techniques can be either cemented, e.g., with Duco cement,[63,69] or soldered[62] to one of the transducer electrodes. Vacuum grease seems to give a sufficient and convenient bond for low temperature operations.[194]

Lead zirconate–titanate transducers have also been used for piezo-transmission measurements.[65] In order to permit transmission of the light, two matched bar-shaped transducers can be mounted into a yoke. The sample is then placed inside the yoke as shown in Fig. 68a. The piezoelectric transducers can be operated in either a resonant or a nonresonant mode. Frequencies of measurement with piezoelectric transducers in the neighborhood of 1 kc/sec and as high as 130 kc/sec[63] have been reported. In all cases extreme care must be taken to avoid spurious signals due to changes in the path of the light beam produced by the vibration of the sample. In order to reduce these to a minimum, a very smooth and flat reflecting surface is needed. In spite of all precautions some spurious background signal, usually not strongly dependent on frequency, is always present. If a Kramers–Kronig analysis of the data is to be performed, one must make a judicious guess as to the amount of spurious background and this background must be subtracted from the signal.

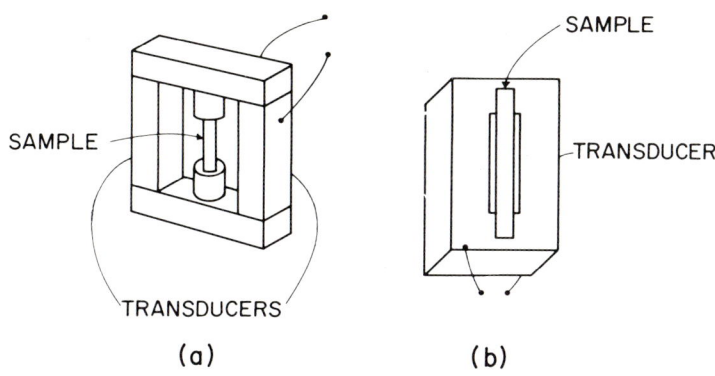

Fig. 68. Yoke-shaped transducer arrangements for piezotransmission measurements. The arrangement in (a) was used by W. E. Engeler, M. Garfinkel, and J. J. Tiemann, *Phys. Rev.* **155**, 693 (1967), while that in (b) was used by A. R. Aggarwal, M. D. Zuteck, and B. Lax, *Phys. Rev. Letters* **19**, 236 (1967).

Gerhardt and co-workers[188,222] have reported piezoreflection measurements on metals and alkali halides with the strain applied by bending the crystal periodically at a frequency of 200 cps. They mention that the signal was completely devoid of spurious uniform background.

Balslev[223] has reported a modification of the standard apparatus for the application of uniaxial stress[67] which permits the superposition of a modulation stress on the static stress. A diagram of this system is shown in Fig. 69a. The sample is held in position by means of bronze blade springs, which provide minimum friction. The modulating stress is applied with a magnet which produces a radial field in the angular gap of the yoke similar to that in a conventional loudspeaker. An ac current is passed through a coil in this gap in order to produce the stress modulation.

Feldman[224] has used a different principle to measure the birefringence induced by a small stress on the reflectivity of a cubic material. He mounts the sample in a small stressing frame and applies a static stress by means of a screw (it should be noted that it is usually difficult to measure the magnitude of such stress). The stressing frame is then rotated about an axis perpendicular to the direction of the stress while linearly polarized light is reflected at nearly normal incidence on the sample (the light beam propagates along the axis of rotation of the frame). A lock-in amplifier is then used to detect the

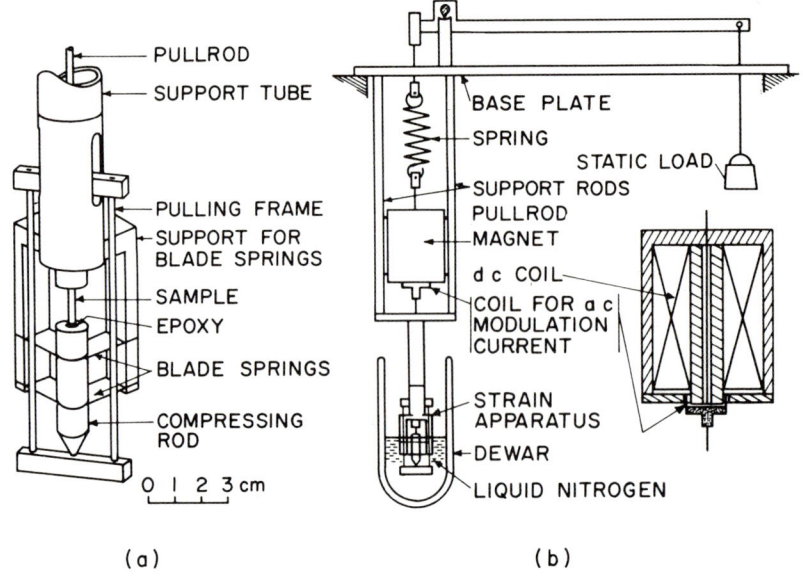

FIG. 69. (a) Sample holder and stressing frame used by I. Balslev *Rev. Sci. Inst.* **38**, 1528 (1967), for modulated stress experiments with a superimposed static stress. The blade springs were made of 0.3 mm bronze. (b) Complete aparatus for optical measurements with a static and a modulated stress.

modulation which is synchronous with the rotation, i.e., $\Delta R_\perp - \Delta R_\parallel$ ($\Delta R_\perp$ and $\Delta R_\parallel$ are the stress induced changes in the reflectivities perpendicular and parallel to the stress direction, respectively). The experiment gives, therefore, less information than a conventional piezoreflectance experiment of the type described earlier, which measures $\Delta R_\perp$ and $\Delta R_\parallel$ independently.

22. Results

a. Semiconductors

Figure 70 shows the piezoreflectance spectra of the E_0 edge (lowest direct edge at $\mathbf{k} = 0$) of Ge and GaAs.[67] The low-energy peaks (1) are similar in shape to those expected for direct interband transitions (the peak in ΔR should be proportional to that in $\Delta \varepsilon_r$ around the E_0 edge) but the high-energy component labeled (2) cannot be explained as caused by interband transitions (for comparison we show the shape of ΔR around E_0 expected for interband transitions according to Fig. 4). In order to explain qualitatively the shapes of Fig. 70 we must invoke exciton effects: the energy of the zero of the signal between (1) and (2) corresponds then to the energy of the exciton

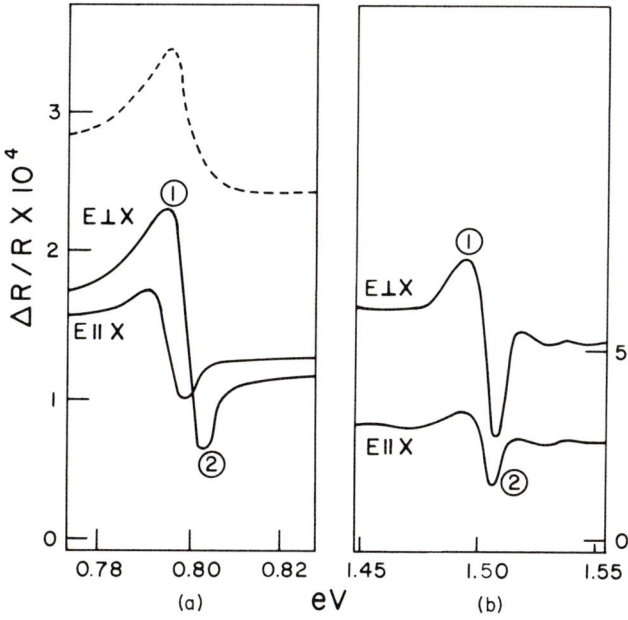

FIG. 70. Piezoreflectance spectra of the lowest direct absorption edge (E_0) of (a) Ge (at 300°K) and (b) GaAs (at 100°K) for [111] stress, as measured by I. Balslev, *Solid State Commun.* **5**, 315 (1967). The samples are under a small (≈ 450 kg cm^{-2}) static compression which can be neglected since splittings are not resolved. The dashed curve represents the shape of $\Delta R/R$ expected for pure interband transitions. The origin of $\Delta R/R$ is arbitrary.

state (Fig. 50). The ratio of the intensities of the peaks for **E** parallel to the stress X to those for **E** perpendicular to X is given by [see Eq. (20.11)]:

$$\frac{\Delta R_{\parallel}/R}{\Delta R_{\perp}/R} = \frac{a(S_{11} + 2S_{12}) - (dS_{44}/2) \, 3^{-1/2}}{a(S_{11} + 2S_{12}) + (dS_{44}/4) \, 3^{-1/2}}. \quad (22.1)$$

Equation (22.1) can be used to determine d since a is known for Ge and for GaAs from hydrostatic pressure measurements.[53] Using the known elastic constants of Ge[225] and GaAs[111] and $a = 7.7$ eV for Ge and $a = 8.9$ eV for GaAs, we find:

$$d = 4.5 \quad \text{for Ge}$$
$$d = 3.7 \quad \text{for GaAs.} \tag{22.2}$$

This value of d for Ge is in reasonable agreement with those obtained by other methods (for a detailed comparison see Chapter VIII). For GaAs, $d = 3.7$ is somewhat smaller than the value generally accepted (~ 5.5 eV).[226]

Figure 71 shows the piezoreflectance spectrum of germanium

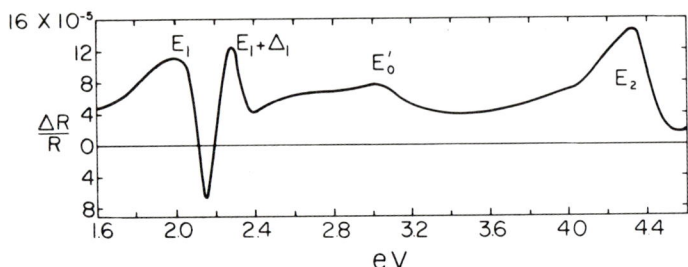

FIG. 71. Piezoreflectance spectrum of germanium at room temperature as reported by W. E. Engeler, H. Fritzsche, M. Garfinkel, and J. J. Tiemann, *Phys. Rev. Letters* **14**, 1069 (1965).

obtained by Engeler *et al.*[62] The measurements were performed with unpolarized light and the orientation of the sample used is not known. Hence a Kramers–Kronig analysis of the data is not possible. Such an analysis would probably resolve conclusively the $E_1 - (E_1 + \Delta_1)$ peaks (Section 9): A knowledge of their existence is required to identify them in Fig. 71. The E_0' peak and the E_2 peak appear clearly at around 3.0 and 4.3 eV, respectively. Measurements with polarized light and oriented samples should yield information about the symmetry of the transitions causing these peaks.

Figure 72 shows the real and imaginary parts of the W_{11}, W_{12}, and W_{44} components of the strain-optical tensor of silicon at room temperature obtained by means of a Kramers–Kronig analysis of the data of Gobeli and Kane.[63]

Figure 73 shows the spectral dependence of $(\Delta R_\perp - \Delta R_\parallel)/R$ obtained by Feldman[224] for silicon stressed along [100] by the rotation method discussed in Section 20 (circles). We have also plotted in Fig. 73 (solid curve) the difference in the conventional piezoreflectance signal of silicon for light polarized with **E** perpendicular and parallel to the [100] stress direction as obtained from Gobeli and Kane's data.[63] The solid curve of Fig. 73 has been normalized so as to agree with Feldman's measurements (circles) at the 3.5 eV peak. The agreement between the $(\Delta R_\perp - \Delta R_\parallel)/R$ spectra obtained by these two different methods is good.

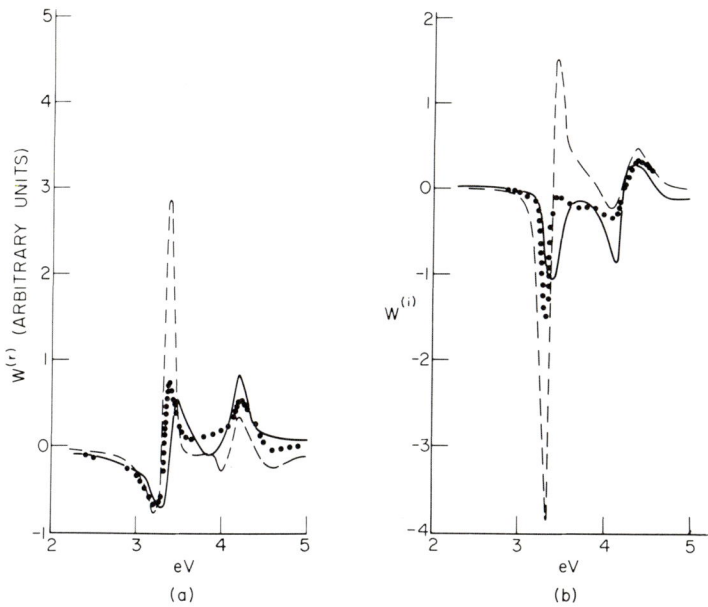

FIG. 72. Real and imaginary parts of the components of the strain-optical tensors of silicon obtained from the data of G. W. Gobeli and E. O. Kane, *Phys. Rev. Letters* **15**, 142 (1965) (at room temperature). (a) Shows $W_{11}^{(r)}$ (dotted line), $W_{12}^{(r)}$ (dashed line), and $W_{44}^{(r)}$ (solid line). (b) Shows $W_{11}^{(i)}$ (dotted line), $W_{12}^{(i)}$ (dashed line), and $W_{44}^{(i)}$ (solid line).

The piezoabsorption spectrum of germanium at room temperature in the region of indirect transitions has been reported by Engeler *et al.*[65] These authors performed measurements on oriented samples with polarized light by the yoke-shaped transducer method described in Section 21. The modulation in the transmitted light produced by the stress is mostly due to the modulation in absorption coefficient, however, a small modulation due to the sample reflectivity R [in this region $R = |(\varepsilon_r^{1/2} - 1)/(\varepsilon_r^{1/2} + 1)|^2$] and the change in thickness d with stress is always present and must be corrected. By differentiating the expression for the transmitted intensity [Eq. (8.1)] we obtain[65]:

$$\frac{\Delta I}{I} = \left[-\frac{R}{1-R} + \frac{2R^2 e^{-2\alpha d}}{1 - R^2 e^{-2\alpha d}} \right] \frac{\Delta R}{R}$$

$$- \left[\frac{1 + R^2 e^{-2\alpha d}}{1 - R^2 e^{-2\alpha d}} \right] (d\,\Delta\alpha + \alpha\,\Delta d) \quad (22.3)$$

$$\Delta R = 4 \frac{n_r - 1}{(n_r + 1)^3} \Delta n_r.$$

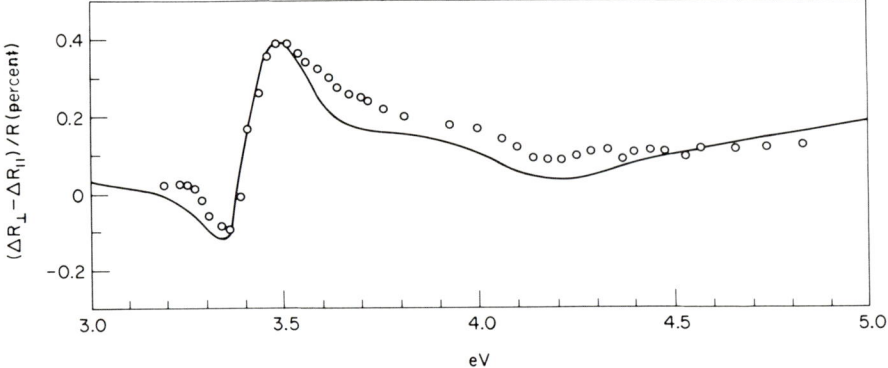

FIG. 73. $(\Delta R_\perp - \Delta R_\parallel)/R$ for silicon with a stress along [100]. The circles were measured by rotation of a stressed sample. [A. Feldman, *Phys. Letters* **25**, 627 (1966). The solid curve has been calculated from the piezoreflectance data of G. W. Gobeli and E. O. Kane, *Phys. Rev. Letters* **15**, 142 (1965).

The dominant term in Eq. (22.3) is that produced by $\Delta\alpha$. The contribution to ΔI produced by ΔR can be either measured directly or evaluated from the measured value of Δn_r for the appropriate polarization direction. The correction due to Δd can be also calculated from the elastic constant if the stress is known (Engeler et al.[65] mounted strain gauges on the samples for this purpose) and hence $\Delta\alpha$ can be calculated from the experimentally measured quantity $\Delta I/I$. From $\Delta\alpha$ for several orientations and polarization directions $W_{11}^{(i)}$, $W_{12}^{(i)}$, and $W_{44}^{(i)}$ can be obtained. These components of the strain-optical tensor obtained for germanium at room temperature are shown in Fig. 74 as a function of photon energy. The structure assigned in the upper part of Fig. 74 to longitudinal acoustical (LA) phonons is of the type shown in Fig. 51. It corresponds therefore to indirect allowed exciton transitions. As we have seen in Section 20d indirect transitions with $\Gamma_{2'}$ as intermediate state are only allowed for LA phonons. These transitions should be dominant in the observed spectrum.

The LO phonon transitions are allowed only with L_3 as an intermediate state. This state gives a large energy denominator and hence

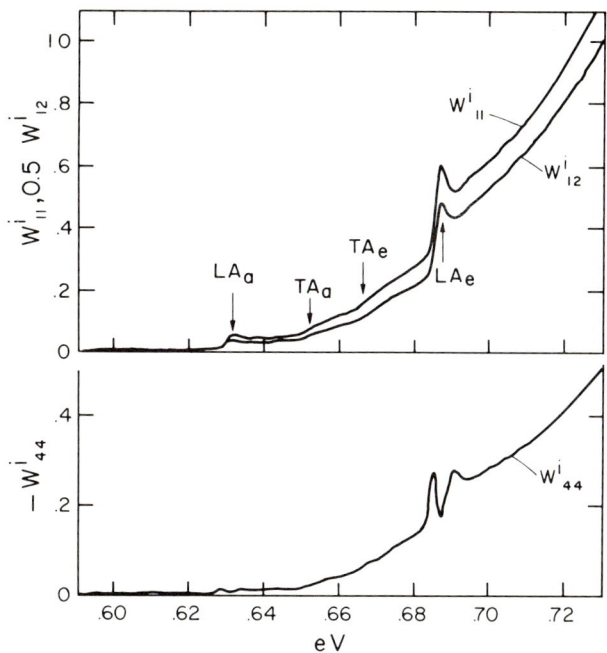

FIG. 74. Imaginary parts of the 2 independent components of the strain optical tensor W of germanium in the region of the indirect gap at room temperature. The peaks and edges associated with the various phonon aided exciton transitions are indicated by arrows. Above the $\omega_g - w_{ex}$ energy (0.659 eV) we find phonon emission aided transitions while below this energy the transitions take place with phonon absorption.

forbidden transitions with $\Gamma_{2'}$ as intermediate state may actually be stronger than allowed ones via L_3. These transitions are expected to be weak and nearly degenerate with the LA transitions (see Table XI): they are not resolved in the spectrum of Fig. 74.

Two other phonons (TA and LO) contribute to the structure in Fig. 74. The corresponding structure is weak and it is difficult to say whether the shape is peak-like (allowed transitions) or threshold-like $\sim (\omega - \omega_g \pm \omega_{phon})^{1/2}$ (forbidden transitions). However, the

TABLE XI. ENERGIES (IN meV) OF THE PHONONS AT THE L POINT FOR GERMANIUM AS OBTAINED BY SEVERAL METHODS INCLUDING PIEZOTRANSMISSION, ELECTROABSORPTION, THERMOTRANSMISSION, AND WAVELENGTH MODULATION

Phonon	Piezotransmission[a]		Indirect transitions				Infrared absorption[f]	Neutron diffraction[g]
	[001]	[011]	Electroabsorption[b]	Tunneling[c]	Transmission[d]	Thermoreflectance[e]		
TA	7.6	7.7	7.3 ± 0.5	7.6	7.6	~10	7.9	8.0
LA	27.7	28.1	27.7 ± 0.1	27.5	27.3	29	26.9	26.7
LO				31.6	30.7		30.4	30.6
TO	35.4	35.9		36.0	36.7		34.5	34.7

[a] M. Garfinkel, J. J. Tiemann, and W. E. Engeler, *Phys. Rev.* **148**, 698 (1966).
[b] A. Frova, P. Handler, F. A. Germano, and D. E. Aspnes, *Phys. Rev.* **145**, 575 (1966).
[c] A. G. Chynoweth, R. A. Logan, and D. E. Thomas, *Phys. Rev.* **125**, 877 (1962).
[d] T. P. McLean "Progress in Semiconductors" (A. F. Gibson, ed.), Vol. 5. Heywood, London, 1960.
[e] Section 19a
[f] S. J. Fray, F. A. Johnson, and R. Jones, *Proc. Phys. Soc.* **76**, 939 (1960).
[g] F. A. Johnson, in "Progress in Semiconductors." (A. F. Gibson, ed.), Vol. 9. Heywood, London, 1965.

positions assigned to the singularity [$\omega_g - \omega_{ex} \pm \omega_{phon}$] in Fig. 74 vary depending on our assumption for the line shape. Since the absorption–emission singularities for a given phonon must lie symmetrically on both sides of $\omega_g - \omega_{ex}$, we have to conclude that the TA singularity is threshold-like (forbidden) while the TO singularity is peak-like (allowed). This is to be expected since the TA transitions are allowed only via the L_3 states [Eq. (20.25)] while the TO transitions are allowed via $L_{3'}$ and have, therefore, a much smaller energy denominator.

The lower part of Fig. 74 shows a rather peculiar structure for the transitions with emission of LA phonons. The reason for this strange doublet-like shape may be the fact that the sample was subjected to a small steady dc stress so as to keep it under tension throughout the whole modulation cycle. This tension should produce splittings[65] in the LA peaks of about 5 meV when W_{44} is measured.

An estimate of the experimental intensities of these LA phonon emission peaks can be easily made from Fig. 74. Since they all have comparable widths, their intensities should be proportional to the height of the steep (low energy) side of the peak for $W_{11}^{(i)}$ and $W_{12}^{(i)}$. The complicated shape of W_{44}^i makes this estimate rather imprecise; it is, however, reasonable to assume that the intensity of this peak lies somewhere between the height of the negative central peak and twice this height. The theoretical relative strengths of these peaks can be estimated from the known deformation potentials of the $\Gamma_{25'}$ and L_1 bands of germanium as indicated in Section 20d. The results of these estimates are compared with the experimental intensities in Table XII.

Indirect transitions have also been reported[227] in the piezoreflectance spectrum of germanium. These transitions should appear extremely weak in any reflection spectrum and it becomes difficult to

TABLE XII. EXPERIMENTAL AND CALCULATED INTENSITIES OF THE PEAKS IN THE STRAIN OPTICAL CONSTANTS OF GERMANIUM FOR INDIRECT, LA PHONON-AIDED TRANSITIONS. THE CALCULATED INTENSITIES HAVE BEEN NORMALIZED SO AS TO BRING W_{11} INTO AGREEMENT WITH THE EXPERIMENTAL W_{12}

Intensities	$W_{11}^{(i)}$	$W_{12}^{(i)}$	$W_{44}^{(i)}$
Experimental	0.28	0.44	0.1 ± 0.1
Calculated	0.12	0.44	0.16

decide whether the corresponding structure is produced by the modulation in the reflectivity or by a transmission modulation of a back reflected component of the beam.

b. Insulators

The piezoreflectance spectrum of KBr and KI has been measured by Gerhardt and Mohler[222] at 90°K by the bending method in the region between 5.7 and 7.6 eV [$E_0 - (E_0 + \Delta_0)$ peaks]. Figure 75 shows the piezoreflectance spectra of KBr (E_0, $E_0 + \Delta_0$ peaks) and KI (E_0 peak only) for a [111] stress, obtained by these authors, together with the normal incidence reflection spectrum. These authors performed a Kramers–Kronig analysis of their data by a method different from that described in Section 13. They added to the

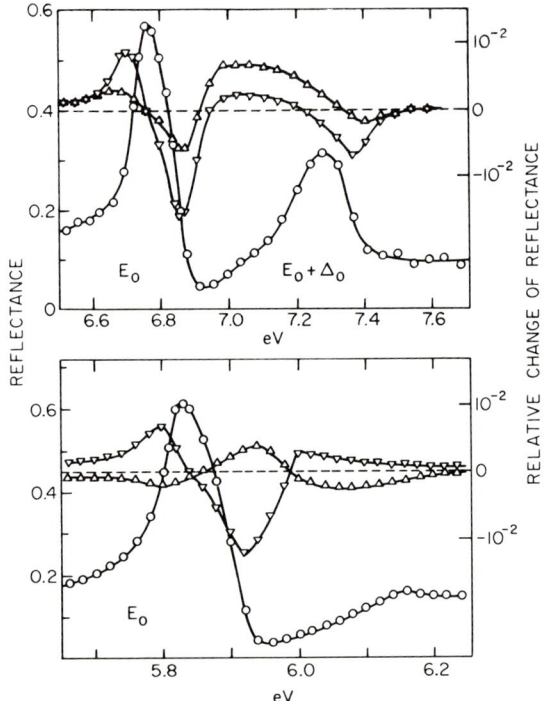

FIG. 75. Normal incidence reflectivity and piezoreflectance spectra of KBr (E_0 and $E_0 + \Delta_0$ peaks) with $e_{111} = 6.5 \times 10^{-4}$ and KI (E_0 peak) with $e_{111} = 4 \times 10^{-4}$ at 90°K for a [111] modulated stress. From U. Gerhardt and E. Mohler, *Phys. Status Solidi* **18**, K45 (1966). $\Delta R/R$: △ △E[111] and ▽ ▽E[11$\bar{2}$], and R : ○○.

reflectivity spectrum a multiple ($\times 10$) of the measured modulation ΔR and found, by means of a conventional Kramers–Kronig analysis of the reflectivity with and without stress, the optical constants (ε_r and ε_i) with and without stress.

Figure 76 shows the results for the KBr spectra of Fig. 75 around E_0. The shift of the E_0 peaks with stress can be easily determined. Inside the peaks of Fig. 76, we see the average points for the various

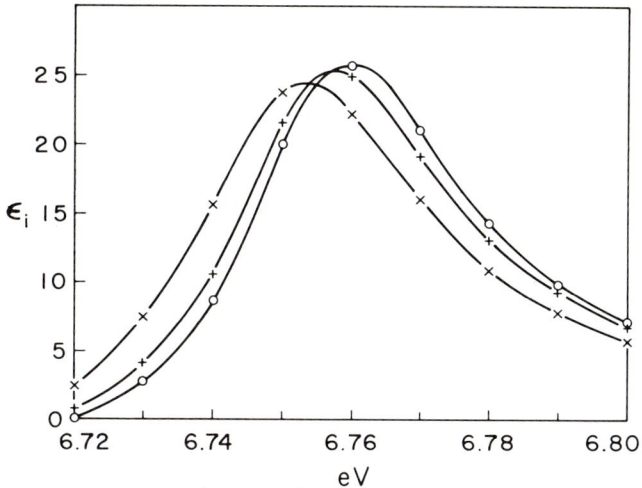

FIG. 76. Effect of stress on the imaginary part of the dielectric constant of KBr around the E_0 peak for light polarized with **E** parallel and perpendicular to the direction of the modulated stress [111] with $e[111] = 6.5 \times 10^{-3}$. The experimental strain was 10 times smaller than that given in this figure: the measured differential effects were multiplied by 10 to make the display possible. $w^{\parallel} = -2.05$ meV and $w^{\perp} = -6.9$ meV indicate the measured energy shifts. The figure shows $E[111](+++)$, $E[11\bar{2}]$ ($\times \times \times$), and $X = 0$ (O O O). U. Gerhard, private communication.

heights. The shifts of these average points with stress are independent of the height chosen and hence shifts can be determined rather accurately. From these shifts we can obtain the shear and the hydrostatic deformation potentials of the E_0 and $E_0 + \Delta_0$ excitons assuming that no plastic deformation occurs while the crystal is vibrated. As discussed in Section 20c the shear deformation potentials of the $E_0 + \Delta_0$ peaks should be zero if the exciton splittings are produced only by the splittings of the corresponding band edges. Under these conditions the $E_0 + \Delta_0$ line should have the same intensity for both polarization directions. For KBr it is difficult to test this condition with accuracy (see Figs. 75 and 76) since the strong E_0 peak interferes with $E_0 + \Delta_0$. In Fig. 75 this condition holds provided one measures the intensity of $E_0 + \Delta_0$ from the 7 eV plateau to the 7.4 eV minimum. In Fig. 76 the ratio of the parallel to the perpendicular intensities for $E_0 + \Delta_0$ is about the same as for E_0!.

We have listed in Table XIII the experimental values of a_{ex}, b_{ex}, and d_{ex} [Eq. (20.22)] derived from these measurements. Our notation differs from that used by Gerhardt and Mohler[222]: the deformation potentials, A, B, and C used in this reference are related to ours by $a_{ex} = A$, $b_{ex} = 2B$, and $d_{ex} = 2C/3^{1/2}$. If the E_0 splittings are equal to the corresponding band edge (Γ_8^-) splittings, the parameters a_{ex}, b_{ex}, and d_{ex} are equal to a, $-\frac{1}{2}b$, and $-\frac{1}{2}d$, respectively, where a, b, and c are the valence band splitting parameters of Eq. (20.6). We have also listed in Table XIII the ratios b_{ex}/a_{ex} and d_{ex}/a_{ex} calculated from Gerhardt and Mohler's values of the deformation potentials and those obtained directly from the ratios of the piezoreflectance line intensities by means of Eqs. (22.1) and (20.11).

c. Metals

The piezoreflectance spectra of polycrystalline silver, copper, and gold have been reported by Garfinkel et al.[178] These measurements were made on polycrystalline thin film evaporated on the transducers. Hence the Kramers–Kronig analysis of the data performed by the authors yields only the changes in ε_r and ε_i averaged for all possible film orientations. Gerhardt et al.[188] measured the piezoreflectance spectrum of single crystal copper for 3 different sample orientations by the bending technique described in Section 21. They derived by Kramers–Kronig analysis the complete piezoreflectance tensor Q (Section 20) and the strain optical tensor W. Figure 77 shows the 3 linear combinations $W_{11}^{(i)} + 2W_{12}^{(i)}$, $W_{44}^{(i)}$, and $W_{11}^{(i)} - W_{12}^{(i)}$ of the imaginary parts of the 3 independent components of the strain optical tensor of copper. These linear combinations correspond to a pure hydrostatic stress ($W_{11} + 2W_{12}$), a [111] shear stress (W_{44}) and a [100] shear stress ($W_{11} - W_{12}$). For comparison we also show in Fig. 77 the ε_i spectrum of copper.

The increased sharpness of the $W^{(i)}$ spectra as compared to ε_i is strikingly apparent in Fig. 77. The structure in the components of

TABLE XIII. DEFORMATION POTENTIALS[a] FOR THE E_0 AND $E_0 + \Delta_0$ EXCITONS OF KBr AND FOR THE E_0 EXCITON OF KI. ALSO LISTED ARE THE RATIOS b_{ex}/a_{ex} AND d_{ex}/a_{ex} OBTAINED FROM THE PRECEDING VALUES OF a_{ex}, b_{ex}, AND d_{ex}, AND THOSE OBTAINED FROM THE RATIO OF INTENSITIES OF THE EXCITON LINES FOR E PARALLEL AND PERPENDICULAR TO THE STRESS

	a_{ex}	b_{ex}	d_{ex}	$b_{ex}/a_{ex}{}^b$	$b_{ex}/a_{ex}{}^c$	$d_{ex}/a_{ex}{}^b$	$d_{ex}/a_{ex}{}^c$
KBr, E_0	-2.22 ± 0.4	0.1 ± 0.04	0.28 ± 0.04	-0.05	-0.04	-0.12	-0.11
KBr, $E_0 + \Delta_0$	-2.13 ± 0.4	± 0.2	± 0.1				
KI, E_0	-2.3	0.08	0.57	-0.03	-0.03	-0.25	-0.25

[a] U. Gerhardt and E. Mohler, *Phys. Status Solidi* **18**, K45 (1966).
[b] Calculated from the preceding values of a_{ex}, b_{ex}, and c_{ex}.
[c] Calculated from the ratio of the intensities of the peaks for E polarized parallel and perpendicular to the stress axis.

$W^{(i)}$ near 2 eV corresponds to the $L_3^{(2h)} \rightarrow E_F$ transitions (near $L_{2'}$) discussed in Section 19b (see Fig. 40). It appears in Fig. 77 that the main contribution to this structure comes from hydrostatic strain (such strain changes the electron density and thus the position of the

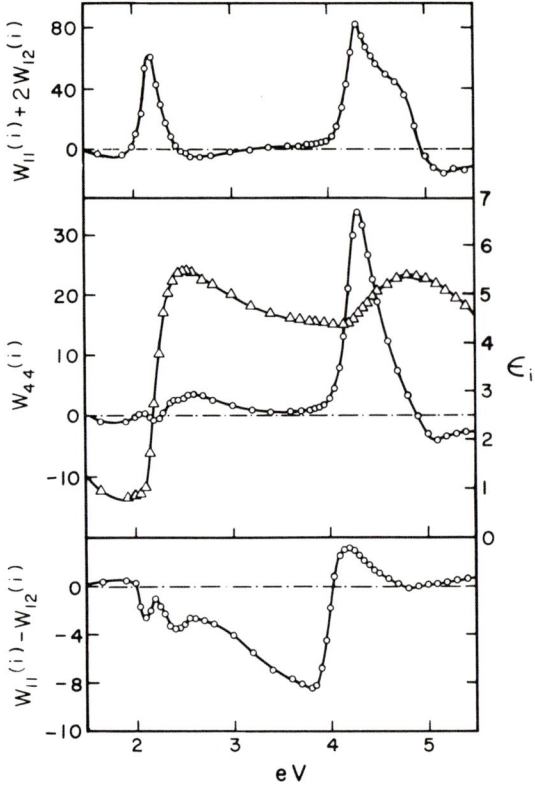

FIG. 77. Imaginary part of the hydrostatic and pure shear components of the piezo-optical tensor of copper at 300°K. The imaginary part of ε has been added for comparison. The triangles represent ε_i and the circles W_{44}. From U. Gerhardt, D. Beaglehole, and R. Sandrock, *Phys. Rev. Letters* **19**, 309 (1967).

Fermi level). Both [111] and [100] shear produce an $L_3^{(2h)} \rightarrow E_F$ (near L_2) signal, which illustrates the fact that these transitions do not occur exactly at L but, according to Fig. 40, in a rather low symmetry region (W-L line) of **k**-space. The structure in $W_{11}^{(i)} - W_{12}^{(i)}$ at 2.4 eV could be tentatively assigned to $\Delta_5 \rightarrow E_F$ (near Δ_1) transitions since it appears strongest in $W_{11}^{(i)} - W_{12}^{(i)}$ (points along equivalent Δ directions split for [100] but not for [111] strain).

The doublet structure between 4 and 5 eV is very similar to that seen in the thermoreflectance and piezoreflectance of gold between 3.5 and 4.5 eV (Fig. 62). The first component of this doublet, the peak at 4.3 eV, does not appear for a [100] pure shear stress. It is therefore reasonable to assign it to $L_3^2 \to L_1^2$ transitions. A peak

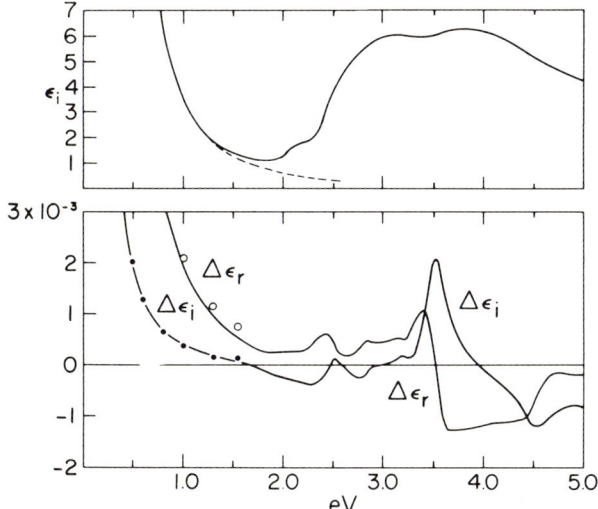

FIG. 78. $\Delta \varepsilon_r$ and $\Delta \varepsilon_i$ obtained from the piezoreflectance spectrum of gold and ε_i obtained by the Kramers–Kronig analysis of the normal incidence reflection (from W. E. Engeler, H. Fritzsche, M. Garfinkel, and J. J. Tiemann, *Phys. Rev. Letters* **14**, 1069 (1965)). All data correspond to room temperature. The dots and the dashed curve represent a calculation of the free electron contribution to the effect (see text). $e_{xx} = e_{yy} = 7 \times 10^{-5}$.

which appears only in $W_{11}^{(i)} - W_{12}^{(i)}$ at 3.9 eV can be assigned to $\Delta_5 \to E_F$ (near Δ_1) transitions. The 4.75 peak, which does not seem to appear for [100] stress, can be tentatively assigned, on the same grounds, to $L_3^1 \to E_F$ (near $L_{2'}$) transitions.

Figure 78 shows the $\Delta \varepsilon_r$ and $\Delta \varepsilon_i$ spectra obtained by the Kramers–Kronig method from the piezoreflectance spectra of polycrystalline gold.[178] Again the higher resolution of the modulation spectra is apparent. The dots and the dashed curve in Fig. 78 were calculated by

Garfinkel et al.[178] from the free electron model [Eq. (3.2)]:

$$\varepsilon_{i_{\text{free}}} = \frac{\omega_p^2 \tau}{(1 + \omega^2 \tau^2)}$$

$$\frac{\Delta \varepsilon_i}{\varepsilon_i} = \frac{2\Delta\omega_p}{\omega_p} + \left(\frac{1 - \omega^2 \tau^2}{1 + \omega^2 \tau^2}\right) \frac{\Delta \tau}{\tau} \qquad (22.4)$$

$$\frac{\Delta \varepsilon_r}{\varepsilon_r - 1} = \frac{2\Delta\omega_p}{\omega_p}.$$

The contribution to the stress modulation due to $\Delta\tau$ was obtained from Eq. (22.4) and the known pressure dependence of the bulk resistivity; that owing to $\Delta\omega_p$ from:

$$\frac{2\Delta\omega_p}{\omega_p} = -\frac{\Delta V}{V}, \qquad (22.5)$$

where ΔV is the applied modulation in the volume. These authors pointed out that significant deviations from the free electron model occur below the lowest interband gap (~ 2.4 eV for gold). They suggested these anomalies may be similar to those observed in the optical spectra of the alkali metals.[39] A tentative assignment of the interband structure in Fig. 78 has already been given in Table VIII.

VII. Electric Field Modulation

23. INTRODUCTION

As mentioned in Section 10 there exists an essential difference between the application of a uniform electric field to a solid and that of a uniaxial or a hydrostatic stress: a solid loses at least part of its translational symmetry operations when the electric field is applied. The contribution of the electric field $\mathscr{E}$ to the one-electron Hamiltonian is $-e\mathbf{r}\cdot\mathscr{E}$ (e is negative for electrons). If the field is small, it is still possible to define a local band structure in the vicinity of the point $\mathbf{r}$. The valence and the conduction band edges of a Van Hove singularity vary with $\mathbf{r}$ as shown in Fig. 43. The optical energy gap disappears for $\mathscr{E}\neq 0$ and therefore an electric field modulation spectrum cannot be considered as the derivative of the conventional spectrum with respect to the energy gap.

In Section 24 we shall discuss the theory of the effect of a uniform electric field on the interband and exciton spectra of solids (insulators and semiconductors): the fields encountered in the usual experimental arrangements are reasonably uniform over many lattice constants (except for very heavily doped semiconductors). Uniformity of the electric field can therefore be safely assumed in the microscopic derivation of the dielectric constant. Since the applied field is not always uniform over a wavelength of the light, spatial dependence of the optical constants (electric field) may have to be taken into account explicitly in the solution of the boundary value problem which corresponds to the propagation of the light.

We shall consider 3 contributions to the effect of the electric field on the optical constants of insulators and semiconductors. In Section 24a we treat the effect of an electric field on direct (allowed and forbidden) interband transitions between one-electron bands. This phenomenon has become known as the Franz-Keldysh effect.[72,73] In Section 24b we treat the effect of a uniform electric field on indirect interband transitions (M_0 edge). In Section 24c we treat the broadening and shift by an electric field of hydrogenic exciton levels associated with M_0 critical points (Stark effect of excitons). In Section 24d we treat effects on the optical constants e.g., band structure, caused by ionic displacements produced by the electric field. Such effects are expected to be largest in ferroelectric and paraelectric materials. We shall confine our discussion to materials of the $SrTiO_3$ (perovskite) family, which have been extensively studied.

The techniques used for electric field modulation fall into two broad categories. If the sample is a good insulator and a poor photoconductor one can apply the field by "brute force." A capacitor-like structure is built with the sample as a dielectric. The field in the bulk of the material is rather uniform (over a wavelength or a penetration depth of the light) and the modulation of the reflection (electroreflectance) or transmission (electroabsorption) can be easily calculated by assuming spatially uniform optical constants. Experiments can be performed with the modulating field either along or perpendicular to the direction of propagation of the light.

For semiconducting materials, a high electric field cannot be set up over a large volume. However, in a reflection modulation experiment the high electric field need only be confined to a depth of about a wavelength (or a penetration depth, whichever is smaller) away from the reflecting surface. Hence, we can use the field at a surface barrier, which can be easily modulated with an external transparent electrode.[61] Under these conditions, the applied field is not completely uniform over the penetration depth of the light (or the wavelength, whichever is smaller) and the solution of Maxwell's equations for spatially varying optical constants may be necessary in order to compute the measured modulation spectra from the microscopic theory of the effect of a *uniform* electric field $\mathscr{E}$ on the optical constants. The surface barrier field is, of necessity, perpendicular to the reflecting surface and, hence, only measurements with the light polarized perpendicular to the modulating field $\mathscr{E}$ are possible at normal incidence. The effects of $\mathscr{E}$ on light polarized with **E** parallel to $\mathscr{E}$ can conceivably be extracted from measurements at oblique incidence.

24. Theory

a. Interband Effects: Direct Transitions

Two different but equivalent methods have been used for the calculation of the dielectric constant in the neighborhood of a direct interband critical point in the presence of a uniform electric field. One of these methods[74,228-231] involves the calculation of the conduction and valence band eigenfunctions in the presence of the electric field and the evaluation of ϵ_i by using the first line of Eq. (6.3). The necessary optical matrix element is calculated from those wave functions. The other method[75,76,232] involves the assumption of parabolic bands around a Van Hove singularity at $\mathbf{k}_0$ and the use of the effective mass approximation. The interband matrix element in the absence of the modulating field is taken either constant (allowed transitions) or linear in $\Delta \mathbf{k} = \mathbf{k} - \mathbf{k}_0$ (first-order forbidden transitions). The matrix element of Eq. (6.3) is then obtained with either Eq. (6.8) [allowed transitions] or Eq. (6.18) [forbidden transitions], where $\phi(r)$ is the solution of the effective mass equation (6.9), with $V(r) = -e\mathbf{r} \cdot \mathscr{E}$. The exciton interaction can, in principle, be easily included in this second formulation by adding to $V(r)$ the electron hole Coulomb potential $-e/\mathbf{r}$ (see Section 24c).

We shall now discuss the first method[74] of calculating ϵ_i in the presence of $\mathscr{E}$. We assume that we are dealing with nondegenerate bands and neglect interband mixing produced by the electric field (this cannot be done, in principle if the bands are degenerate at some points of k-space). An immediate drawback of this assumption is that it breaks down for the valence band of germanium-like materials at $\mathbf{k} = 0$ (a calculation including this coupling has yet to be given). We expand the eigenfunction of a band in the presence of an electric field as a linear combination of Bloch functions of that band (no interband coupling). We assume that $\mathscr{E}$ is along the direction of *some* reciprocal lattice vector. Under these conditions the translational symmetry perpendicular to $\mathscr{E}$ is preserved and the electric field only mixes wave functions whose $\mathbf{k}_\perp$ is the same, therefore:

$$\phi_{v,n}(\mathbf{k}_\perp, \mathbf{r}) = \int A_{v,n}(\mathbf{k}) \psi_n(\mathbf{k}, \mathbf{r}) \, dk_\parallel . \qquad (24.1)$$

This integral is extended to all nonequivalent values of $k_\parallel$, the component of $\mathbf{k}$ along the $\mathscr{E}$ direction, i.e., to $-\tfrac{1}{2}K \leq k_\parallel \leq \tfrac{1}{2}K$, where $\mathbf{K}$ is the shortest reciprocal lattice vector along $\mathscr{E}$. The index v denotes the quantum number which is to replace the mixed parallel components of $\mathbf{k}$, $k_\parallel$ and n labels the band under consideration. The mixing coefficients $A_{v,n}(\mathbf{k})$ are the solutions of the equation [229,231]

$$[\omega_n(\mathbf{k}) + i\mathscr{E} \, \partial/\partial k_\parallel] A_{v,n}(\mathbf{k}) = W_{v,n}(\mathbf{k}_\perp) A_{v,n}(\mathbf{k}). \qquad (24.2)$$

In Eq. (24.2) $\omega_n(\mathbf{k})$ is the energy of the corresponding Bloch state shifted by the diagonal matrix element of the electric field Hamiltonian. This shift vanishes if parity is a good quantum number for the extrema under consideration; it is otherwise linear in $\mathscr{E}$. We shall disregard this linear shift since it is negligible for small fields in comparison with the Franz-Keldysh effect (proportional to $\mathscr{E}^{1/3}$) which results from this calculation.

The solution of Eq. (24.2) is:

$$A_{v,n}(\mathbf{k}) = K^{-1/2} \exp(1/i\mathscr{E}) \int_0^{k_\|} [W_{v,n}(\mathbf{k}_\perp) - \omega_n(\mathbf{k}_\perp, k_\|')] \, dk_\|'. \quad (24.3)$$

The factor $\mathbf{K}^{-1/2}$ in Eq. (24.3) is a normalization constant. The energy eigenvalue $W_{v,n}(\mathbf{k}_\perp)$ is determined by the condition:

$$A_{v,n}(\mathbf{k} + \mathbf{K}) = A_{v,n}(\mathbf{k}), \quad (24.4)$$

which expresses the periodicity of the Bloch functions in $\mathbf{k}$-space. By replacing Eq. (24.4) into Eq. (24.3) we obtain:

$$W_{v,n}(\mathbf{k}_\perp) = -(2\pi v \mathscr{E}/K) + \bar{\omega}_n(\mathbf{k}_\perp) \quad (24.5)$$

with

$$\bar{\omega}_n = \frac{1}{K} \int_{-K/2}^{K/2} dk_\|' \, \omega_n(\mathbf{k}_\perp, k_\|'),$$

and $v = 0, 1, \ldots, N$, so as to keep the wave packet of Eq. (24.1) within the crystal (N is the number of equivalent crystal planes perpendicular to $\mathscr{E}$). We assume the length of the crystal along this direction to be unity, therefore $N = K/2\pi$.

The component of $\mathbf{k}$ perpendicular to $\mathscr{E}$ is preserved for electric dipole transitions. The optical matrix element is[†]:

$$\hat{\mathbf{n}} \cdot \mathbf{P}_{v'c, vv} = \langle v'c | \hat{\mathbf{n}} \cdot \mathbf{p} | vv \rangle = \int \phi_{v',c}^*(\mathbf{k}_\perp, \mathbf{r}) \hat{\mathbf{n}} \cdot \mathbf{p} \phi_{v,v}(\mathbf{k}_\perp, \mathbf{r}) \, dV_\mathbf{r}$$

$$= \int_{-K/2}^{K/2} A_{v',c}^*(\mathbf{k}) A_{v,v}(\mathbf{k}) \hat{\mathbf{n}} \cdot \mathbf{p}_{cv}(\mathbf{k}) \, dk_\|, \quad (24.6)$$

where $\mathbf{p}_{cv}(\mathbf{k})$ is the matrix element of $\mathbf{p}$ between the Bloch functions of the conduction and valence band of crystal momentum $\mathbf{k}$. The first

[†] Equation (6.3) has been written down for a cubic material. We must use here the more general expression in which the average matrix element $\frac{1}{3}|\langle|\mathbf{p}|\rangle|^2$ is replaced by $|\langle|\hat{\mathbf{n}} \cdot \mathbf{p}|\rangle|^2$, where $\hat{\mathbf{n}}$ is the unit vector along the direction of $\mathbf{E}$.

row of Eq. (6.3) must be summed over $\mathbf{k}_\perp$, v, and v'. By replacing Eq. (24.3) into Eq. (24.6) we find:

$$\mathbf{P}_{v'c,\,vv} = K^{-1} \int_{-K/2}^{K/2} \mathbf{p}_{cv}(\mathbf{k}) \exp\left\{\frac{1}{i\mathscr{E}} \int_0^{k_\parallel} \left[\frac{2\pi\mathscr{E}}{K}(v'-v)\right.\right.$$

$$\left.\left. + \bar{\omega}_c - \bar{\omega}_v + \omega_v - \omega_c\right] dk_\parallel'\right\} dk_\parallel. \qquad (24.7)$$

The imaginary part of the dielectric constant is, according to Eq. (6.3):

$$\varepsilon_i(\omega) = \frac{4\pi^2}{\omega^2} \sum_{vv'} \int_s (ds_\perp/2\pi^2)\,|\hat{\mathbf{n}} \cdot \mathbf{P}_{v'c,\,vv}|^2 \delta(W_{v'c} - W_{vv} - \omega), \qquad (24.8)$$

where $ds_\perp$ is the element of surface in $\mathbf{k}$-space perpendicular to $\mathscr{E}$. Both the matrix element $p_{v'c,vv}$ and the δ function in the energy of Eq. (6.3) are functions of the $(v-v')$ combination and not of v and v' separately. Hence we can apply to the $|\hat{\mathbf{n}} \cdot \mathbf{P}|^2 \cdot \delta$ product of Eq. (24.8) the expression [231,233a]

$$\sum_{vv'} f(v-v') = N \sum_{l=-\infty}^{+\infty} \int_{-\infty}^{+\infty} d\xi\, f(\xi) e^{-2\pi i l\xi}. \qquad (24.9)$$

By so doing we obtain:

$$\varepsilon_i(\omega) = \frac{2\pi K}{\omega^2} \sum_{l=-\infty}^{+\infty} \int_s \frac{ds_\perp}{2\pi^2} \int_{-\infty}^{+\infty} d\xi\, |\hat{\mathbf{n}} \cdot \mathbf{P}_{0c,\,\xi v}|^2$$

$$\times \delta\left(\frac{2\pi\mathscr{E}}{K}\xi + \bar{\omega}_c - \bar{\omega}_v - \omega\right) \cdot e^{2\pi i l\xi}$$

$$= -\frac{K^2}{\omega^2\mathscr{E}} \left\{ \int_s \frac{ds_\perp}{2\pi^2} |\hat{\mathbf{n}} \cdot \mathbf{P}_{0c,\,\xi_0 v}|^2 + 2\sum_{l=1}^{\infty} \frac{ds_\perp}{2\pi^2} |\hat{\mathbf{n}} \cdot \mathbf{P}_{0c,\,\xi_0 v}|^2 \cos(2\pi l\xi_0) \right\}, \qquad (24.10)$$

where:

$$\xi_0 = -(K/2\pi\mathscr{E})(\bar{\omega}_c - \bar{\omega}_v - \omega).$$

Each of the terms in the summation of the last line of Eq. (24.10) represents a rapidly oscillating function of ω. These are the so-called Stark oscillations.[228,233b] These terms should give no contribution to ε_i for small fields ($2\pi\mathscr{E}/K \ll \omega$): The Stark oscillations† in interband

---

† Oscillations in the absorption spectrum of CdS, somewhat similar to Stark oscillations, have been reported by Vavilov et al.[233b] These authors, however, ruled out the explanation in terms of Stark steps because of the expected lifetime broadening.[233c]

optical spectra have yet to be observed. We shall neglect these terms and write:

$$\varepsilon_i(\omega, \mathscr{E}) = -\frac{K^2}{2\pi^2 \omega^2 \mathscr{E}} \int ds_\perp |\hat{\mathbf{n}} \cdot \mathbf{P}_{0c,\xi_0 v}|^2, \qquad (24.11)$$

with $\mathbf{P}_{0c,\xi_0 v}$ given in Eq. (24.7). By replacing Eq. (24.7) into Eq. (24.11) we find:

$$\varepsilon_i(\omega, \mathscr{E}) = -\frac{1}{2\pi^2 \omega^2 \mathscr{E}} \int_{BZ} dk^3 \int_{-K/2}^{K/2} \hat{\mathbf{n}} \cdot \mathbf{p}_{cv}(\mathbf{k}_\perp, k_\parallel) \hat{\mathbf{n}} \cdot \mathbf{p}_{cv}^*(\mathbf{k}_\perp, q) \, dq$$

$$\times \exp\left\{\frac{1}{i\mathscr{E}} \int_q^{k_\parallel} dk_\parallel' [\omega_c(\mathbf{k}_\perp, k') - \omega_v(\mathbf{k}_\perp, k') - \omega]\right\}.$$

(24.12)

With a change in the variable of integration, Eq. (24.12) can also be written in the form[231]:

$$\varepsilon_i(\omega, \mathscr{E}) = \frac{1}{\pi^2 \omega^2} \int_{BZ} dk^3 \int_{-K}^{+K} ds \, \hat{\mathbf{n}} \cdot \mathbf{p}_{cv}(\mathbf{k} + s\mathscr{E}) \hat{\mathbf{n}} \cdot \mathbf{p}_{cv}^*(\mathbf{k} - s\mathscr{E})$$

$$\times \exp\left\{-i \int_{-s}^{s} dt \, [\omega_c(\mathbf{k} + \mathscr{E}t) - \omega_v(\mathbf{k} - \mathscr{E}t) - \omega]\right\}.$$

(24.13)

Equation (24.13) is valid for any shape of the energy bands (except for the exclusion of degeneracy). In order to proceed further with Eqs. (24.12) or (24.13), we must have the energy bands ($\omega_c - \omega_v$) and the interband matrix elements $\mathbf{p}_{cv}$ as a function of $\mathbf{k}$. Since it is reasonable to assume that the most interesting contributions to $\varepsilon_i(\omega, \mathscr{E})$ come from the vicinity of Van Hove singularities, we can confine the power series expansion of $\omega_c - \omega_v$ around a given singularity to quadratic terms in the components of $\mathbf{k} - \mathbf{k}_0$. We can also assume, for allowed transitions, that $\mathbf{p}_{cv}$ is constant and thus can be taken out of the integrals. Let us consider, for instance, an M_0 critical point at $\mathbf{k} = 0$ with isotropic reduced mass μ:

$$\omega_c - \omega_v = \omega_g + (k^2/2\mu). \qquad (24.14)$$

The dielectric constant ε_i becomes (Eq. 24.12):

$$\varepsilon_i(\omega, \mathscr{E}) = -\frac{|\mathbf{p}_{cv} \cdot \hat{\mathbf{n}}|^2}{2\pi^2 \omega^2 \mathscr{E}} \int_{BZ} ds_\perp$$

$$\times \left| \int_{-K/2}^{K/2} dk_\parallel \exp\left\{ \frac{1}{i\mathscr{E}} \int_0^{k_\parallel} \left[\omega_g + \frac{k_\perp^2 + k_\parallel'^2}{2\mu} - \omega \right] dk_\parallel' \right\} \right|^2$$

$$= -\frac{|\mathbf{p}_{cv} \cdot \hat{\mathbf{n}}|^2}{2\pi^2 \omega^2 \mathscr{E}} \int_{BZ} ds_\perp$$

$$\times \left| \int_{-K/2}^{K/2} dk_\parallel \exp\left\{ \frac{1}{i\mathscr{E}} \left[\left(\omega_g + \frac{k_\perp^2}{2\mu} - \omega \right) k_\parallel + \frac{k_\parallel^3}{6\mu} \right] \right\} \right|^2$$

(24.15)

For small fields we can extend the limits of integration in Eq. (24.15) to $\pm\infty$, after changing the variable of integration $k_\parallel$ to the dimensionless variable $k_\parallel/(2\mu\mathscr{E})^{1/3}$. The integral over $k_\parallel$ of Eq. (24.15) can then be easily reduced to the Airy function regular at infinity $Ai(x)$, given by[234]:

$$\text{Ai}(x) = \frac{1}{2\pi} \int_{-\infty}^{\infty} ds \, \exp(i[s^3/3 + sx]). \qquad (24.16)$$

We find:

$$\int_{-K/2}^{K/2} dk_\parallel \exp \frac{i}{\mathscr{E}} \left[\left(\omega_g + \frac{k_\perp^2}{2\mu} - \omega \right) k_\parallel + \frac{k_\parallel^3}{6\mu} \right] \simeq \frac{2\pi\mathscr{E}}{\theta} \text{Ai}\left(-\frac{\Omega}{\theta} \right) \qquad (24.17)$$

with $\theta = (\mathscr{E}^2/2\mu)^{1/3}$ and $\Omega = \omega - \omega_g - (k_\perp^2/2\mu)$. By substituting Eq. (24.17) into Eq. (24.15) we find:

$$\varepsilon_i(\omega, \mathscr{E}) = \frac{2|\mathbf{p}_{cv} \cdot \hat{\mathbf{n}}|^2 \mathscr{E}}{\theta^2 \omega^2} \int_{BZ} ds_\perp \left| \text{Ai}\left(-\frac{\Omega}{\theta} \right) \right|^2$$

$$= \frac{2\pi}{\omega^2} (2\mu)^{3/2} |\mathbf{p}_{cv} \cdot \hat{\mathbf{n}}|^2 \theta^{1/2}$$

$$\times \left[\text{Ai}'^2\left(\frac{\omega_g - \omega}{\theta} \right) - \frac{\omega_g - \omega}{\theta} \text{Ai}^2\left(\frac{\omega_g - \omega}{\theta} \right) \right], \qquad (24.18)$$

where use has been made of the relationship derived in Appendix I:

$$\int_t^\infty d\omega \, \text{Ai}^2(\omega) = t\text{Ai}^2(t) - \text{Ai}'(t). \qquad (24.19)$$

The type of calculation performed above is typical of any calculation of the interband electrooptic effect: one always obtains integrals over **k**-space (or energy) of squares of Airy functions. A general method to evaluate these integrals has been given by Aspnes.[76] Using the asymptotic expressions[234] of Ai(x):

$$\text{Ai}(x) \underset{x \to -\infty}{\to} \pi^{-1/2}(-x)^{-1/4} \sin[\tfrac{2}{3}(-x)^{3/2} + \tfrac{1}{4}\pi],$$
$$\text{Ai}(x) \underset{x \to \infty}{\to} \tfrac{1}{2}\pi^{-1/2}x^{-1/4} \exp(-\tfrac{2}{3}x^{3/2}), \tag{24.20}$$

we find

$$\varepsilon_i \to 0 \quad \text{for} \quad (\omega - \omega_g)/\theta \to -\infty$$

and

$$\varepsilon_i(\omega, \mathscr{E}) \to \frac{2(2\mu)^{3/2}}{\omega^2} |\mathbf{p}_{cv} \cdot \hat{\mathbf{n}}|^2 (\omega - \omega_g)^{1/2}, \tag{24.21}$$

for $(\omega - \omega_g)/\theta \to \infty$.

Equation (24.21) is the dielectric constant $\varepsilon_i(\omega)$ in the absence of an electric field [Eqs. (4.3) and (4.4)]. Thus we see that the electric field modifies ε_i only near the Van Hove singularity. The modification introduced in ε_i by the electric field is usually written as[76]:

$$\Delta\varepsilon_i(\omega, \mathscr{E}) = \frac{B\theta^{1/2}}{\omega^2} F\left(\frac{\omega_g - \omega}{\theta}\right), \tag{24.22}$$

where

$$B = 2|\mathbf{p}_{cv} \cdot \hat{\mathbf{n}}|^2 (2\mu)^{3/2}$$

and

$$F(\eta) = \pi[\text{Ai}'^2(\eta) - \eta \,\text{Ai}^2(\eta)] - (-\eta)^{1/2} H(-\eta),$$

where $H(\eta)$ is the unit step function. The modification in ε_r produced by the electric field can be calculated from $\Delta\varepsilon_i$ with Eq. (1.1):

$$\Delta\varepsilon_r(\omega, \mathscr{E}) = \frac{2B\theta^{1/2}}{\pi} P \int_0^\infty \frac{d\omega'}{\omega'(\omega'^2 - \omega^2)} F\left(\frac{\omega_g - \omega'}{\theta}\right)$$
$$\simeq \frac{2B\theta^{1/2}}{\pi} P \int_{-\infty}^\infty \frac{d\omega'}{\omega'(\omega'^2 - \omega^2)} F\left(\frac{\omega_g - \omega'}{\theta}\right). \tag{24.23}$$

The lower limit of integration has been extended to $-\infty$ in Eq. (24.23) because $F[(\omega_g - \omega')/\theta] \simeq 0$ for $\omega' < 0$ if the field is small ($\omega_g/\theta \gg 1$). Equation (24.23) is then easily evaluated as a sum of residues of F for $\omega' = 0$ and $\omega' = \pm\omega$.

The function $\mathscr{F}(x) = \text{Ai}^2(x) + i\,\text{Ai}(x)\,\text{Bi}(x)$ has the integral representation (see Appendix I):

$$\text{Ai}^2(x) + i\,\text{Ai}(x)\,\text{Bi}(x)$$
$$= \tfrac{1}{2}\pi^{-3/2} \int_0^\infty s^{-1/2}\, ds\, \exp[i(1/12)s^3 + ixs + i(\pi/4)]. \quad (24.24)$$

Equation (24.24) describes a function of x which can be analytically continued in the upper half-plane. Its integral along the half-circle of Fig. 2 tends to zero as the radius tends to infinity. Therefore, using the contour of Fig. 2, we derive:

$$\text{Ai}^2(x) = \frac{1}{\pi} P \int_{-\infty}^\infty \frac{dx'}{x' - x}\, \text{Ai}(x')\,\text{Bi}(x'),$$
$$\text{Ai}(x)\,\text{Bi}(x) = \frac{1}{\pi} P \int_{-\infty}^\infty \frac{dx'}{x - x'}\, \text{Ai}^2(x'). \quad (24.25)$$

Similar expressions are obtained for the function $\text{Ai}'^2(x) + i\,\text{Ai}'(x)\,\text{Bi}'(x)$. Equation (24.23) can be decomposed into integrals of the type of Eqs. (24.25) by using the method of partial fractions, since F contains the functions Ai^2 and Ai'^2. Following this prescription we find[232]:

$$\Delta\varepsilon_r(\omega, \mathscr{E}) = \frac{B\theta^{1/2}}{\omega^2} \left[G\!\left(\frac{\omega_g - \omega}{\theta}\right) + G\!\left(\frac{\omega_g + \omega}{\theta}\right) - 2G\!\left(\frac{\omega}{\theta}\right) \right], \quad (24.26)$$

with:

$$G(\eta) = \pi[\text{Ai}'(\eta)\,\text{Bi}'(\eta) - \eta\,\text{Ai}(\eta)\,\text{Bi}(\eta)] + \eta^{1/2} H(\eta).$$

The terms with the step function $\eta^{1/2} H(\eta)$ in Eq. (24.26) arise from the real part of the dielectric constant in the absence of an electric field. The 3 terms in Eq. (24.26) correspond to the 3 terms obtained in Eq. (4.11a) for $\mathscr{E} = 0$. Since usually $\omega_g/\theta \gg 1$, the terms involving $G[(\omega_g + \omega)/\theta]$ and $G(\omega_g/\theta)$ can be dropped: they correspond to practically zero applied field and hence they are negligible. The modification introduced by the field in ε_r is large only near ω_g and, for small fields, i.e., for $\omega_g/\theta \gg 1$, is given by:

$$\Delta\varepsilon_r(\omega, \mathscr{E}) = \frac{B\theta^{1/2}}{\omega^2} G\!\left(\frac{\omega_g - \omega}{\theta}\right). \quad (24.27)$$

The small-field approximation is not as restrictive as it may seem. First of all the condition $\omega_g/\theta \gg 1$ is fulfilled in any experiment involving nondegenerate ($\omega_g \neq 0$) bands. Also, the extension of the limits of integration to infinity in Eqs. (24.18) and (24.15) is only justified in the small-field limit and, therefore, there is no reason to keep the terms in $(\omega_g + \omega)/\theta$ and (ω_g/θ) in Eq. (24.26). The electrooptic functions $F(\eta)$ and $G(\eta)$ defined above are shown in Fig. 79.

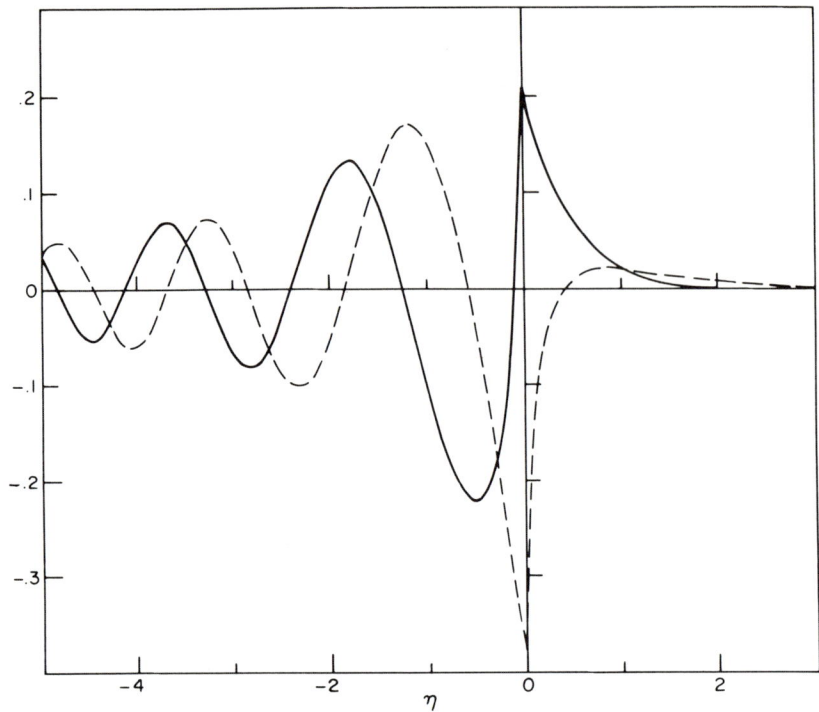

FIG. 79. Three-dimensional electrooptic functions $F(\eta)$ (solid line) and $G(\eta)$ (dashed line), according to D. Aspnes, *Phys. Rev.* **147**, 554 (1966); **153**, 972 (1967).

We shall now discuss the method of Tharmalingham[75] and Aspnes[76,232] for the evaluation of $\Delta\varepsilon_i$, which is based on the solution of the effective mass equation of the electron hole pair near a Van Hove singularity in the presence of a uniform electric field. This equation is:

$$\left\{\frac{1}{\mu_1}\frac{\partial^2}{\partial r_1^2} + \frac{1}{\mu_2}\frac{\partial^2}{\partial r_2^2} + \frac{1}{\mu_3}\frac{\partial^2}{\partial r_3^2} - 2\mathscr{E}\cdot\mathbf{r} + 2\mathbf{W}\right\}\phi(\mathbf{r}) = 0, \qquad (24.28)$$

referred to the principal axes of the effective reduced mass tensor. The solution $\phi(\mathbf{r})$ of Eq. (24.28) can be written as a product of 3 functions $\phi(r_i)$ each of which fulfills the equation:

$$\left[\frac{1}{\mu_i}\frac{\partial^2}{\partial r_i^2} - 2\mathscr{E}_i r_i + 2W_i\right]\phi(r_i) = 0. \tag{24.29}$$

The eigenvalue of Eq. (24.28) is $W = W_1 + W_2 + W_3$. If the component of the field along the ith direction is zero, the solution of the corresponding Eq. (24.29) is a plane wave $\phi(r_i) \propto \exp(ik_i r_i)$. If $\mathscr{E}_i$ is not zero, Eq. (24.29) can be reduced to the equation:

$$d^2\phi(\xi)/d\xi^2 = \pm \xi \phi(\xi) \tag{24.30}$$

by means of the change in variables

$$\xi_i = W_i/\theta_i - r_i(2\mu_i\mathscr{E}_i)^{1/3}, \tag{24.31}$$

where $\theta_i = [\mathscr{E}_i^2/2|\mu_i|]^{1/3}$ [see Eq. (24.17)]. The minus sign in Eq. (24.30) appears for a positive μ_i and the plus sign for a negative μ_i. The solution of Eq. (24.30) regular at infinity is[234]:

$$\phi(\xi_i) = C_i \, \text{Ai}\,(\pm \xi_i). \tag{24.32}$$

The normalization constant C_i is chosen so as to have the scalar product of two wave functions $\phi_{i,1}(r_i)$ and $\phi_{i,2}(r_i)$, which belong to eigenvalues W_{i1} and W_{i2}, equal to a delta function of $(W_{i1} - W_{i2})$:

$$\begin{aligned}\delta(W_{i1} - W_{i2}) &= C_i^2 \int_{-\infty}^{\infty} dr_i \, \text{Ai}\left(\pm \frac{r_i \mathscr{E}_i}{\theta_i} \pm \frac{W_{i1}}{\theta_i}\right) \text{Ai}\left(\pm \frac{r_i \mathscr{E}_i}{\theta_i} \pm \frac{W_{i2}}{\theta_i}\right) \\ &= C_i^2 \frac{\theta_i}{|\mathscr{E}_i|} \int_{-\infty}^{\infty} dx \, \text{Ai}\left(x \pm \frac{W_{i1}}{\theta_i}\right) \text{Ai}\left(x \pm \frac{W_{i2}}{\theta_i}\right) \\ &= C_i^2 \frac{\theta_i}{|\mathscr{E}_i|} \delta\left(\frac{W_{i1} - W_{i2}}{\theta_i}\right) \\ &= C_i^2 \frac{\theta_i^2}{|\mathscr{E}_i|} \delta(W_{i1} - W_{i2}). \end{aligned} \tag{24.33}$$

(See Appendix I). Hence, the normalization constant C_i is:

$$C_i = |\mathscr{E}_i|^{1/2}/\theta_i. \tag{24.34}$$

The imaginary part of the dielectric constant $\varepsilon_i(\omega, \mathscr{E})$ can now be obtained with the first line of Eq. (6.3). For allowed transitions, the matrix element is [Eq. (6.8)]:

$$\phi(0)\mathbf{p}_{cv} = \frac{\mathbf{p}_{cv}|\mathscr{E}_1\mathscr{E}_2\mathscr{E}_3|^{1/2}}{\theta_1\theta_2\theta_3} \operatorname{Ai}\left(\pm\frac{W_1}{\theta_1}\right)\operatorname{Ai}\left(\pm\frac{W_2}{\theta_2}\right)\operatorname{Ai}\left(\pm\frac{W_3}{\theta_3}\right), \quad (24.35)$$

under the assumption that the electric field $\mathscr{E}$ is not along one of the principal axes of the effective mass. If $\mathscr{E}$ is along one of these axes, we must drop all terms which do not refer to this axis in Eq. (24.35). By replacing Eq. (24.35) into Eq. (6.3), we find the general expression:

$$\varepsilon_i(\omega, \mathscr{E}) = \frac{4\pi^2|\mathscr{E}_1\mathscr{E}_2\mathscr{E}_3||\mathbf{p}_{cv}\cdot\hat{\mathbf{n}}|^2}{\omega^2\theta_1^2\theta_2^2\theta_3^2}\int_{-\infty}^{\infty}dW_1\,dW_2\,dW_3$$

$$\times\left[\operatorname{Ai}^2\left(\frac{\pm W_1}{\theta_1}\right)\operatorname{Ai}^2\left(\frac{\pm W_2}{\theta_2}\right)\operatorname{Ai}^2\left(\frac{\pm W_3}{\theta_3}\right)\right]$$

$$\times\delta(\omega_g + W_1 + W_2 + W_3 - \omega), \quad (24.36)$$

where the $+$ and $-$ signs in front of the W_i correspond to positive or negative masses respectively. Equation (24.36) can be evaluated for any type of critical point.[76] As an example, let us consider an M_0 critical point with $\mathscr{E}$ along one of the principal axes ($\mathscr{E}_2 = \mathscr{E}_3 = 0$). Equation (24.36) must be replaced by:

$$\varepsilon_i(\omega, \mathscr{E}) = \frac{4\pi^2|\mathbf{p}_{cv}\cdot\hat{\mathbf{n}}|^2|\mathscr{E}_1|}{\omega^2\theta_1^2}\int_{-\infty}^{\infty}N_d(\omega - \omega_g - W_1)\operatorname{Ai}^2\left(-\frac{W_1}{\theta_1}\right)dW_1, \quad (24.37)$$

where N_d is the density of states for the two-dimensional free electron bands given by:

$$N_d(\omega) = (\mu_2\mu_3)^{1/2}/\pi \quad \text{for } \omega > \omega_g,$$
$$N_d(\omega) = 0 \quad \text{for } \omega < 0. \quad (24.38)$$

Hence, according to Eq. (24.19):

$$\varepsilon_i(\omega, \mathscr{E}) = \frac{4\pi|\mathbf{p}_{cv}\cdot\hat{\mathbf{n}}|^2|\mathscr{E}_1||\mu_2\mu_3|^{1/2}}{\omega^2\theta_1^2}\int_{\omega-\omega_g}^{-\infty}-\operatorname{Ai}^2\left(-\frac{W_1}{\theta_1}\right)dW_1$$

$$= \frac{2\pi}{\omega^2}(8\mu_1\mu_2\mu_3)^{1/2}|\mathbf{p}_{cv}\cdot\hat{\mathbf{n}}|^2$$

$$\times\theta_1^{1/2}\left[\operatorname{Ai}'^2\left(\frac{\omega_g-\omega}{\theta_1}\right) - \left(\frac{\omega_g'-\omega}{\theta_1}\right)\operatorname{Ai}^2\left(\frac{\omega_g-\omega}{\theta_1}\right)\right]. \quad (24.39)$$

Equation (24.39) is analogous to Eq. (24.18) derived earlier for isotropic masses, with the mass μ replaced by $(\mu_1 \mu_2 \mu_3)^{1/3}$ in the density of states factor $\mu^{3/2}$ and by μ_1 in the parameter θ_1 which gives the effect of the electric field: the effect of $\mathscr{E}$ on ε_i will be larger the smaller the reduced mass along $\mathscr{E}$ is.

The integral of Eq. (24.36) can also be found for the field $\mathscr{E}$ in an arbitrary direction through a somewhat more laborious process.[232] If one defines the average reduced effective mass μ:

$$\frac{1}{\mu} = \frac{1}{|\mathscr{E}|^2}\left[\frac{\mathscr{E}_1^2}{\mu_1} + \frac{\mathscr{E}_2^2}{\mu_2} + \frac{\mathscr{E}_3^2}{\mu_3}\right], \tag{24.40}$$

$\Delta\varepsilon_r$ and $\Delta\varepsilon_i$ are given by Eqs. (24.22) and (24.27) with μ replaced by the density of states effective mass $(\mu_1 \mu_2 \mu_3)^{1/3}$ in B and by Eq. (24.40) in θ.

The derivation given for $\varepsilon_i(\omega, \mathscr{E})$ near an M_0 critical point can be easily adapted to an M_3 critical point: the sign of the argument of the Airy function in Eq. (24.37) must be changed so as to take into account the *negative* effective masses. The density of states of Eq. (23.38) must also be reflected about the $\omega = \omega_g$ line and, therefore, one obtains for $\varepsilon_i(\omega, \mathscr{E})$ near an M_3 singularity the result obtained for an M_0 singularity [Eq. (24.39)] reflected about the $\omega = \omega_g$ line. These simple reflection arguments can be carried over to the *real* part of the dielectric constant provided one takes into acccount the fact that the energy denominator in the Kramers–Kronig relation [Eq. (1.1)] produces a reversal in the sign [235] of $\varepsilon_r - 1$ when reflecting about $\omega = \omega_g$. Therefore, the change induced by the electric field in ε_r near an M_3 singularity is:

$$\Delta\varepsilon_r(\omega, \mathscr{E}) = -\frac{B\theta^{1/2}}{\omega^2} G\left[\frac{\omega - \omega_g}{\theta}\right], \tag{24.41}$$

with the density of states effective mass $|\mu_1 \mu_2 \mu_3|^{1/3}$ used in B and the effective mass $|\mu|$

$$\frac{1}{|\mu|} = \frac{1}{|\mathscr{E}|^2}\left|\frac{\mathscr{E}_1^2}{\mu_1} + \frac{\mathscr{E}_2^2}{\mu_2} + \frac{\mathscr{E}_3^2}{\mu_3}\right| \tag{24.42}$$

in the expression for θ[Eq. (24.17)].

A similar derivation leads to $\Delta\varepsilon_i$ and $\Delta\varepsilon_r$ for an M_1 critical point[76,236] provided $\mathscr{E}$ is along the negative mass direction and for an M_2 critical point if $\mathscr{E}$ is along one of the positive mass directions. In these cases the two-dimensional energy bands for **k** perpendicular to $\mathscr{E}$ have either a minimum (M_1) or a maximum (M_2) at the critical point. The density of States of these two-dimensional bands is that of Eqs. (24.38) for M_1 or that of Eqs. (24.38) reflected with respect to $\omega = \omega_g$ for M_2. An integral similar to Eq. (24.37), with proper reflections in N_d and changes in sign in the argument of the Airy function is obtained. The results are[76,232]:

$$\begin{matrix} M_1 \\ \mathscr{E}_1 \neq 0 \\ \mathscr{E}_2 = \mathscr{E}_3 = 0 \\ \mu_1 < 0 \end{matrix} \begin{cases} \Delta\varepsilon_i = -\dfrac{B\theta^{1/2}}{\omega^2} F\left(\dfrac{\omega - \omega_g}{\theta}\right) \\ \Delta\varepsilon_r = \dfrac{B\theta^{1/2}}{\omega^2} G\left(\dfrac{\omega - \omega_g}{\theta}\right) \end{cases}$$

$$\begin{matrix} M_2 \\ \mathscr{E}_1 \neq 0 \\ \mathscr{E}_2 = \mathscr{E}_3 = 0 \\ \mu_1 > 0 \end{matrix} \begin{cases} \Delta\varepsilon_i = -\dfrac{B\theta^{1/2}}{\omega^2} F\left(\dfrac{\omega_g - \omega}{\theta}\right) \\ \Delta\varepsilon_r = -\dfrac{B\theta^{1/2}}{\omega^2} G\left(\dfrac{\omega_g - \omega}{\theta}\right) \end{cases}$$

(24.43)

with $B = 4|\mathbf{p}_{cv} \cdot \hat{\mathbf{n}}|^2 |2\mu_1 \mu_2 \mu_3|^{1/2}$ and $\theta = (\mathscr{E}^2/2|\mu_1|)^{1/3}$. The cases of an M_1 critical point with $\mathscr{E}$ along the positive mass direction and that of an M_2 critical point with $\mathscr{E}$ along the negative mass direction yield saddle points for the two-dimensional bands with **k** perpendicular to $\mathscr{E}$. The density of states for these two-dimensional bands has a logarithmic singularity at the critical point [Eq. (4.9)]. An evaluation of $\Delta\varepsilon_i$ and $\Delta\varepsilon_r$ yields in these cases a permutation of the functions F and G in the expressions for $\Delta\varepsilon_r$ and $\Delta\varepsilon_i$[76,232]:

$$\begin{matrix} M_1 \\ \mathscr{E}_1 \neq 0 \\ \mathscr{E}_2 = \mathscr{E}_3 = 0 \\ \mu_1 > 0 \end{matrix} \begin{cases} \Delta\varepsilon_i = \dfrac{B\theta^{1/2}}{\omega^2} G\left(\dfrac{\omega_g - \omega}{\theta}\right) \\ \Delta\varepsilon_r = -\dfrac{B\theta^{1/2}}{\omega^2} F\left(\dfrac{\omega_g - \omega}{\theta}\right) \end{cases}$$

$$\begin{matrix} M_2 \\ \mathscr{E}_1 \neq 0 \\ \mathscr{E}_2 = \mathscr{E}_3 = 0 \\ \mu_1 < 0 \end{matrix} \begin{cases} \Delta\varepsilon_i = \dfrac{B\theta^{1/2}}{\omega^2} G\left(\dfrac{\omega - \omega_g}{\theta}\right) \\ \Delta\varepsilon_r = \dfrac{B\theta^{1/2}}{\omega^2} F\left(\dfrac{\omega - \omega_g}{\theta}\right) \end{cases}$$

(24.44)

While we have presented Eqs. (24.43) and (24.44) as being valid only for $\mathscr{E}$ along a principal direction of the effective mass tensor, it can be shown that they are valid in general, provided one uses in the expression of θ instead of μ_1 the reduced mass μ of Eq. (24.40). The condition $\mu_1 \gtrless 0$ must be replaced by $\mu \gtrless 0$. The line shapes of $\Delta\varepsilon_r$ and $\Delta\varepsilon_i$ for any type of direct allowed three-dimensional critical point are shown in Fig. 80.

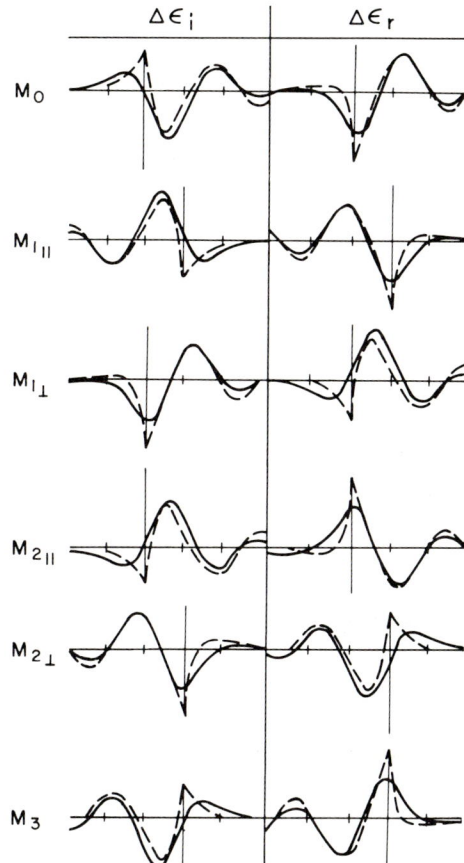

FIG. 80. Schematic representation of $\Delta\varepsilon_r$ and $\Delta\varepsilon_i$ for all possible field configurations and three-dimensional direct allowed critical points. The dashed curves are unbroadened. The solid curves correspond to a broadening parameter $\Gamma = \omega_c/\theta = 0.5$. The vertical scale has been expanded by a factor of 4 in the broadened case.

It is interesting to note that owing to the dependence of the argument of the Airy functions on the sign of the reduced mass along the direction of the electric field, one finds, in all cases discussed, oscillations in $\Delta\varepsilon_r$ and $\Delta\varepsilon_i$ *above* $\omega_g(\omega > \omega_g)$ if μ [Eq. (24.40)] is positive and *below* ω_g if μ is negative. This conclusion also holds for electrons in two-[230] and one-dimensional bands in the presence of an electric field [in these cases one has to use only 2 or 1 terms in the sum of Eq. (24.40)]. The two dimensional case may be of interest for the treatment of three dimensional bands with a very large reduced mass in one direction (flat bands such as those encountered in layer-type crystal structures), while the one-dimensional case is useful for the treatment of bands flat along 2 directions of **k**-space and for the case of a solid in the presence of parallel uniform electric and magnetic fields. (See Chapter VIII.)

The results obtained for $\Delta\varepsilon_r$ and $\Delta\varepsilon_i$ near three-dimensional critical points and those which can be obtained for two-[230] and one-dimensional singularities have been listed in Table XIV. The one-dimensional electrooptic functions $F_1(\eta)$ and $G_1(\eta)$ are defined as follows:

$$F_1(\eta) = 2\pi \, \text{Ai}^2(\eta) - H(-\eta)(-\eta)^{-1/2}$$
$$G_1(\eta) = 2\pi \, \text{Ai}(\eta) \, \text{Bi}(\eta) - H(\eta)\eta^{-1/2} \qquad (24.45)$$

The two-dimensional electrooptic functions are:

$$F_2(\eta) = \text{Ai}_1(\eta) - H(-\eta)$$
$$G_2(\eta) = \text{Gi}_1(\eta) + \pi^{-1} \ln|\eta| \qquad (24.46)$$

The functions $\text{Ai}_1(\eta)$ and $\text{Gi}_1(\eta)$ are defined by:

$$\text{Gi}(\eta) = \pi^{-1} \, \text{Im} \int_0^\infty ds \, \exp i[s^3/3 + s\eta]$$

$$\text{Ai}_1(\eta) = \int_0^\infty \text{Ai}(\eta + x) \, dx \qquad (24.47)$$

$$\text{Gi}_1(\eta) = \int_0^\infty \text{Gi}(\eta + x) \, dx.$$

The electrooptic functions F_2, G_2, F_1, and G_1 are plotted in Figs. 81 and 82.

The effect of Lorentzian broadening on the three-dimensional electrooptic functions discussed above, has been considered by Aspnes.[232] An alternative treatment of broadening has been given by Enderlein[237]; a numerical calculation of several particular cases has been also given by Seraphin and Bottka.[149] The calculation of ε_i can be performed with Eq. (6.3) provided one replaces the δ function by its Lorentzian broadened counterpart [see also Eq. (7.2b)]:

$$\omega_\tau/\pi^{-1}[(\omega - \omega_j)^2 + \omega_\tau^2], \qquad (24.48)$$

24. THEORY

TABLE XIV. CHANGES INDUCED IN THE REAL AND IMAGINARY PARTS OF THE DIELECTRIC CONSTANT BY A UNIFORM ELECTRIC FIELD $\mathscr{E}$ IN THE NEIGHBORHOOD OF THREE-, TWO-, AND ONE-DIMENSIONAL CRITICAL POINTS. THE RESULTS ARE EXPRESSED IN TERMS OF THE ELECTROOPTIC FUNCTIONS F, G, F_2, G_2, F_1, AND G_1 DEFINED IN THE TEXT

Dimensions	Critical point	Sign of effective mass μ	$\Delta \varepsilon_r$	$\Delta \varepsilon_i$
3	M_0	+	$\dfrac{B\theta^{1/2}}{\omega^2} G\left(\dfrac{\omega_g - \omega}{\theta}\right)$	$\dfrac{B\theta^{1/2}}{\omega^2} F\left(\dfrac{\omega_g - \omega}{\theta}\right)$
	M_1	−	$\dfrac{B\theta^{1/2}}{\omega^2} G\left(\dfrac{\omega - \omega_g}{\theta}\right)$	$-\dfrac{B\theta^{1/2}}{\omega^2} F\left(\dfrac{\omega - \omega_g}{\theta}\right)$
	M_1	+	$-\dfrac{B\theta^{1/2}}{\omega^2} F\left(\dfrac{\omega_g - \omega}{\theta}\right)$	$\dfrac{B\theta^{1/2}}{\omega^2} G\left(\dfrac{\omega_g - \omega}{\theta}\right)$
	M_2	+	$-\dfrac{B\theta^{1/2}}{\omega^2} G\left(\dfrac{\omega_g - \omega}{\theta}\right)$	$-\dfrac{B\theta^{1/2}}{\omega^2} F\left(\dfrac{\omega_g - \omega}{\theta}\right)$
	M_2	−	$\dfrac{B\theta^{1/2}}{\omega^2} F\left(\dfrac{\omega - \omega_g}{\theta}\right)$	$\dfrac{B\theta^{1/2}}{\omega^2} G\left(\dfrac{\omega - \omega_g}{\theta}\right)$
	M_3	−	$-\dfrac{B\theta^{1/2}}{\omega^2} G\left(\dfrac{\omega - \omega_g}{\theta}\right)$	$\dfrac{B\theta^{1/2}}{\omega^2} F\left(\dfrac{\omega - \omega_g}{\theta}\right)$
2	Minimum	+	$\dfrac{4\pi(m_1 m_2)^{1/2}}{\omega^2} G_2\left(\dfrac{\omega_g - \omega}{\theta}\right)$	$\dfrac{4\pi(m_1 m_2)^{1/2}}{\omega^2} F_2\left(\dfrac{\omega_g - \omega}{\theta}\right)$
	Saddle point	−	$-\dfrac{4\pi(m_1 m_2)^{1/2}}{\omega^2} F_2\left(\dfrac{\omega - \omega_g}{\theta}\right)$	$\dfrac{4\pi(m_1 m_2)^{1/2}}{\omega^2} G_2\left(\dfrac{\omega - \omega_g}{\theta}\right)$
	Saddle point	+	$\dfrac{4\pi(m_1 m_2)^{1/2}}{\omega^2} F_2\left(\dfrac{\omega_g - \omega}{\theta}\right)$	$\dfrac{4\pi(m_1 m_2)^{1/2}}{\omega^2} G_2\left(\dfrac{\omega_g - \omega}{\theta}\right)$
	Maximum	−	$-\dfrac{4\pi(m_1 m_2)^{1/2}}{\omega^2} G_2\left(\dfrac{\omega - \omega_g}{\theta}\right)$	$\dfrac{4\pi(m_1 m_2)^{1/2}}{\omega^2} F_2\left(\dfrac{\omega - \omega_g}{\theta}\right)$
1	Minimum	+	$\dfrac{2\pi}{\omega^2}\left(\dfrac{2m}{\theta}\right)^{1/2} G_1\left(\dfrac{\omega_g - \omega}{\theta}\right)$	$\dfrac{2\pi}{\omega^2}\left(\dfrac{2m}{\theta}\right)^{1/2} F_1\left(\dfrac{\omega_g - \omega}{\theta}\right)$
	Maximum	−	$-\dfrac{2\pi}{\omega^2}\left|\dfrac{2m}{\theta}\right|^{1/2} G_1\left(\dfrac{\omega - \omega_g}{\theta}\right)$	$\dfrac{2\pi}{\omega^2}\left|\dfrac{2m}{\theta}\right|^{1/2} F_1\left(\dfrac{\omega - \omega_g}{\theta}\right)$

where ω_τ represents the energy broadening, or collision frequency. Hence, the electric field change in ε_i in the presence of Lorentzian broadening, may be obtained from the unbroadened change $\Delta\varepsilon_i (\omega, \mathscr{E})$ by using:

$$\Delta\varepsilon_i(\omega, \mathscr{E}, \eta) = \pi^{-1} \int_{-\infty}^{\infty} \frac{\Delta\varepsilon_i(\omega, \mathscr{E})\omega_\tau}{(\omega - \omega')^2 + \omega_\tau^2} d\omega'. \qquad (24.49)$$

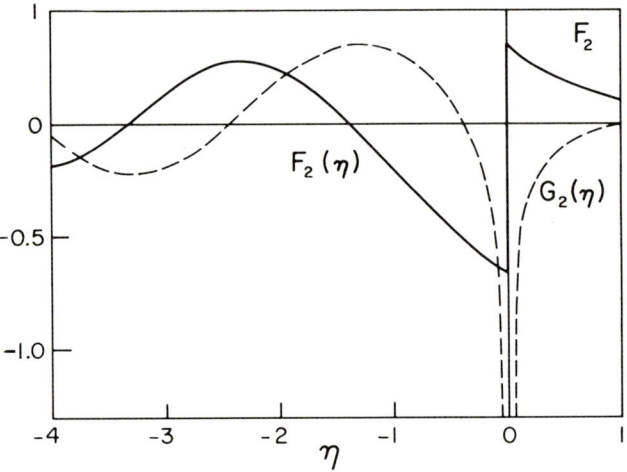

FIG. 81. Two-dimensional electrooptic functions $F_2(\eta)$ and $G_2(\eta)$ defined in Eq. (24.46). Computed by D. E. Aspnes.

Since both $\Delta\varepsilon_i(\omega, \mathscr{E}, \omega_c)$ and $\Delta\varepsilon_i(\omega, \mathscr{E})$ are the imaginary parts of functions analytic in the upper half plane of complex ω and since these functions fulfill the necessary asymptotic behavior for $|\omega| \to \infty$, $\Delta\varepsilon(\omega, \mathscr{E})$ and $\Delta\varepsilon(\omega, \mathscr{E}, \omega_\tau)$ also fulfill Eq. (24.49). A contour integration of Eq. (24.49) yields:

$$\Delta\varepsilon(\omega, \mathscr{E}, \omega_\tau) = \Delta\varepsilon(\omega + i\omega_\tau, \mathscr{E}). \qquad (24.50)$$

Equation (24.50) signifies that the broadened electrooptic functions can be obtained by replacing ω by $\omega + i\omega_\tau$ into the unbroadened ones. They involve Airy functions of complex argument. Figures 83a,b show the broadened functions $F(\eta, \Gamma)$ and $G(\eta, \Gamma)$ by which one has to replace F and G in Table XIV in order to obtain $\Delta\varepsilon$ for three-dimensional critical points. The dimensionless broadening parameter Γ is equal to ω_τ/θ.

We have so far discussed the case of direct *allowed* transitions, i.e., we have assumed that the interband matrix element is a constant independent of **k**. Under these conditions we have found that the electrooptic effect is not an explicit function of the relative orientation of **n̂** (the polarization vector of the radiation) and $\mathscr{E}$: it depends on **n̂** through the matrix element of **n̂ · p** and on $\mathscr{E}$ through the reduced

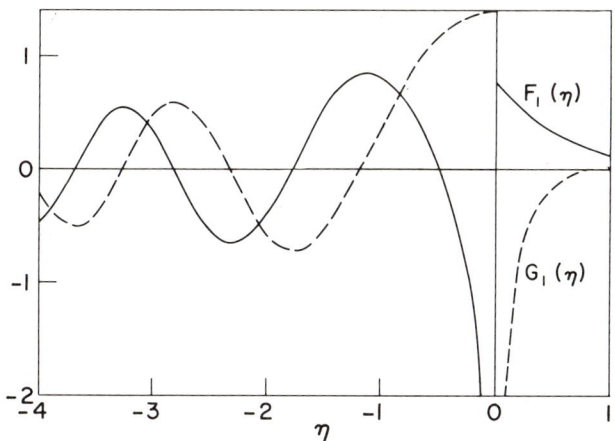

FIG. 82. One-dimensional electrooptic functions $F_1(\eta)$ and $G_1(\eta)$ defined in Eq. (24.45). Computed by D. E. Aspnes.

effective mass. If **n̂ · p** is independent of the direction of **n̂**, e.g., $\Gamma_7 \to \Gamma_1$ or $E_0 + \Delta_0$ transitions of zincblende, the electrooptic effect is isotropic. These conclusions do not hold when higher-order terms in **k** are included in the expansion of **n̂ · p**. As an example, we shall discuss the case of first-order forbidden transitions with the electric field along one of the crystal axes ($\mathscr{E}_1 \neq 0$, $\mathscr{E}_2 = \mathscr{E}_3 = 0$).[75] The square of the matrix element to be used in Eq. (6.3) is [Eq. (6.18)]:

$$A^2 |\mathbf{\hat{n}} \cdot \nabla\phi(x_1)e^{i(k_2 x_2 + k_3 x_3)}|^2_{x_1 = 0}, \tag{24.51}$$

where it has been assumed, for simplicity, that the tensor **A** reduces to a scalar A (spherical symmetry). The function $\phi(x_1)$ is given in Eq. (24.32). Equation (24.51) yields different matrix elements depending on whether **n̂** is parallel ($\varepsilon_\parallel$) or perpendicular ($\varepsilon_\perp$) to $\mathscr{E}$ and an anisotropy appears in contrast to the case of allowed transitions. A

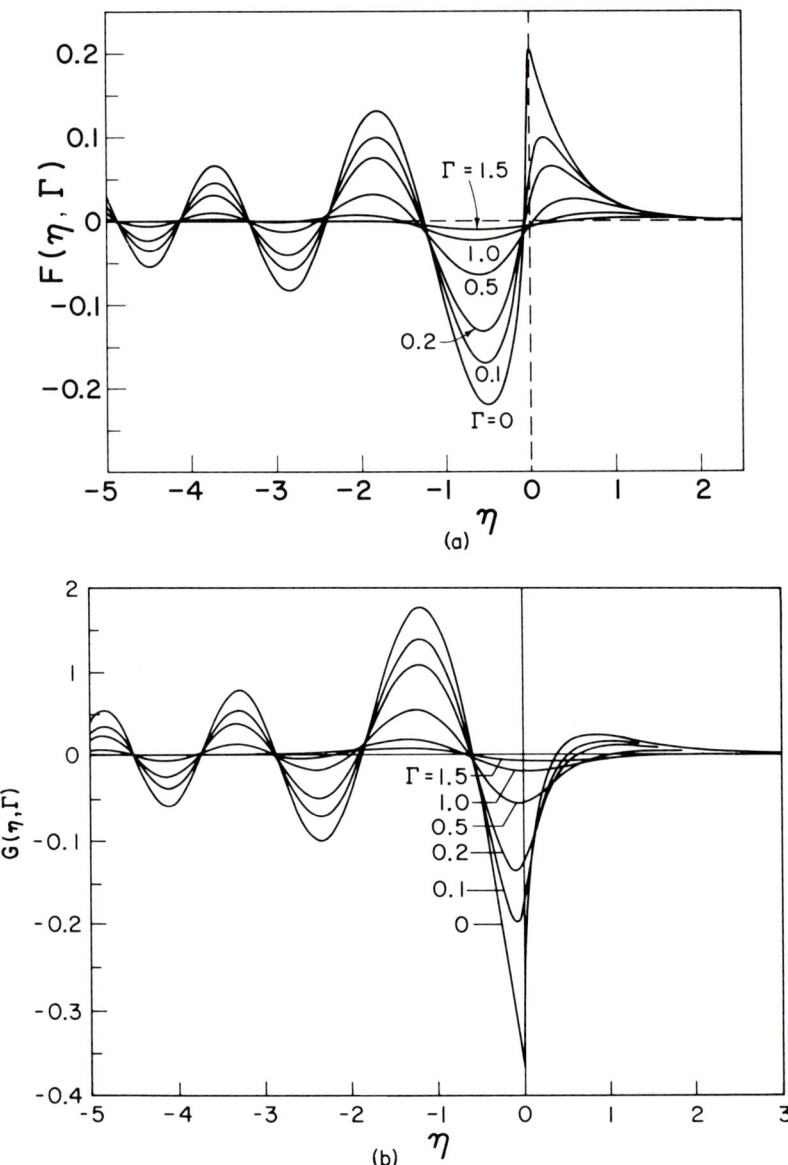

Fig. 83. Electrooptic functions $F(\eta, \Gamma)$ $G(\eta, \Gamma)$ needed to obtain the electrooptic effect of a broadened direct allowed three-dimensional edge (see Table XIV), for several values of the broadening parameter $\Gamma = \omega_\tau/\theta$.

straightforward calculation[75] for isotropic bands near an M_0 singularity yields:

$$\varepsilon_{\parallel i}(\omega, \mathscr{E}) = \frac{D\theta^{3/2}}{\omega^2} \int_{(\omega_g - \omega)/\theta}^{\infty} |Ai'(t)|^2 \, dt$$

$$= \frac{D\theta^{3/2}}{3\omega^2} \left\{ \left(\frac{\omega_g - \omega}{\theta}\right)^2 \left[Ai\left(\frac{\omega_g - \omega}{\theta}\right)\right]^2 \right.$$

$$- \left(\frac{\omega_g - \omega}{\theta}\right) \left[Ai'\left(\frac{\omega_g - \omega}{\theta}\right)\right]^2$$

$$\left. - 2 Ai\left(\frac{\omega_g - \omega}{\theta}\right) Ai'\left(\frac{\omega_g - \omega}{\theta}\right) \right\}$$

(24.52)

$$\varepsilon_{\perp i}(\omega, \mathscr{E}) = \frac{D\theta^{3/2}}{2\omega^2} \int_{(\omega_g - \omega)/\theta}^{\infty} \left(t - \frac{\omega_g - \omega}{\theta}\right)^2 Ai^2(t) \, dt$$

$$= \frac{D\theta^{3/2}}{3\omega^2} \left\{ \left(\frac{\omega_g - \omega}{\theta}\right)^2 \left[Ai\left(\frac{\omega_g - \omega}{\theta}\right)\right]^2 \right.$$

$$- \left(\frac{\omega_g - \omega}{\theta}\right) \left[Ai'\left(\frac{\omega_g - \omega}{\theta}\right)\right]^2$$

$$\left. - \frac{1}{2} Ai\left(\frac{\omega_g - \omega}{\theta}\right) Ai'\left(\frac{\omega_g - \omega}{\theta}\right) \right\},$$

where $D = 8(2^{1/2})\pi|A|^2\mu^{5/2}$. The changes in ε are obtained by subtracting from Eqs. (24.52) the zero field dielectric constant:

$$\varepsilon_r = -(D/3\pi\omega^2)(\omega_g - \omega)^{3/2}$$
$$\varepsilon_i = (D/3\pi\omega^2)(\omega - \omega_g)^{3/2}.$$

(24.53)

Figure 84 shows the functions $\Delta\varepsilon_{\parallel i}$ and $\Delta\varepsilon_{\perp i}$ obtained from Eqs. (24.52) and (24.53), divided by the factor $D\theta^{3/2}/3\omega^2$. An anisotropy $\Delta\varepsilon_{\parallel}/\Delta\varepsilon_{\perp} = 4$ appears at $\omega = \omega_g$.

The general case of forbidden transitions for parabolic bands and arbitrary masses has been treated by Aspnes.[238] His calculation includes a constant term and terms linear in the components of **k** in the expansion of $\mathbf{p}_{cv}$. He finds for M_0 and M_3 critical points:

$$\Delta \varepsilon_i = \frac{2(2\mu)^{3/2}\theta^{1/2}}{\omega^2} \left\{ |\mathscr{P}_{\hat{\mathbf{n}}}|^2_M F(\beta) + \pi 2^{1/2} |\mu| \theta^{1/2} \operatorname{Im}(\mathscr{P}_{\hat{\mathbf{n}}} \hat{\mathbf{n}} \cdot \nabla_{\mathbf{k}} \mathscr{P}_{\hat{\mathbf{n}}}^*)_M \operatorname{Ai}^2(\beta) \right.$$

$$+ \theta \left(\sum_i \mu_i \left| \frac{\partial \mathscr{P}_{\hat{\mathbf{n}}}}{\partial k_i} \right|^2_M \right) [\pi \{ \tfrac{2}{3} \beta^2 \operatorname{Ai}^2(\beta) - \tfrac{2}{3}\beta \operatorname{Ai}'^2(\beta) - \tfrac{1}{3} \operatorname{Ai}(\beta) \operatorname{Ai}'(\beta) \}$$

$$+ \tfrac{2}{3}\beta(-\beta)^{1/2} H(-\beta)] - \theta |\mu| |\hat{\mathbf{n}} \cdot \nabla_{\mathbf{k}} \mathscr{P}_{\hat{\mathbf{n}}}|^2_M \pi \operatorname{Ai}(\beta) \operatorname{Ai}'(\beta) \Bigg\}, \quad (24.54)$$

where $\mathscr{P}_{\hat{\mathbf{n}}}$ is equal to $p_{cv} \cdot \hat{\mathbf{n}}$, $\beta = (\omega_g - \omega)/\theta$ for an M_0 critical point and to $(\omega - \omega_g)/\theta$ for an M_3 critical point. The constants μ and θ are given in Eqs. (24.40) and (24.43), respectively. The subindex M indicates the value at the critical point.

For M_1 and M_2 critical points one must also distinguish between the "parallel" configuration ($\mu < 0$ for M_1 and $\mu > 0$ for M_2 critical

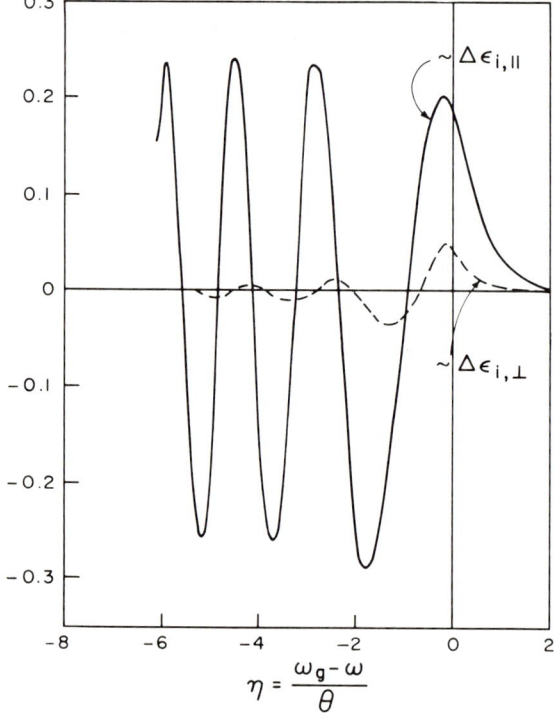

FIG. 84. Functions which give the changes induced in ε_i by an electric field near a direct forbidden edge. Notice the strong anisotropy of the effect.

points) and the opposite "perpendicular" or "transverse" configuration ($\mu > 0$ for M_1 and $\mu < 0$ for M_2 critical points). For the parallel configuration, one finds:

$$\Delta\varepsilon_{i_{M_1\text{para}}} = -\Delta\varepsilon_{i_{M_3}} \\ \Delta\varepsilon_{i_{M_2\text{para}}} = -\Delta\varepsilon_{i_{M_0}}. \tag{24.55}$$

A calculation of the "transverse" case near an $M_1(\mu > 0)$ or an M_2 ($\mu < 0$) edge yields:

$$\Delta\varepsilon_i = \frac{2(2\mu)^{3/2}\theta^{1/2}}{\omega^2} \Big\{ |\mathscr{P}_{\hat{n}}|_M^2 \, G(\beta) + \pi 2^{1/2} |\mu| \, \theta^{1/2} \, \text{Im}(\mathscr{P}_{\hat{n}} \hat{n} \cdot \nabla_k \mathscr{P}_{\hat{n}})_M \\ \times \text{Ai}(\beta) \, \text{Bi}(\beta) + \theta\Big(\sum_i \mu_i \left|\frac{\partial \mathscr{P}_{\hat{n}}}{\partial k_i}\right|_M^2\Big)[\pi\{\tfrac{2}{3}\beta^2 \, \text{Ai}(\beta) \, \text{Bi}(\beta) \\ - \tfrac{2}{3}\beta\text{Ai}'(\beta)\text{Bi}'(\beta) - \tfrac{1}{6}[\text{Ai}(\beta)\text{Bi}'(\beta) + \text{Ai}'(\beta) \, \text{Bi}(\beta)]\} \\ - \tfrac{2}{3}\beta(\beta)^{1/2}H(\beta)] \\ - \theta |\mu| \, |\hat{n} \cdot \nabla\mathscr{P}_{\hat{n}}|_M^2 \tfrac{1}{2}\pi[\text{Ai}(\beta)\text{Bi}'(\beta) + \text{Ai}'(\beta) \, \text{Bi}(\beta)] \Big\}, \tag{24.56}$$

with $\beta = (\omega_g - \omega)/\theta$ for an M_1 edge and $\beta = (\omega - \omega_g)/\theta$ for an M_2 edge. It is worth mentioning that Eq. (24.56) can be obtained by replacing Ai Ai by Ai Bi into Eq. (24.54) in a symmetrical manner [(Ai Ai') → $\tfrac{1}{2}$(Ai Bi' + Ai' Bi)]. A similar type of substitution can be used to obtain $\Delta\varepsilon_r$ from Eqs. (24.54), (24.55), and (24.56). The resulting equations will not be given here.

It is interesting to consider the ratio of the electrooptic effect in the allowed case [e.g., Eq. (24.18)] for an M_0 edge, and the forbidden case, e.g., Eq. (24.52). At $\omega = \omega_g$, we find:

$$\frac{\Delta\varepsilon_{\|i}(\omega = \omega_g, \text{forbidden})}{\Delta\varepsilon_i(\omega = \omega_g, \text{allowed})} = \frac{7|A|^2\mu}{|\mathbf{p}_{cv}|^2} \theta. \tag{24.57}$$

The magnitude of the coefficient of the linear terms in $\mathbf{k}$ of $\mathscr{P}_{\hat{n}}$ can be expected to be at most of the order of $|\mathbf{p}_{cv}|$ times $a/2\pi$ (a = lattice constant) and $(2\pi/a)^2 \times \mu^{-1}$ is of the order of the band widths. Therefore, the ratio in Eq. (24.57) is of the order of θ divided by the sum of the widths of valence and conduction bands; it is small for most practical values of the modulating field.

In the formulas given in the preceding pages, the applied modulating field appears in the argument of the Airy functions and in the multiplication factors as a fractional power. Hence a complicated dependence of the electrooptic effect on the orientation of $\mathscr{E}$ is expected.

The argument of the Airy function contains through θ the mass along the direction of the electric field. For M_1 and M_2 edges, the sign of this mass depends on the direction of $\mathscr{E}$ and, therefore, as the orientation of $\mathscr{E}$ with respect to the mass axes is varied the position of the oscillations with respect to the gap should vary: for $\mathscr{E}$ inside the positive mass cone oscillations will occur *above* ω_g and for $\mathscr{E}$ outside this cone they will occur *below* ω_g. The multiplicity of critical points off $\mathbf{k} = 0$ imposed by the crystal point group makes it possible to have both, oscillations above and below ω_g. Let us consider, for instance, a cubic material, e.g., silicon, with M_1 critical points along [100] and all equivalent directions. The [100] constant energy surfaces have symmetry of revolution around [100]: the mass along the axis of revolution must be the negative one. For $\mathscr{E}$ along [111], the contribution of all equivalent ellipsoids to the electrooptic effect is the same; its relative intensity is given by the matrix element $|\mathbf{p}_{cv} \cdot \hat{\mathbf{n}}|^2$ (under the assumption of allowed transitions and $\mathbf{k}$-independent matrix element). Therefore, the effect has, in this case, the same symmetry as the dielectric constant for $\mathscr{E} = 0$ (isotropic). If $\mathscr{E}$ is along [100], however, the [100] valleys give oscillations *below* ω_g while the [010] and [001] valleys give oscillations *above* ω_g. The line shapes of these 2 contributions are also different since for the [100] valleys $\Delta\varepsilon_i$ is given by F functions while for [010] and [001] valleys it is given by G functions. The relative contribution of each valley to the overall effect depends on $\mathbf{p}_{cv}$: if, as is sometimes the case, the transitions are only allowed for $\hat{\mathbf{n}}$ perpendicular to the axis of the valley, only the [010] and [001] valleys contribute (oscillations *above* ω_g) for $\hat{\mathbf{n}}$ parallel to $\mathscr{E}$ while both, the [100] and the [010] valleys contribute for $\hat{\mathbf{n}}$ parallel to [001] (oscillations *above* and *below* ω_g). If in this last case one of the masses, e.g., the longitudinal one, is *large* the oscillations above ω_g may, however, not be observable at moderate values of the field.

For [111] valleys, the electrooptic effect of *allowed* transitions is independent of $\hat{\mathbf{n}}$ when $\mathscr{E}$ is along [100] and strongly anisotropic ($\hat{\mathbf{n}}$-dependent) for $\mathscr{E}$ along [111]: in this case oscillations result both above and below ω_g for M_1 and M_2 critical points. As an illustration, we show in Fig. 85 the electroreflectance spectrum of the E_1 transitions of GaAs (see Fig. 29) calculated by Rössler and Bottka[239] under the assumption of a matrix element of $\mathbf{p}$ independent of the polarization of **E**. This figure demonstrates the existence of oscillations above and below ω_g for $\mathscr{E} \parallel [111]$. However the oscillations below ω_g are very closely spaced because of the large longitudinal mass: broadening is likely to wipe out these oscillations altogether. For $\mathscr{E}$ along [110], one only sees oscillations above ω_g since the masses of all valleys along the $\mathscr{E}$ direction are positive. There are, however, two sets of equivalent valleys which give different masses along $\mathscr{E}$: $\{\pm[111], \pm[11\bar{1}]\}$ and $\{\pm[1\bar{1}1], \pm[1\bar{1}\bar{1}]\}$. Hence, two distinct periods of oscillation appear for $\omega > \omega_g$.

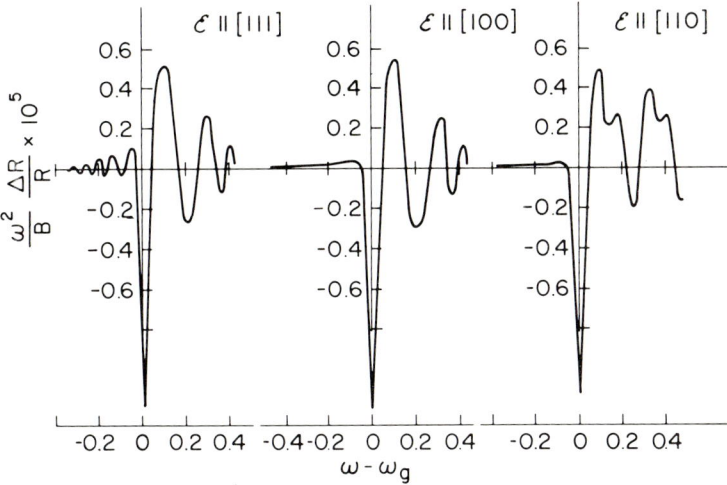

FIG. 85. Electroreflectance spectra of the $E_1 - (E_1 + \Delta_1)$ transitions in GaAs with M_1 critical point at L or Λ calculated by U. Rössler and N. Bottka, *Solid State Commun.* **5**, 939 (1968). $\mathscr{E} = 6 \times 10^4$ V cm^{-1}.

b. *Interband Effects: Indirect Transitions*

The effect of an electric field on an allowed indirect edge can be also calculated[240] by solving the corresponding effective mass equation for the relative coordinate of the electron hole pair and thus calculating $\phi(\mathbf{r} = 0)$ [we assume constant matrix elements of **p** for the optical transition]. Derivations which parallel the other methods of Section 24a and yield the same results as the one given here, have also been published.[241,242]

We shall consider transitions between a valence band maximum and a conduction band minimum at different points of **k**-space: transitions between other types of extrema are usually superimposed on larger backgrounds and hence not observed. The difference in **k** between the electron and the hole extrema is $\mathbf{K}_0$. We assume, for simplicity, that $\mathscr{E}$ is along one of the principal axes of the electron hole reduced mass [Eq. (6.11) with tensorial masses]. Generalization to an arbitrary direction can be done in a manner similar to that used in Section 24a. It amounts to the use of average effective masses and does not change the electrooptic functions involved. The energy of the final exciton state is the eigenvalue of Eq. (24.29) for the r_1 coordinate ($\mathscr{E}_1 \neq 0$), plus the kinetic energy ω_r of the relative motion perpendicular to $\mathscr{E}$:

$$2\omega_r = (k_2^2/\mu_2) + (k_3^2/\mu_3), \qquad (24.58)$$

plus the energy ω_{cm} of the center of mass motion of the electron hole pair[17]

$$2\omega_{cm} = \mathbf{K} \cdot \mathbf{M}^{-1} \cdot \mathbf{K}. \qquad (24.59)$$

In Eqs. (24.58) and (24.59) **k** is the crystal momentum of the relative motion of the electron and the hole

$$\mathbf{k} = (\mathbf{k}_e - \mathbf{k}_h - \mathbf{K}_0) \qquad (24.60)$$

and **K** the center of mass momentum

$$\mathbf{K} = \tfrac{1}{2}(\mathbf{k}_e + \mathbf{k}_h - \mathbf{K}_0). \qquad (24.61)$$

The tensor M is the sum of the effective mass tensors of the electron and the hole. The transition probability, proportional to $|\phi(0)|^2$ with $\phi(0)$ given by Eq. (24.32), must be summed for all possible values of k_2, k_3, and **K** subject to the energy conservation condition:

$$\omega = W_1 + \omega_g + \omega_r + \omega_{cm} \pm \omega_{phon}, \qquad (24.62)$$

with the plus sign for phonon emission and the minus sign for phonon absorption. The density of states for the energy of relative motion perpendicular to $\mathscr{E}$ is the step function of Eq. (24.38) while that for the center of mass motion is proportional to $\omega_{cm}^{1/2}$. The imaginary part of the dielectric constant $\varepsilon_i(\omega, \varepsilon)$ is therefore proportional to:

$$I(\omega_g - \omega \pm \omega_{phon})$$
$$= \int_{\omega_g - \omega \pm \omega_{phon}}^{\infty} \theta^{-1/2}(W - \omega_g + \omega \pm \omega_{phon})^{1/2} \, dW \int_{W/\theta}^{\infty} |\text{Ai}(t)|^2 \, dt. \qquad (24.63)$$

Evaluation of the integral of Eq. (24.63) by the methods of Aspnes[76] yields a dielectric constant ε_i for allowed indirect transitions proportional to:

$$I(\eta) = (2^{1/3}\pi/16)[\text{Ai}(\eta) + \eta \text{Ai}'(\eta) + \eta^2 \, \text{Ai}_1(\eta)] \qquad (24.64)$$

with $\eta = (\omega_g - \omega \pm \omega_{phon})/\theta$. The change in ε_i induced by $\mathscr{E}$ is proportional to $\mathscr{E}^{4/3}$ and to the electrooptic function:

$$F_i(\eta) = \mathrm{Ai}(\eta) + \eta \mathrm{Ai}'(\eta) + \eta^2 \, \mathrm{Ai}_1(\eta) - \eta^2 H(-\eta). \quad (24.65)$$

The function $F_i(\eta)$ is shown in Fig. 86. The real part of the dielectric constant associated with an indirect edge is readily obtained by replacing $\mathrm{Ai}(\eta)$ by $\mathrm{Gi}(\eta)$ in Eq. (24.64). Because of their weak nature, indirect transitions are usually observed only in transmission and, therefore, only $\Delta\varepsilon_i$ is of relevance for these spectra.†

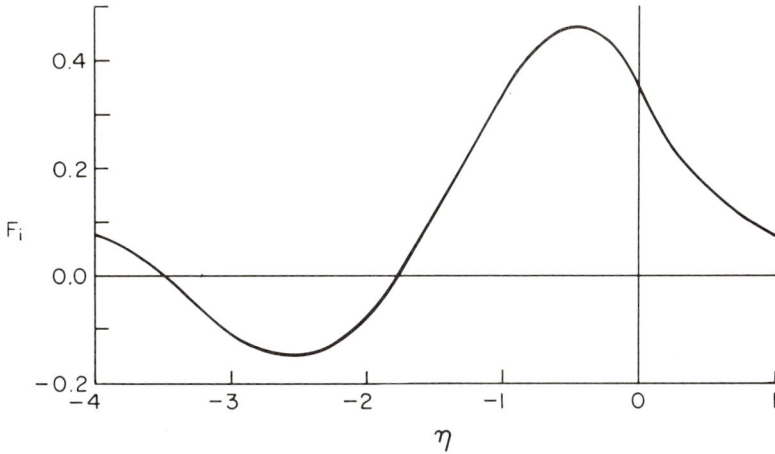

FIG. 86. Electrooptic function $F_i(\eta)$. This function determines $\Delta\varepsilon_i(\omega, \mathscr{E})$ for an allowed indirect edge. Computed by D. E. Aspnes.

c. Excitons[79]

In the preceding sections, we have treated the dielectric constant in the neighborhood of an energy gap in the presence of an electric field neglecting exciton effects, i.e., the Coulomb interaction between the electron and the hole. Such interaction can be included in the effective mass equation [Eq. (24.28)] and the corresponding $\varepsilon_i(\omega, \mathscr{E})$ can, in principle, be obtained from Eq. (6.3). The optical matrix elements are calculated from the wave functions of the effective mass

---

† Reports of indirect transitions observed weakly in electroreflectance spectra have appeared in the literature[243]; it is not clear, however, whether this observation is due to structure in the reflectivity or to electroabsorption structure.

equation for $\mathbf{r}=0$ [Eq. (6.8), allowed transitions] or from their gradient also at $\mathbf{r}=0$ [Eq. (6.18), forbidden transitions].

As indicated in Section 6c, the exciton potential may not be as simple as a Coulomb potential. This is particularly true in the neighborhood of M_1 and M_2 critical points. We shall however, treat here only the case of an M_0 critical point with pure Coulomb interaction.[79] The treatment can be easily extended to M_3 critical points. We shall also assume isotropic effective masses. No theoretical treatment of the electrooptic effect with exciton interaction for M_1 and M_2 critical point has appeared to date.

The effective mass equation can be conveniently written by using as unit of energy *twice* the binding energy of the ground state of the exciton $2\omega_{\mathrm{ex}} = \mu/\varepsilon_0^2$ and as unit of length the corresponding Bohr radius $a_{\mathrm{B}} = \varepsilon_0/\mu$. We find:

$$[\tfrac{1}{2}\nabla^2 - ez + (1/r)]\phi(\mathbf{r}) = 0. \qquad (24.66)$$

The field e is along the z direction and is measured in reduced units:

$$e = \mathscr{E} a_{\mathrm{B}}/2\omega_{\mathrm{ex}}. \qquad (24.67)$$

Notice that the reduced field e is one-fourth of the ratio of the difference in potential due to $\mathscr{E}$ across the diameter of exciton orbit to the binding energy of the exciton. Ionization of the exciton in the external field is expected to occur when this ratio is of the order of one ($e \simeq \tfrac{1}{4}$). It is convenient to define a new parameter $n = (-2W)^{1/2}$; in the $e = 0$ case, the hydrogenic bound states correspond to integer values of n. We shall confine our discussion to the energy region in the vicinity of these bound states. Equation (24.66) can be separated by changing to parabolic coordinates[244]:

$$x = (\xi\eta)^{1/2}\cos\phi, \quad y = (\xi\eta)^{1/2}\sin\phi, \quad z = \tfrac{1}{2}(\xi-\eta). \qquad (24.68)$$

Due to the symmetry of revolution around z, the wave function must have a factor $e^{im\phi}$ with m an integer. The wave function can be written:

$$\phi(r) = (2/\pi)^{1/2}\chi_1(\rho_1)\chi_2(\rho_2)e^{im\phi}/(\eta^2\rho_1\rho_2)^{1/2}, \qquad (24.69)$$

where $\rho_1 = \xi/n$ and $\rho_2 = \eta/n$. The functions χ_1 and χ_2 fulfill the one-dimensional equations:

$$d^2\chi_i/d\rho_i^2 + [\tfrac{1}{4} - U_i(\rho_i)]\chi_i = 0,$$

where

$$U_i(\rho_i) = (m^2 - 1)\rho_i^2 - [n_i + \tfrac{1}{2}(1 + |m|)]/\rho_i - (-1)^i \mathscr{E} n^3 \rho_i/4 \quad (24.70)$$

and $n_1 + n_2 = n - 1 - |m|$. The differential equations (24.70) cannot be solved in terms of standard tabulated functions. Such a solution is possible in the absence of modulating field ($e = 0$) and also for $e \neq 0$ in the absence of Coulomb interaction. This consideration lead Duke and Alferieff[79] to approximate the U_i functions of Eqs. (24.70) by the pure Coulomb case ($e = 0$) for ρ_i smaller than a cutoff ρ_i^0 and by the pure uniform electric field case for ρ_i larger than this cutoff. The solution of the pure Coulomb case for $\rho_i^0 \geq \rho_i$ is obtained in terms of the Kummer functions[245] (the solutions *do not* have to be regular at infinity) $M(a, b, x)$:

$$\chi_i(\rho_i) \propto n^{1/2} \rho_i^{(1+|m|)/2} \exp(-\rho_i/2) \, M(-n_i, 1 + |m|, \rho_i) \quad \text{for } \rho_i \leq \rho_i^0.$$
(24.71)

The solutions for $\rho_i > \rho_i^0$ (no Coulomb potential) are linear combinations of the two Airy functions $\text{Ai}(\rho_i)$ and $\text{Bi}(\rho_i)$; for $i = 1$, however, due to the fact that $U_1 \to +\infty$ for $\rho_1 \to +\infty$, χ_1 must tend to zero for $\rho_1 \to \infty$ and hence only $\text{Ai}(\rho_1)$ contributes to χ_1. Finding the solution of Eq. (24.70) with this model potential is, therefore, reduced to matching (continuity of χ_i and first derivatives) the Kummer function solution and the sums of Airy functions at $\rho_i = \rho_i^0$. A certain degree of arbitrariness is left in the choice of ρ_i^0. It is reasonable to chose ρ_i^0 as those values of ρ_i for which the Coulomb and the externally applied potentials are equal. For the details of the wave function matching the reader is referred to Duke and Alferieff.[79]

Regularity of $\phi(\mathbf{r})$ for $r \to 0$ requires $\phi(0) = 0$ unless $m = 0$. Hence, for allowed transitions, we only have to find the solutions with $m = 0$. In this case and in the absence of the modulating field ($e = 0$), positive integer values of n give the discrete hydrogenic series of bound states as discussed in Section 6b. The quantum numbers n_1 and n_2 are, in this case, also positive integers: the nth level is n-fold degenerate for $m = 0$. For $e \neq 0$ the wave function matching described above yields the quantization of n_1 which, however, will not be an integer any longer. The other quantum number n_2 is not quantized: the energy spectrum becomes continuous. Nevertheless, continuity for $e \to 0$ suggests that for small values of e broadened peaks will occur in the vicinity of the hydrogenic lines. Each hydrogenic line corresponding to a given n should split into n components in the presence of the electric field. For large electric fields the Coulomb potential becomes negligible and a smooth spectrum of the type discussed in the previous section is obtained below ω_g. Figure 87 shows the contribution of the lowest eigenvalue of n_1 to ε_i in the neighborhood of the lowest exciton peak (allowed transitions). It is clear that the exciton

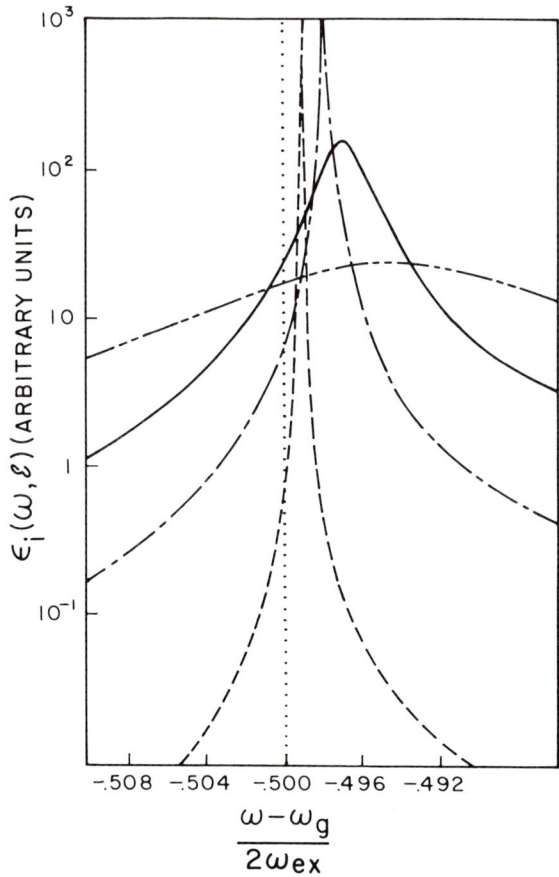

FIG. 87. ε_i in the vicinity of the first exciton peak (in arbitrary units) for various values of the reduced electric field $\varepsilon = \mathscr{E}a_B/2\omega_{ex}$ for $\varepsilon = 0.01$ (···), $\varepsilon = 0.05$ (- - -), $\varepsilon = 0.075$ (— · —), $\varepsilon = 0.10$ (—), and $\varepsilon = 0.15$ (— - - —). Only contributions from the lowest eigenvalue of n have been included. From C. B. Duke and M. E. Alferieff, *Phys. Rev.* **145**, 583 (1966).

structure disappears for ε in the neighborhood of 0.15. Since ε is one quarter of the ratio of the potential difference across the $n = 1$ exciton orbit to the binding energy of the exciton, we have confirmed the qualitative conclusion reached earlier: this ratio must be about 1 in order to destroy the exciton with the electric field.

Figure 88 shows the contribution to ε_i of the 2 lowest eigenvalues of n_1 near the $n = 2$ state of the unperturbed exciton. These peaks

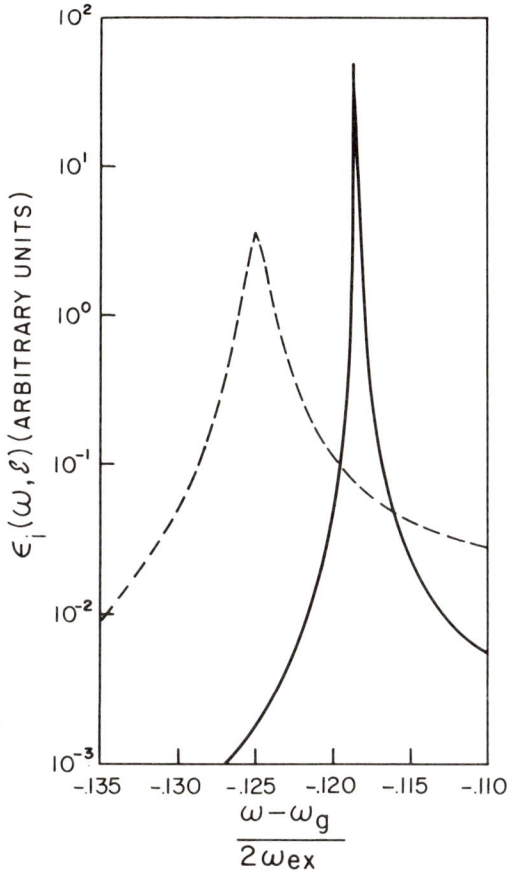

FIG. 88. Contributions of the 2 lowest eigenvalues of n_1 to ε_i: $\varepsilon = 0.01$, $i = 1$ (—) and $\mathscr{E} = 0.01$ $i = 0$ (- - -) (in arbitrary units) in the neighborhood of the $n = 2$ exciton peak for the 2 lowest eigenvalues of n_1. The splitting of the $n = 2$ states by the electric field is clearly noticeable. From C. B. Duke and M. E. Alferieff, *Phys. Rev.* **145**, 583 (1966).

disappear completely[79] for $\mathscr{e} = 0.05$. The splitting of the $n = 2$ excitons for $\mathscr{e} \neq 0$, caused by the 2 possible eigenvalues of n_1, appears clearly in this figure.

The case of excitons associated with forbidden interband transitions has not been treated[79] in as much detail as that of allowed transitions. Equations (24.68) and (24.69) indicate that the selection rules depend on the direction of polarization of the incident light, i.e., on whether its electric field is parallel (E ∥ $\mathscr{E}$) or perpendicular (E ⊥ $\mathscr{E}$) to the modulating electric field. For the E ∥ $\mathscr{E}$ configuration, we find $m = 0$ and for the E ⊥ $\mathscr{E}$ configuration $m = \pm 1$ as the states to which transitions are allowed. For E ∥ $\mathscr{E}$ and $n = 2\kappa + 1$ (κ an integer) the state with $n_1 = n_2 = \kappa$, forbidden in the absence of an electric field, becomes allowed in its presence. Hence the $n = 1$ state, forbidden for "forbidden" interband transitions, e.g., the lowest exciton state of Cu_2O (see Eq. (6.19)) becomes allowed in the presence of the electric field. For $\mathscr{E} = 0$ electric dipole transitions are only allowed to p-states of the "forbidden exciton." Since for $n = 1$ we only have an s-state, such transitions are forbidden. The electric field mixes some p-like states to the s, $n = 1$ state and, hence, the transitions become allowed. Such electric field induced transitions should be readily observable in a modulation spectrum. An example (Cu_2O) will be presented in section 26b.

As shown in Figs. 87 and 88, the effect of the electric field on an exciton spectrum is to broaden the exciton lines and to shift them to higher energies. The calculations described here[79] have been performed neglecting the line width of the exciton for $\mathscr{E} = 0$. However, it is reasonable to assume that the conclusions obtained here for $\mathscr{E} \neq 0$ remain qualitatively valid when the zero-field broadening is included. The field should increase the line width and shift the peaks to higher energies. Depending on whether the shift is larger than the broadening or vice versa, the shapes of the differential spectra for ε_r and ε_i should qualitatively be either those of Eqs. (15.2) or (17.3).

The effect of an electric field on an indirect exciton edge [Eq. (6.41)] has not been treated theoretically. However, the qualitative conclusions discussed above are expected to hold: the differential field modulation spectra should be a combination of the corresponding shift (Fig. 51) and broadening spectra (Table VI). It is difficult, however, to decide with qualitative arguments what relative weights should be given to each one of these contributions.

d. Paraelectric and Ferroelectric Crystals

Paraelectric and ferroelectric crystals are known to exhibit very large electrooptic phenomena[246,247] which cannot be accounted for in terms of the interband electrooptic effect (Franz–Keldysh) of Sections 24a–b. On the one hand the large magnitude ($\sim 25\%$) of the electric field induced changes in optical constants would require quite unreasonable values of the effective masses† (or matrix elements $\mathbf{p}_{cv}$) if it were caused by Franz–Keldysh effects.[248] Also, the line shapes observed are not of the Franz–Keldysh type: no oscillations are seen in the rather broad spectra, which resemble more closely stress modulation than Franz–Keldysh-type spectra. Considerable shifts in the positions of the critical points with $\mathscr{E}$ are seen.[249] These shifts are usually proportional to the square of the polarization $\mathbf{P}$, although shifts proportional to $\mathbf{P}$ have been reported; this fact suggests that one is actually observing a change in the band structure produced by the shifts in the positions of the constituent atoms associated with the polarization $\mathbf{P}$.

An applied electric field is expected to change the lattice constants of the material (electrostriction, piezoelectricity) and the internal coordinates of the atoms in the unit cell. Hence any electric field modulation spectrum is expected to have, beside the explicit (Franz–Keldysh) electric field contribution discussed in the previous subsections, contributions of the type described in Chapter VI. In most materials these contributions can be shown to be (from the known electrostriction and piezoelectric coefficients) considerably smaller than the explicit electrooptic effects (piezoelectric and electrostrictive contributions are proportional to $\mathscr{E}$ and $\mathscr{E}^2$, respectively, while unbroadened Franz–Keldysh contributions are, for direct allowed three-dimensional transitions, proportional to $\mathscr{E}^{1/3}$). In ferroelectric and paraelectric materials, however, large changes in the atomic positions can be effected by an applied field. The associated changes in lattice constants seem to be responsible for the observed electrooptic effect in some ferroelectric materials, e.g., SbSI,[246] while the change in the internal coordinates seems to be responsible for the effects observed in perovskite-type paraelectric and ferroelectric materials, e.g.,

---

† The authors of reference 248 have attempted to fit the electroreflectance spectrum of rutile (TiO_2) to F- and G-type electrooptic functions. However, the values of the constants B [Eq. (24.22)] required to fit the amplitude of the data would demand completely unreasonable values of either $\mathbf{m}^*$ or $\mathbf{p}_{cv}$.

BaTiO$_3$, SrTiO$_3$, KTaO$_3$.[249] This change in internal coordinates produced by the electric field is similar to the change undergone in the absence of an electric field when the temperature of the material is brought below the ferroelectric Curie temperature.[250] It seems, however, that atomic displacements cannot account for the large electrooptic effects observed in Rutile (TiO$_2$).[248] It is worthwhile pointing out that these effects, in contrast to the Franz–Keldysh effects, should relax for ac applied modulating fields at frequencies at which the ions no longer follow the applied field.

We shall concentrate our discussion on the materials with perovskite, e.g., SrTiO$_3$, BaTiO$_3$ above 120°C) and related structures (BaTiO$_3$ below 120°C, TiO$_2$) since these materials have commanded considerable theoretical and experimental attention. The structure of perovskite is simple cubic with 5 atoms/unit cell distributed as indicated in Fig. 89a. The corresponding Brillouin zone, with the standard notation for high-symmetry points and lines is shown in Fig. 89b. The

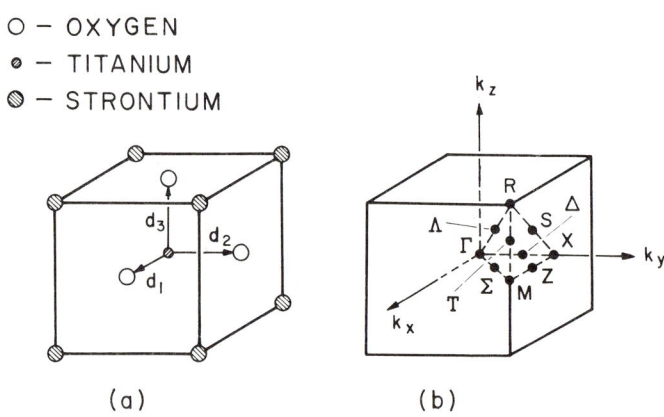

FIG. 89. (a) Unit cell of perovskite structure SrTiO$_3$; (b) Brillouin zone for the simple cubic structure of SrTiO$_3$ [from A. H. Kahn and A. J. Leyendecker, *Phys. Rev.* **135** A132 (1964)].

band structure of SrTiO$_3$ has been calculated by Kahn and Leyendecker[251] (KL) by the tight binding (linear combination of atomic orbitals, LCAO) method. The valence band at $\mathbf{k} = 0$ arises in the KL model from the p valence orbitals of the 3 oxygen atoms in the unit cell (9 states which split under the cubic crystal field into 3 groups of threefold states of symmetries Γ_{15}, Γ_{15}, Γ_{25}). The $\mathbf{k} = 0$ conduction bands arise from the 5 d orbitals of the titanium ($\Gamma_{25'}$, Γ_{12}). The overlap integrals between these p and d orbitals determine the amount of

p–d mixing away from $\mathbf{k}=0$. The presence of Sr atoms does not influence the calculation. These atoms produce energy levels too far removed from the energy gap to have any influence on the bands discussed above: no essential difference is expected to exist between the band structure of $SrTiO_3$ and that of cubic (above 120°C) $BaTiO_3$.

An adjustable parameter in the KL calculation is the ionicity of the compound which affects the band structure primarily through the electrostatic or Madelung energy of the point ions; the ionicity is adjusted so as to obtain a smallest energy gap for the material close to that observed experimentally (~ 3 eV). This occurs for a negative charge on the oxygen ions of about 1.68 (ionicity $\sim 85\%$).

The KL band structure of $SrTiO_3$ is shown in Fig. 90. The lowest

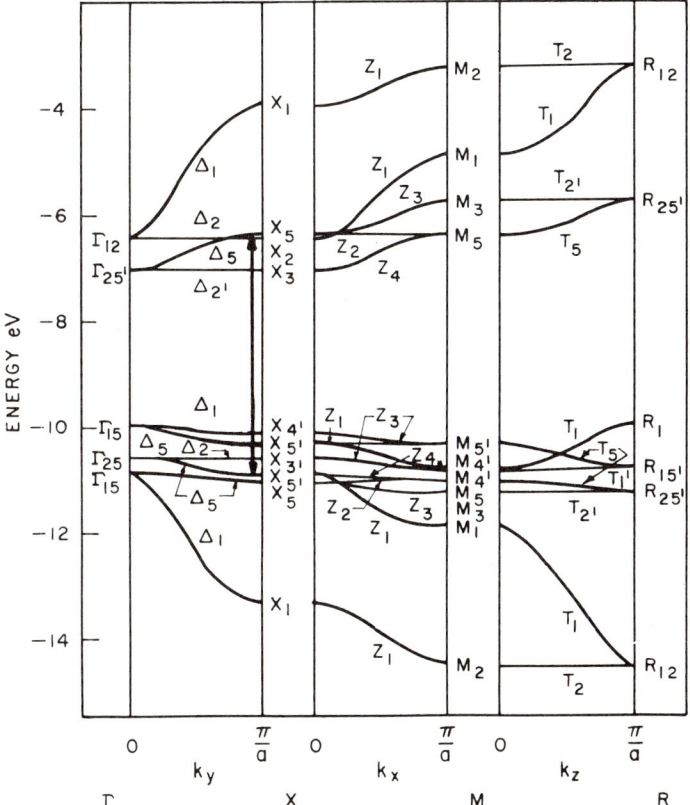

FIG. 90. Band structure of $SrTiO_3$ along several high-symmetry lines [from A. H. Kahn and A. J. Leyendecker, *Phys. Rev.* **135**, A132 (1964)].

direct gap seems to occur at $\mathbf{k}=0$ although the possibility of its occurring at the X point cannot be ruled out. While the $\Gamma_{15} - X_3$ conduction band is flat, a small perturbation, not included in the KL calculation, probably lowers the X point and gives a lowest X_3 minimum. It has been suggested[252] that the main dispersion mechanism in this material (that which is responsible for the long wavelength electronic polarizability) is the transitions between the $X_{5'}$ and X_5 bands (connected by an arrow in Fig. 90), which are flat along $X - M(Z_4 \to Z_2)$.

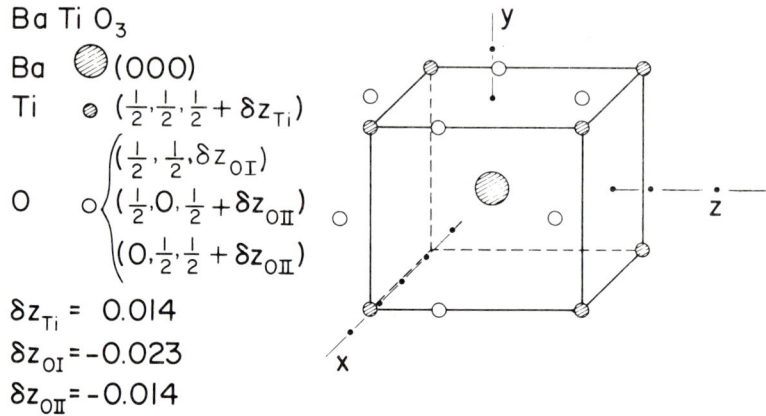

FIG. 91. Unit cell of tetragonal ferroelectric $BaTiO_3$. The crystal point group is C_{4v}. The displacements of the oxygen and titanium atoms have been exaggerated so as to make then clearly visible. The unit cell parameters are taken from B. C. Frazer, H. R. Danner, and R. Pepinsky, *Phys. Rev.* **100**, 745 (1955).

The application of an electric field along [001] to $SrTiO_3$ is expected to produce an effect similar to that of the transition to tetragonal ferroelectric $BaTiO_3$ (see Fig. 91): the Ti and O sublattices should shift with respect to each other so as to establish a net polarization **P**. Several numerical calculations of the effect of these displacements on the band structure have appeared[249,253,254] plus a group theoretical analysis which includes also the effect of [111] fields.[255] Zook and Casselman[253] have calculated the effect on the band structure of a shift $\Delta\rho$ of the oxygen atom from the center of the O − Ti − O chains along the polarization axis z. They neglected the effect on the matrix elements of changes in the bond angles (see Fig. 91). Under these assumptions, only the Z_2 levels on faces perpendicular to the direction of the shift $\Delta\rho$ are affected by the distortion; all the Z_4 levels and the Z_2 levels on faces parallel to $\Delta\rho$ are not changed. The $Z_4 \to Z_2$ transitions split into 2 sets allowed only for light polarization parallel and perpendicular to the displacement $\Delta\rho$, respectively. Brews[254] has shown that the change in bond angles has a nonnegligible effect on

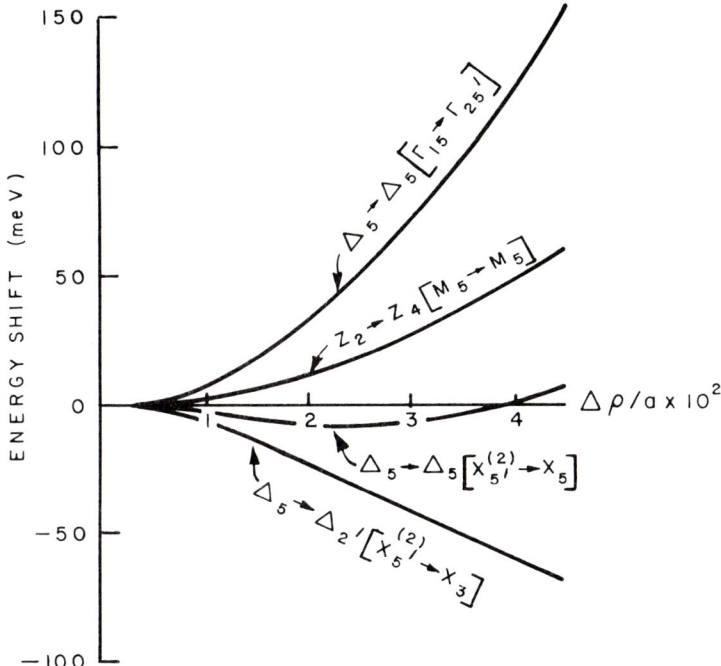

FIG. 92. Dependence of several energy gaps of $SrTiO_3$ on relative displacement $\Delta\rho/a$ of the Ti and O atoms along the direction of the polarization P.

the band structure of the distorted material and made a calculation of the band structure of $BaTiO_3$ in the ferroelectric phase on the basis of the undistorted KL band structure of $SrTiO_3$. We show in Fig. 92 the motion of several of the transitions of Fig. 90 as a function of $\Delta\rho/a$ (a = edge of the unit cell). The polarization **P** is proportional to $\Delta\rho$; the proportionality coefficient is readily obtained from the atomic charges. Figure 92 shows a quadratic variation of gaps with polarization for small values of $\Delta\rho$, i.e., small **P**. For moderately large values of **P** (of the order of those found in ferroelectric $BaTiO_3$) large deviations from the quadratic behavior are found for some gaps ($X_{5'} \to X_3$

becomes almost linear with P). The $X_{5'} \rightarrow X_5$ gap has an initial shift with **P** whose sign is opposite to that calculated by Zook and Casselman.

Gähwiller[249] has pointed out that the Madelung constants of the 3 oxygen atoms in the unit cell become different when a distortion from the cubic structure is introduced. This fact produces, regardless of changes in overlap integrals, splittings of the cubic degeneracies which can be easily calculated by calculating the Madelung constants of the perturbed lattice. Gähwiller[249] has performed a perturbation calculation of the effect of the ionic displacements on the Madelung constants of the various atoms and obtained increases in band gaps quadratic in $\Delta\rho$ and of the same order as those in Fig. 92: for the displacement $\Delta\rho$ of $BaTiO_3$ ($\Delta\rho/a = 0.037$) he obtained $\Delta\omega_g^\perp = 0.13$ eV and $\Delta\omega_g^\parallel = 0.18$ eV. While some doubt can be cast upon this calculation because of the assumption of constant ionicity, it is clear that the effect of changes in Madelung constants with **P** has to be added to the calculations of Zook and Casselman[253] and Brews.[254]

25. Experimental Techniques

a. Insulators

Uniform electric fields can be easily applied to materials with high resistivity in the capacitor configuration of Fig. 93a. The material to be measured must have high resistivity (typically $\rho \gtrsim 10^8$ Ω-cm) and also extremely low photoconductivity (to check the absence of photoconductivity it is appropriate to perform measurements at 2 different light levels). This configuration, with transparent electrodes, is quite useful in transmission modulation studies[247] with the modulating field along the direction of propagation. Semitransparent evaporated metallic layers and also semiconducting SnO_2 deposits (see Section 25b) can be used as electrodes. Another convenient method is to place the sample between two electrolytic cells and use the electrolytes as electrodes[247] (see Fig. 93b): aqueous electrolytes, while opaque in the infrared ($\lambda > 1.2$ μ) are transparent in the ultraviolet to about 1800 Å (see Section 25c). The sample can be sealed to the cell with either rubber O-rings or with wax. Contact is made to each cell with an inert electrode, e.g., platinum.

While the method described above is quite useful for transmission measurements, in reflection measurements the sample is usually much thicker than the light penetration depth. Under these conditions, a small amount of photoconductivity, nearly always present, is sufficient to short-circuit the field in the region which determines the reflection of the light and hence no modulation is obtained. This difficulty can be overcome by performing measurements with $\mathscr{E}$ perpendicular to

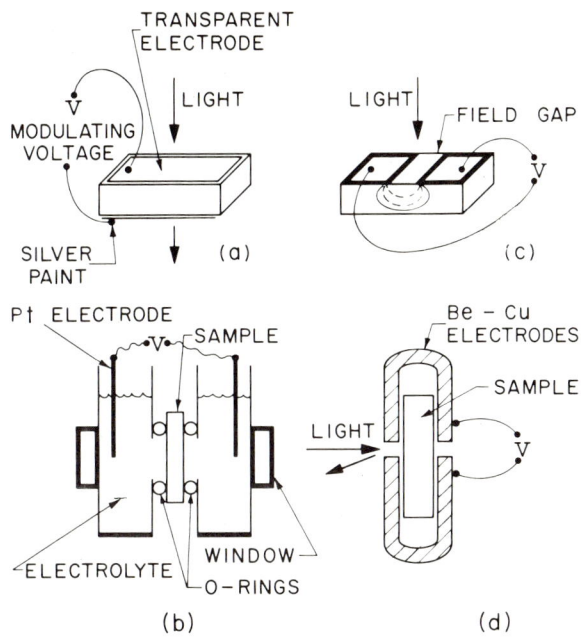

Fig. 93. Configurations for optical modulation experiments with high resistivity, low photoconductivity samples. (a) Capacitor configuration with semitransparent electrodes. (b) Capacitor configuration with electrolytic electrodes. (c) Asymmetric coplanar electrode configuration. (d) Symmetric coplanar electrode configuration.

the direction of propagation, using as a reflecting surface one of the sample surfaces parallel to $\mathscr{E}$. From the point of view of sample capacitance and ease of sample preparation, it is more convenient, however, to use the configurations of Fig. 93c–d. The increase in conductivity now takes place uniformly between the contacts if the gap is uniformly illuminated and, hence, no short-circuiting of the field takes place. The electrodes can be either evaporated metal (gold) films[256,257] (Fig. 93c) or slip-on metal caps[258] (Fig. 93d). In these configurations, if the electrodes are assumed to extend to infinity along the direction of the gap, the potential distribution can be considered to be two-dimensional. A diagram of the field lines is given in Fig. 93c. The field becomes infinite at the contact edges in the field gap but is reasonably uniform at the reflecting surface away from the contacts (near the center of the gap). For uniform illumination of the gap, the effect observed will be that of an average field which is close to the field at the surface in the middle of the gap (actually slightly higher). We shall assume the absence of surface layers and surface fields when no voltage is applied to the electrodes; such fields (normal to the surface) could change the direction and magnitude of the modulating field.

Typical widths of the field gap of Figs. 93c–d are between 0.2 and 1 mm. The maximum fields attainable are limited in general by breakdown across the gap, either in the sample or in the outside medium (arcing); once breakdown occurs, the samples are usually destroyed. The breakdown field can be lowered, when measurements at low temperatures are performed, by immersing the sample in liquid nitrogen. Noise due to bubbles in the nitrogen can be kept to a minimum by keeping dewar and sample extremely clean and free of ice and dust. Slow bubbling of helium through the nitrogen also helps to prevent unwanted bubbling.[259] Measurements are usually performed with fields around 10^4 V cm^{-1}, hence, the ac voltage applied to the sample must be of the order of 1000 V. In order to avoid the unnecessary complication of averaging time varying fields it is convenient to use square wave modulating voltages. If deformation of the square wave by the sample capacitance is to be avoided, the square wave generator must be capable of delivering fairly high currents (low output impedance).

The arrangements of Figs. 93c–d have the advantages over those of Figs. 93a–b that the reflecting surface is free, and hence one is not limited by the transparency of the front electrode. Also, the coplanar electrode arrangements permit measurements with light polarized parallel and perpendicular to $\mathscr{E}$ and hence yield more information than the capacitor arrangements (at least for normal incidence).

A simple square wave power supply which operates in the 100–40,000 cps range with a maximum voltage of 3000 V at 300 mA can be built[259] with a Fairchild UA710A high-speed comparator to which a small sine-wave of the desired frequency, e.g., the reference signal of the lock-in amplifier, is fed. The output of the comparator is applied to the grid of an RCA 8122 power tetrode whose plate is capacitively coupled to a biasing network. In this manner a square wave of arbitrary base line is obtained.

In order to find the field distribution in the configuration of Figs. 93c–d one must solve the corresponding boundary value problem. This problem is simplified considerably if one assumes that the conductivity of the material is large enough so that the field, normal to the surface of the material, is zero except at the contacts. It is then possible to map conformally the two-dimensional configurations of Figs. 93c–d onto a plane parallel capacitor configuration (Fig. 93a). This conformal mapping has been studied in detail[260] for all configurations of Fig. 94. We shall not discuss the antisymmetric cases here

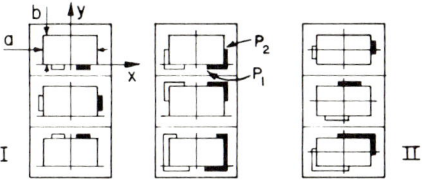

FIG. 94. Two-dimensional electrode configuration whose field distribution is studied by W. E. Wilhelm, *Z. Angew. Math. Mech.* **45**, 121 (1965); S. Oberlander and W. E. Wilhelm, *Phys. Status. Solidi* **12**, 569 (1965). We shall only make use of the symmetric configurations (I).

since the configurations of Figs. 93c–d are symmetric. The x and y components of the field are obtained from the complex potential χ:

$$\mathscr{E}_x = -\mathrm{Re}\, d\chi/dz; \qquad \mathscr{E}_y = \mathrm{Im}\, d\chi/dz. \tag{25.1}$$

The function $d\chi/dz$ is:

$$\frac{d\chi}{dz} = -\alpha \frac{B}{A} \frac{cn\left(\frac{z}{A}, k\right) dn\left(\frac{z}{A}, k\right)}{\left(\left[1 - \frac{1}{\alpha^2} sn^2\left(\frac{z}{A}, k\right)\right]\left[1 - \frac{h^2}{\alpha^2} sn^2\left(\frac{z}{A}, k\right)\right]\right)^{-1/2}} \tag{25.2}$$

where cn, dn, and sn are elliptic functions[261], k and A are obtained from the sample dimensions a and b through

$$a = 2AK(k) = 2AK; \qquad b = 2AK(1 - k^2)^{1/2} = 2AK',$$

where $K(k)$ is the complete elliptic integral of the first kind of modulus k, α is the elliptic function[261] $\zeta = sn\,(z/A, k)$ evaluated at the contact edge $P_1[\zeta(P_1)]$, while $h = \zeta(P_1)/\zeta(P_2)$. The parameter B is given by:

$$B = V/2K(k),$$

where V is the applied voltage.

The somewhat formidable Eq. (25.2) can be used to calculate the field distribution of the configurations of Figs. 93c–d. In Fig. 95 we

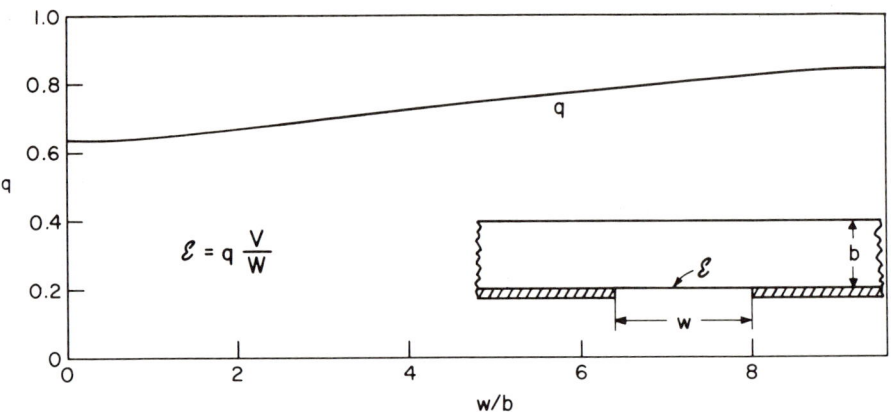

FIG. 95. Correction factor for the evaluation of the field at the center of the field gap of the sketched configuration [D. E. Aspnes, Private communication].

plot the correction factor q which must be used to calculate the field at the center of the gap of Fig. 93c ($a \to \infty$) with the expression:

$$\mathscr{E} = qV/w, \tag{25.3}$$

where w is the width of the field gap. The average effective field for uniform illumination is expected to be slightly higher than that in Eq. (25.3).

b. Semiconductors: Field Effect

The Seraphin method of optical modulation,[61] based on the field effect, is shown schematically in Fig. 96. A field is applied with a transparent electrode to the surface of the material to be measured. This field penetrates into the material to a depth which is typically of the order of 10^4 Å for semiconductors. Since this depth varies with the carrier concentration N of the semiconductor approximately as

$N^{1/2}$, a wide range of carrier concentrations produces penetration depths whose order of magnitude is 10^4 Å. Light is reflected on the surface of the sample after traversing the transparent electrode and the insulating layer. The depth of sample which contributes to the reflected intensity is of the order of either λ or $1/\alpha$, which ever is smaller (α is the absorption coefficient). Hence the penetration depth

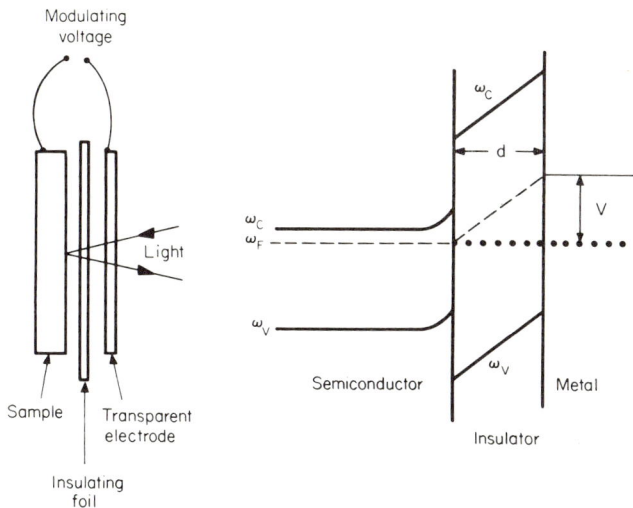

FIG. 96. Schematic diagram of the sample arrangement used in reflection modulation experiments by the field effect method. Also, energy band diagram of the arrangement in the absence of surface states.

of the light is of the order of that of the applied electric field and reflection modulation (electroreflectance) becomes possible. The method can also be used for transmission modulation (electro-absorption) provided the sample is not much thicker than the field penetration depth. The conventional energy band diagram of the system is also shown in Fig. 96 under the assumption of no surface states and a metallic field electrode (penetration depth ≈ 0 for the electrode). Under this condition the field at the surface of the sample (but inside the sample) is $\mathscr{E} = V/\varepsilon_0 d$ where ε_0 is the static dielectric constant of the intrinsic material. V is positive when the field electrode is positive. The electrostatic potential $\phi(x)$ inside the sample can be found by integrating Poisson's equation:

$$d^2\phi/dx^2 = -[4\pi\rho(x)/\varepsilon_0]. \tag{25.4a}$$

We have assumed that the sample extends to infinity in the 2 dimensions perpendicular to x. The charge density $\rho(x)$ is determined by the density of states (energy bands + impurity levels) and the appropriate Fermi–Dirac occupation functions. The Fermi level ω_F (electrochemical potential) must be constant throughout the sample in equilibrium conditions while the origin of energies is shifted by the position-dependent electrostatic potential $\phi(x)$ [Fermi–Thomas approximation, valid only if $\phi(x)$ varies little within a lattice constant]. By multiplying both sides of Eq. (25.4a) by $2\, d\phi/dx$, we obtain:

$$d(d\phi/dx)^2 = -(8\pi\rho(\phi)/\varepsilon_0)\, d\phi,$$

or
(25.4b)

$$(d\phi/dx)^2 = -(8\pi/\varepsilon_0) \int \rho(\phi)\, d\phi + \text{constant}.$$

Equation (25.4b) can be used to obtain the field $(d\phi/dx)_{x=0}$ at the surface of the sample. This field must be related to the applied voltage in the absence of surface states through:

$$[d\phi/dx]_{x=0} = V/\varepsilon_0 d. \tag{25.5}$$

This condition determines the constant of integration in Eq. (25.4b). In order to proceed any further we have to introduce the specific impurity levels and energy bands in Eq. (25.4b). A very illuminating simplification is possible, however, in the case of a semiconductor whose majority carrier concentration N is far from that of the intrinsic material [$\omega_{FB} - \omega_{FI} \gg kT$ where ω_{FB} and ω_{FI} represent the Fermi energies of the bulk of the extrinsic material and that of the intrinsic material, respectively] when the band bending at the surface is such that the surface is nearly intrinsic (depletion layer). The space charge near the surface can then the approximated by a constant $\rho(x) = \pm eN$ (+ for n-type, − for p-type bulk) up to a certain penetration depth δ and by zero beyond that point. We take $\phi(\delta) = 0$. From Eq. (25.4b) we obtain:

$$\phi = \pm(2\pi/\varepsilon_0)eN(x-\delta)^2$$
$$\mathscr{E} = [d\phi/dx]_{x=0} = \pm(4\pi/\varepsilon_0)eN\delta, \tag{25.6}$$

hence

$$\delta = \varepsilon_0|\mathscr{E}|/4\pi eN.$$

A surface barrier with the potential distribution of Eqs. (25.6) is called a Schottky barrier. The total charge density in the surface layer is $\phi = \pm eN\delta$, hence we can define the capacitance C per unit area as:

$$C_s = \left|\frac{Q}{\phi(x=0)}\right| = \frac{\varepsilon_0}{2\pi\delta} = \left(\frac{eN\varepsilon_0}{2\pi|\phi(x=0)|}\right)^{1/2}; \qquad (25.7)$$

this capacitance is in series with the capacitance of the field electrode of Fig. 96. Since d is usually much larger than δ, the contribution of C_s to the total capacitance of the experimental cell is negligible.

Equations (25.4b) can be put in a reasonably simple form even if the previous approximation does not hold, provided Boltzman statistics applies and there are no localized states within the gap. In this case we can write for the electron and hole concentrations $n(x)$ and $p(x)$:

$$n(x) = n_1 e^{u_x}; \qquad p(x) = n_1 e^{-u_x}, \qquad (25.8)$$

where n_1 is the intrinsic electron concentration, and u_x is $[\omega_F(x) - \omega_{Fi}]/kT$, the difference between the Fermi energy at x and that of the intrinsic material in units of kT. It is convenient to write $u_x = u_B + y$, where u_B is the bulk value of u_x and $y = e\phi/kT$. The charge density $\rho(x)$ needed in Eqs. (25.4b) is:

$$\rho(x) = e[p(x) - n(x) - p_B(x) + n_B(x)], \qquad (25.9)$$

where the subindex B indicates the bulk values of n and p. By replacing Eqs. (25.9) and (25.8) into Eq. (25.4b) we obtain[262]:

$$\frac{d\phi}{dx} = \pm \left(\frac{8\pi kTn_1}{\varepsilon_0}\right)^{1/2} [e^{-u_B}(e^{-y}-1) + e^{u_B}(e^{+y}-1) + (e^{-u_B} - e^{u_B})y]^{1/2}$$

$$= \frac{kT}{e\mathscr{L}} F(y, u_B), \qquad (25.10)$$

where $\mathscr{L}$ is the "Debye screening length" of intrinsic material

$$\mathscr{L} = \left(\frac{\varepsilon_0 kT}{8\pi e^2 n_1}\right)^{1/2}, \qquad (25.11)$$

and F the so-called space-charge function. The $+$ sign in Eq. (25.10) corresponds to upward bending bands. The function F is plotted in Fig. 97 as a function of $u_x = y + u_B$ with u_B as a parameter[263]. Numerical calculations of $d\phi/dx$ with Eqs. (25.4b) involving Fermi statistics have been performed by Seiwatz and Green.[264]

If the first two terms in Eq. (25.10) are neglected (as can be done when the surface is much more intrinsic than the bulk), we find the Schottky barrier discussed above (depletion layer). An enrichment layer is obtained when the bands are bent downward at the surface in n-type material [the second term dominates the right-hand side of Eq. (25.10)] and upward in p-type material [the first term in the

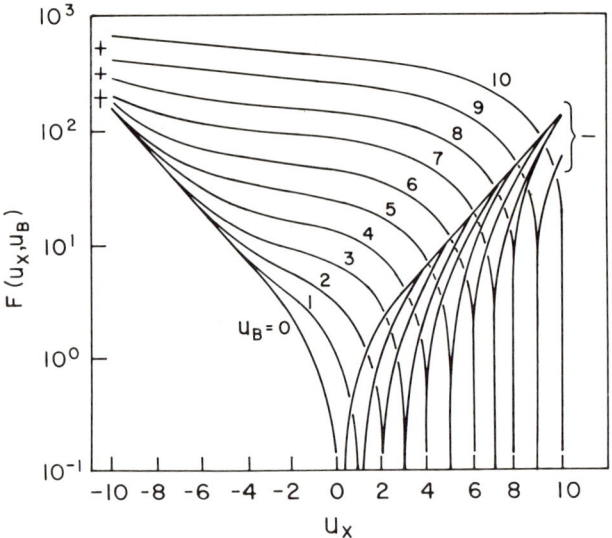

FIG. 97. The space–charge function $F(u_x, u_B)$. From R. H. Kingston and S. Neustadter, *J. Appl. Phys.* **29**, 1034 (1955).

radical of Eq. (25.10) is dominant]. An inversion layer is obtained when the magnitude of the surface potential is increased beyond that which makes the surface intrinsic (depletion layer).

The previous considerations also apply when surface states are present, provided ϕ refers to the potential at the bulk side of the surface states. The charge in the surface states is determined by $\phi(x = 0)$ through the density of surface states, the occupation factor, and the condition of constant electrochemical potential. The field for $x = 0$ inside the sample is not given, however, by Eq. (25.5): the charge in the surface states screens the externally applied field. The field $\mathscr{E}_i$ inside the sample at $x = 0$ is:

$$\mathscr{E}_i = -(d\phi/dx)_{x=0} = [\mathscr{E}_0 - 4\pi Q_{ss}]/\varepsilon_0$$
$$= [(V/d) - 4\pi Q_{ss}]/\varepsilon_0, \qquad (25.12)$$

where Q_{ss} is the charge per unit area in the surface states.

Two types of surface states are present in semiconductors[265]: the slow states, which have time constants of the order of seconds to hours, and the fast states, with time constants from 10^{-3} to 10^{-8} sec. The slow states are usually associated with oxide layers and foreign adsorbed materials; their density is of the order of 10^{13} cm^{-2}, higher than that of fast states. Since a surface density of states of the order of 10^{13} cm^{-2} corresponds to a volume density of states of about 10^{20} cm^{-3}, when such large densities of slow states are present, the Fermi level at the surfaces is likely to be clamped by the surface states regardless of applied external field; a small variation in the surface potential is sufficient to release enough surface charge to completely screen very large surface fields. Owing to the slow nature of these states, the surface potential can be changed by varying the external voltage for a length of time of the order of the time constant of the states (seconds–hours). These time constants usually become longer at low temperatures and therefore it is sometimes possible at 77°K to vary the surface potential for a period of time from hours to days while it is not possible to do it at room temperature. Slow states do not follow modulating voltages of the usual frequencies (100–10,000 cps) but fast state do. Since the density of fast states is smaller (10^{11} cm^{-2}) than that of slow states, it is often possible to achieve significant ac modulation in the surface potential while it is not possible to bias it in a steady manner except, maybe, at low temperatures. The positions of the Fermi level at the surface of a number of group IV and III–V semiconductors at a semiconductor-metal contact have been shown to be clamped by surface states.[266] This does not seem to be the case for II-VI semiconductors.[266]

The height of the surface barrier, and hence the magnitude of the modulating field, was not known in most of the optical field modulation experiments performed so far. It is possible, however, through careful measurements of the field effect conductance, to determine the fields at the surface for the maximum and minimum of the modulating voltage and also the quiescent bias point. The change in conductance parallel to the surface induced by the modulating voltage on the sample of Fig. 96 is due to the change in the charge induced in the space–charge region Q_{sc} and to the surface charge. The surface mobility is usually smaller than the bulk mobility and the surface conductivity can be neglected. Thus the modulation in the surface charge determines the modulation in the surface field. Standard field effect techniques[267,268] can be used to determine the surface barrier height from the dependence of the conductance modulation on applied voltage. As an example, we show in Fig. 98, the surface barrier heights for a germanium sample (*n*-type, 30 Ωcm at room temperature) at 208°K as a function of external voltage (dc bias) as reported by Seraphin.[267] The modulating voltage is sinusoidal and equal to 400 V peak-to-peak. We show in this figure the curve

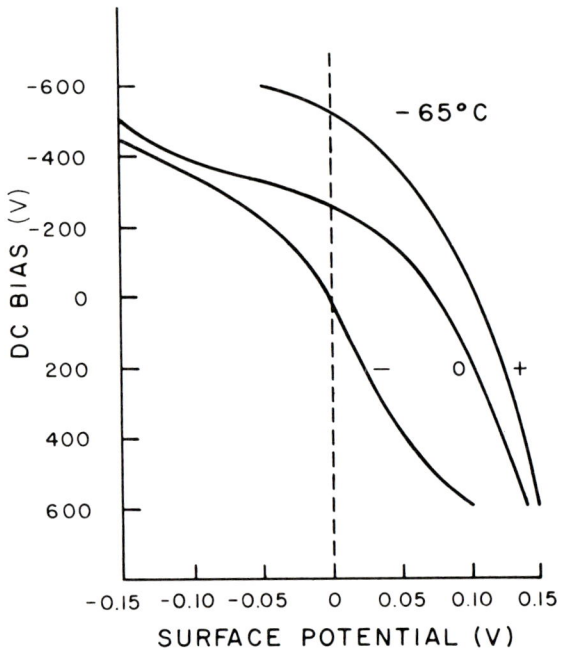

Fig. 98. Variation of surface potential with dc bias applied to the field electrode for a 30 Ωcm (at room temperature) *n*-type germanium sample. The + and − curves correspond to the positive and negative cycles of the modulating voltage, respectively. The 0 curve to no modulation. From B. O. Seraphin, *Surface Sci.* **8**, 399 (1967).

corresponding to the quiescent point and also the 2 curves which correspond to the 2 peak values of the modulation. Throughout the range of bias of Fig. 98, the effect of modulating voltage is somewhat larger than the effect of an equal change in the dc bias, in agreement with the discussion above.

As opposed to the cases discussed in Section 25a, the modulating field varies quite considerably within a penetration depth (or wavelength) of the light when the space–charge barrier technique is used.

It is possible, of course, to reduce this variation to a minimum by using low carrier concentration material at the expense of the modulation amplitude. We shall now treat† the effect of the spatial non-uniformity of the electric field on the measured modulation spectra under the assumption of a one-dimensional geometry. The electric field **E** of the light is obtained as a function of depth x by solving the wave equation:

$$\left[\frac{d^2}{dx^2} + \frac{\omega^2}{c^2}\varepsilon(x)\right]\mathbf{E}(x) = 0. \tag{25.13}$$

We replace in Eq. (25.13) $\mathbf{E}(x)$ by a WKB-type solutions:

$$\mathbf{E}(x) = \hat{\mathbf{z}}E_0 \exp[-ik_0 x + i\phi(x) - i\omega t]; \quad \text{with } k_0 = \frac{\omega}{c}\varepsilon^{1/2} \tag{25.14}$$

where $\mathbf{E}(x)$ has been taken to be polarized along z and $\hat{\mathbf{z}}$ is the unit vector along this direction. We assume that $\phi(x)$ is a slowly varying function of x. Substituting Eq. (25.14) into Eq. (25.13) and writing $\varepsilon(x) = \varepsilon + \Delta\varepsilon(x)$ [ε is the dielectric constant for $\mathscr{E} = 0$ and $|\Delta\varepsilon(x)| \ll |\varepsilon|$] we find for $\phi(x)$ the equation:

$$i\phi'' + 2\phi'k_0 - (\phi')^2 = -(\omega^2/c^2)\Delta\varepsilon(x), \tag{25.15}$$

where the primes indicate derivatives with respect to x. Approximate solutions to Eq. (25.15) can be readily found if either ϕ'' or $(\phi')^2$ or both are negligible with respect to $\phi'k_0$. Neglecting both ϕ'' and $(\phi')^2$ in Eq. (25.15) we find:

$$\phi' \approx \tfrac{1}{2}k_0 \Delta\varepsilon/\varepsilon, \tag{25.16}$$

and therefore ϕ'' and $(\phi')^2$ are negligible provided:

for $(\phi')^2$: $\tfrac{1}{4}|\Delta\varepsilon/\varepsilon| \ll 1$
for ϕ'': $|\Delta\varepsilon'| \ll 2|k_0 \Delta\varepsilon|$.

The first condition above is always satisfied in the usual experimental situations: it was an assumption underlying the theoretical calculations of Section 24b. The second condition implies that the change of $\Delta\varepsilon$ in a wavelength or penetration depth (whatever smaller) must be much smaller than the maximum $\Delta\varepsilon$; this condition is only fulfilled if the field penetration depth is large compared with the wavelength

† This treatment follows unpublished work by D. E. Aspnes.

or the penetration depth of the light (whatever is smaller). If this is the case, Eq. (25.16) yields for the effective propagation vector of the light:

$$k(x) = k_0 \left[1 + \frac{1}{2} \frac{\Delta\varepsilon(x)}{\varepsilon} \right] = \frac{\omega}{c} [\varepsilon + \Delta\varepsilon(x)]^{1/2} = \frac{\omega}{c} [\varepsilon(x)]^{1/2}. \quad (25.17)$$

If normal incidence electroreflectance is measured under conditions for which Eq. (25.17) is valid, the reflection coefficient is given by Eq. (8.2) with the complex refractive index which corresponds to the surface field: $n_r + in_i = [\varepsilon + \Delta\varepsilon(x=0)]^{1/2}$. Thus the measured spectra are essentially the same as those for uniform fields.

It is also interesting to consider the opposite situation, in which $\Delta\varepsilon$ varies very rapidly in a wavelength (or penetration depth). This is easy to accomplish experimentally for highly doped materials or for strong surface enrichment layers. It is sometimes possible to go from the case of slow to that of fast variation of $\Delta\varepsilon$ by varying either the doping, the modulation amplitude, or the bias field. Also, the fulfillment of these conditions depends on the wavelength under consideration and it is possible to go from one extreme to the other in a given experiment while sweeping over the spectral range.

For the case of fast variation of $\Delta\varepsilon$, we obtain from Eq. (25.15) by neglecting $(\phi')^2$ but not ϕ'':

$$\phi'(x) \approx ik_0^2 \exp\left[2ik_0 x \int_{-\infty}^{x} dx' \exp[-2ix'k_0] \Delta\varepsilon(x')/\varepsilon_0 \right]$$

$$\approx ik_0^2 \int_{-\delta}^{x} dx' \, \Delta\varepsilon(x')/\varepsilon_0, \quad (25.18)$$

where $x = -\delta$ is the point at which the field vanishes (penetration depth of $\mathscr{E}$).

By imposing the appropriate boundary conditions, the normal incidence reflection coefficient is found to be, when Eq. (25.18) holds:

$$R = \frac{\left| k_0 - ik_0^2 \int_{-\delta}^{0} dx' \, (\Delta\varepsilon(x')/\varepsilon_0) - k_a \right|^2}{\left| k_0 - ik_0^2 \int_{-\delta}^{0} dx' \, (\Delta\varepsilon(x')/\varepsilon_0) + k_a \right|^2}, \quad (25.19)$$

where k_a is the propagation constant for the medium surrounding the sample. Equation (24.19) yields the same result as that produced by a uniform change in the real and imaginary parts of ε:

$$\Delta\varepsilon_r \to 2\,\text{Im}\,k_0 \int_{-\delta}^{0} dx'\,\Delta\varepsilon(x'), \qquad \Delta\varepsilon_i \to -2\,\text{Re}\,k_0 \int_{-\delta}^{0} dx'\,\Delta\varepsilon(x'). \qquad (25.20)$$

hence, in this case, the effect of changes in the real and imaginary parts of ε become mixed: the coefficients β_r and β_i of Eq. (8.9) do not correspond to the effects of separate changes in ε_r and ε_i. If, for instance, k_0 is mostly real, Eqs. (25.20) indicates that the roles of $\Delta\varepsilon_r$ and $\Delta\varepsilon_i$ are interchanged. Hence it is possible to affect, rather drastically, the shape of observed spectra when going from a long to a short field penetration depth, i.e., when going from a depletion to an enrichment surface layer. One must keep in mind that a Kramers–Kronig analysis of electroreflectance spectra in the short penetration depth limit yields for $\Delta\varepsilon_r$ and $\Delta\varepsilon_i$ the effective values of Eqs. (25.20). The average values of $\Delta\varepsilon_r(\mathscr{E})$ and $\Delta\varepsilon_i(\mathscr{E})$ which appear in the integrals of Eqs. (25.20) can be calculated by replacing in the theory of Section 24b the electrooptic functions, e.g., F and G, by their spatial averages. Aspnes has evaluated the averages $\bar{F}$ and $\bar{G}$ of F and G (unbroadened) for the case of a barrier with a field which is a linear function of position (Schottky barrier). The results are shown in Fig. 99 (as a function

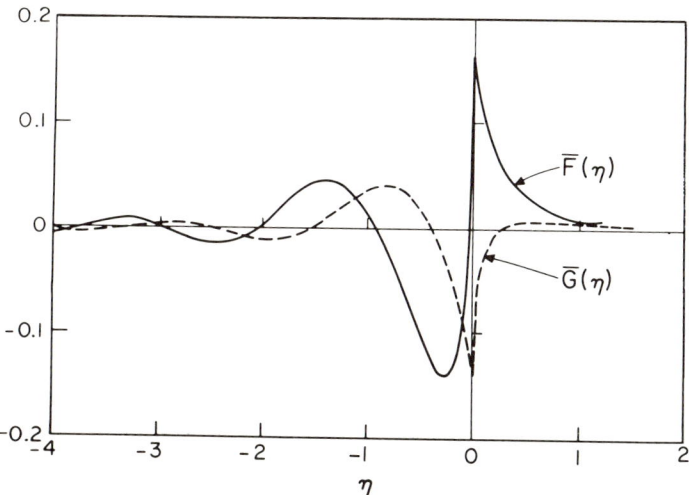

FIG. 99. Functions $\bar{F}(\eta)$ and $\bar{G}(\eta)$, obtained by averaging $F(\eta)$ and $G(\eta)$ over a Schottky barrier. Computed by D. E. Aspnes.

of the η which corresponds to the maximum field) and should be compared with the curves of F and G given in Fig. 79. The strength of F and G is reduced only slightly by the averaging for $\eta = 0$ ($\omega = \omega_g$) but the oscillations for $\omega > \omega_g$ decay rather rapidly, as would be expected as a result of the field averaging.

It has been already pointed out that space–charge barrier measurements at normal incidence do not permit studies with the modulating field along the direction of the electric field of the light. This drawback could, conceivably, be partially eliminated with measurements at oblique incidence: 1 of the 2 possible normal modes of polarization (**E** in the plane of incidence) contains a component of **E** along $\mathscr{E}$. This is the polarization for which the reflection cancels ($R_\| = 0$) at the Brewster angle if $n_i = 0$: the experimentally measured quantity $\Delta R/R$ becomes infinite under these circumstances. If $n_i \neq 0$, but small, one can still define a "pseudo-Brewster" angle (Section 8e) at which $R_\|$ is very small and thus $\Delta R_\|/R$ is very large. One may therefore be inclined to conclude that the sensitivity of electroreflectance measurements is greatly increased at the pseudo-Brewster angle for the $R_\|$ polarization. A careful examination of the $n_i = 0$ case shows, however, that not only $R_\| = 0$ at the Brewster angle but also $\Delta R_\| = 0$ while the ratio $\Delta R_\|/R_\|$ becomes infinite. The signal-to-noise ratio is determined by $\Delta R/R^{1/2}$, which is *finite* and of the order of unity at the Brewster angle. A detailed calculation shows that the signal-to-noise ratio of measurements at the Brewster (and also pseudo-Brewster) angle can sometimes be higher than that for normal incidence.

We shall now discuss the various types of "dry" sample packages used for space–charge barrier optical modulation experiments. The closely related electrolytic methods will be discussed in the next subsection. As shown in Fig. 96, the main components of the measurement package are the sample, the dielectric layer, and the transparent electrode. Both dielectric and electrode must, of course, be transparent in the measurement region and they also should not produce spurious electroreflectance signals. Seraphin and co-workers[61,269] used Saran wrap as a dielectric and SnO_2-coated quartz as a transparant electrode in their original electroreflectance arrangements. The quartz can be easily coated by placing it in a tube furnace at about 600°C and spraying into the tube a solution of $SnCl_2$ with a stream of oxygen. Care must be taken to prevent mechanical vibrations in the package from producing sizable spurious modulations. This can be done by optical matching of the electrode with a mixture of Canada balsam and oil.[191]

Such packages are usable to about 4 eV in the ultraviolet and 0.5 eV in the infrared. A number of other arrangements have been described by other authors since Seraphin's original work. Shaklee[270] has used 0.0004 in thick Mylar film as the insulator and a semitransparent gold layer deposited on it as the field electrode. The film so prepared is glued to the sample with "Duco" cement diluted in amyl acetate. Thin evaporated nickel films (65% transmission) have also been used as transparent electrodes[271] with a photoresist material of thickness about 1 μ as the dielectric.† Another interesting technique is that of Ludeke and Paul.[272] These authors used a vacuum deposited silica layer (evaporated with an electron gun); an SnO_2-film, deposited on the silica layer in the usual manner, was used as the conducting electrode.

Two somewhat less conventional electroreflectance methods are also based on the space–charge-layer principle. The method of Wang et al.[273] does not require the use of the insulator transparent electrode sandwich and hence its wavelength range is, in principle, unlimited. Modulation of the field at the space–charge layer is obtained by shining strong chopped light of photon energy higher than the band gap, onto the surface to be measured. Photocarriers produced by this light modulate the surface field. The effect of this modulation on the optical constants (reflectivity) is then measured with an independent monochromatic source.†† The other unconventional method of surface–barrier electroreflectance [274] is based on the anisotropy of the effect of a surface field on the optical properties of a cubic material for a surface other than (100) or (111). This anisotropy can be measured by rotating the sample while linearly polarized light is reflected on it at normal incidence. The component of the reflected signal synchronous with the rotation (double frequency) measures the *difference* between the electroreflectances for the 2 normal modes contained in the sample surface. This technique has been named rotoreflectance.[274]

Modulation of a space–charge layer can be conveniently obtained at the boundary of a *p–n* junction.[78,275] Surface states are not present in homogeneous junctions and hence their undesirable shielding effects are avoided. The junction region, however, must be accessible to the light and, therefore, this method is only useful when the light penetration depth is larger than the distance from the sample surface at which the *p–n* boundary lies. Typical boundary depths are larger than

---

† Kodak Photo Resist, alternate layers of KTFR and KPR material.
†† This method has become known as *photoreflectance*.

1 μ (see Fig. 100) and hence the method is not very convenient for electroreflectance measurements; it has been extensively used for electrotransmission. Let us assume an abrupt junction with Schottky-type space–charge and potential distributions [Eq. (25.6)]. The correctness of this assumption can be tested by measuring the junction

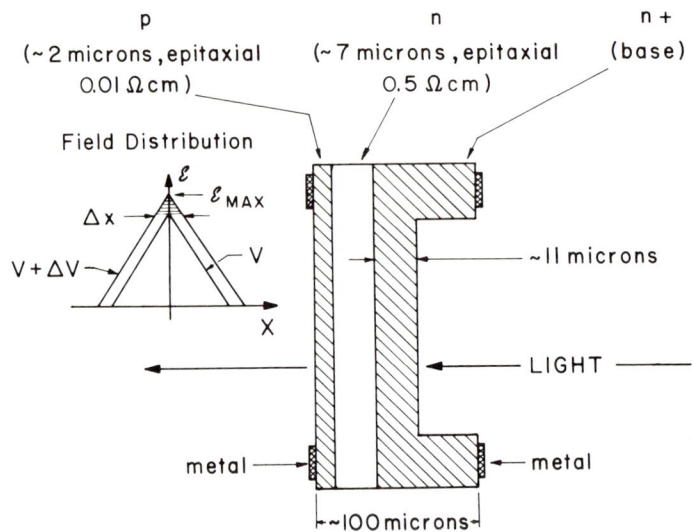

FIG. 100. *p–n* junction configuration used by A. Frova, P. Handler, F. A. Germano, and D. E. Aspnes, *Phys. Rev.* **145**, 575 (1966), for electrotransmission measurements in germanium and silicon; also, field profile for this configuration.

capacitance, which should be given by Eq. (25.7) with ϕ equal to the built-in voltage plus the applied voltage V. Hence, for a Schottky-type junction, $1/C^2$ should be a linear function of V. Junctions prepared in the manner of Fig. 100 fulfill this condition quite well. The field profile as a function of depth X is also shown in Fig. 100 for applied voltages V and $V + \Delta V$. Under the assumption of a small variation of ε in a wavelength (junction width $\gg \lambda$, this condition may not hold very well in practice) the decrease in transmitted intensity when going from bias V to $V + \Delta V$ is produced by the shaded part of the diagram near the maximum field $\mathscr{E}_{\max}$. Hence we obtain:

$$\Delta I/I = -[\alpha(\omega, \mathscr{E}_{\max}) - \alpha(\omega, 0)] \Delta x$$
$$= [\alpha(\omega, \mathscr{E}_{\max}) - \alpha(\omega, 0)] \Delta V/\mathscr{E}_{\max}. \qquad (25.21)$$

Therefore, the measurement of $\Delta I/I$ yields directly the increase in absorption coefficient ($\propto \varepsilon_i$) produced by the field $\mathscr{E}_{max}$. Frova et al.[78] have calculated the correction factor to Eq. (25.21) for a practical case in which the junction capacitance C varies as $V^{-0.45}$. They find the correction to be quite small but not completely negligible.

c. Semiconductors: Electrolytic Method

The possibility of space–charge layer modulation at a semiconductor–electrolyte interface has been known for a long time.[276] In particular, it has been known that a sizeable change in the conductivity (field effect) of a thin semiconductor electrode can be obtained by changing the voltage applied to an electrolytic cell in which the semiconductor is one of the electrodes. These changes must be due to changes in the space–charge, i.e., in the electric fields at the semiconductor–electrolyte interface. These field changes can be conveniently utilized for optical modulation purposes at wavelengths in the vicinity of an energy gap of the semiconductor, as was first recognized by Williams.[277]

The electrochemical properties of semiconductor–electrolyte interfaces have been the object of considerable interest. A number of review articles have been published on this subject.[278,279] A semiconductor-electrolyte interface can be viewed, for our purposes, in a manner somewhat similar to a p–n junction: the charge carriers have different natures in the 2 conductors in contact (semiconductor–electrolyte, p-type–n-type semiconductors). In equilibrium (no current flowing) the electrochemical potentials of the electrons in the solid and in the electrolyte must be equal. When the semiconductor is immersed in the electrolyte, electrons flow from 1 material to the other until the electrochemical potentials are equal and the space–charge layer at the surface of the semiconductor is modified or established (if none existed before). A surface charge layer will also be present at the electrolyte side of the interface (Gouy layer). Both charge layers obey the theory described in Section 25b. A schematic diagram of the semiconductor–electrolyte system and the variation of the energy with the coordinate normal to the interface is shown in Fig. 101. The ions cannot get closer to the surface of the semiconductor than their radius and hence there is a thin layer near that surface which does not obey the equations of electrostatics of continuous media. This is the so-called Helmholtz layer and the potential distribution in it is difficult to estimate. The thickness of the space–charge or Gouy layer in the electrolyte (the Debye length) can be made small (~5 Å) in high-conductivity electrolytes. This is convenient from the point of view of optical modulation experiments since then the capacitance of this layer represents a short circuit to an ac voltage, as compared with the

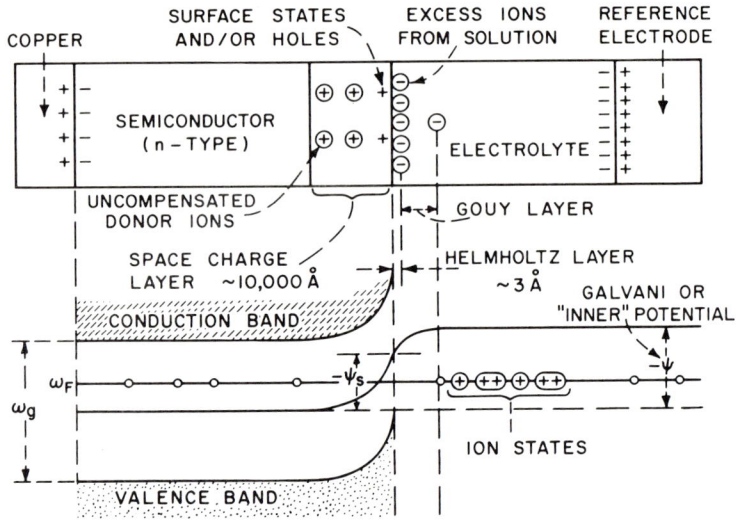

FIG. 101. Schematic representation of a semiconductor electrolyte interface and energy level diagram. From J. F. Dewald, in "Semiconductors" (N. B. Hannay, ed.), p. 727. Reinhold, New York, 1960.

capacitance of the space–charge region in the semiconductor. The impedance of the Gouy layer can always be made negligible to an ac voltage with aqueous electrolytes except for very heavily doped semiconductors (~10^{20} carriers/cm^3); under those circumstances, electrooptical effects in the electrolyte ought to be negligible. Non-aqueous electrolytes (in particular the organic electrolytes to be discussed later) usually have lower ion densities and hence the Gouy layer may produce a drop in the modulating voltage.

The I–V (current density versus voltage) curve of a semiconductor–electrolyte junction, like that of a p–n junction, is very nonlinear. Figure 102 shows the I–V curves of n- and p-type germanium in an 0.1 N aqueous solution of KOH.[280] We see that the current is, in both cases, particularly small for a negatively biased germanium electrode with applied voltages up to about 1.5 V. It is very convenient for optical modulation purposes to work in this region. First, a small current density guarantees a small voltage drop in the bulk of the semiconductor and the electrolyte; most of the applied voltage reaches the junction. Second, high currents have as a result strong electrochemical reactions which produce undesired effects. A large negative voltage applied to the semiconductor produces evolution of hydrogen

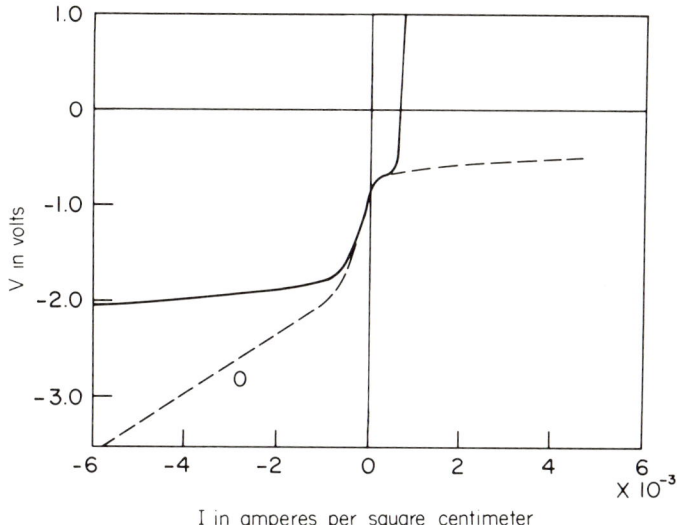

FIG. 102. I–V curves of n- (solid line) and p-type (dashed line) germanium in the darkness according to W. H. Brattain and C. G. B. Garrett *Bell System Tech. J.* **34**, 129 (1955).

gas with its associated noise in the optical reflection. A large positive voltage produces cathodic dissolution of the germanium. This last effect is more serious in compound semiconductors since the rate of dissolution is usually different for the different constituents. It is worthwhile noting that in the small current region of Fig. 102 the I–V curve is neither strongly dependent on the sign of the charge carriers nor on the amount of doping; it depends only on the nature of the electrolyte and of the intrinsic semiconductor. This is due to the fact that the electrochemical reactions at the interface are determined only by the carrier concentrations at the semiconductor surface and not by the bulk. As the applied voltage becomes large the surface concentrations depend on the bulk concentrations and a dependence on the properties of the bulk appears. For n-type material, the semiconductor-to-electrolyte current is mostly electronic. A bias which makes the semiconductor positive (positive bias) produces a decrease in the number of electrons at the surface and hence the saturation in the current seen in Fig. 102. No such saturation occurs for a p-type sample since the number of holes at the surface is increased by a positive bias. Typical operating conditions for germanium are shown in Fig. 103. A dc bias is applied to the cell so as to bring the operating

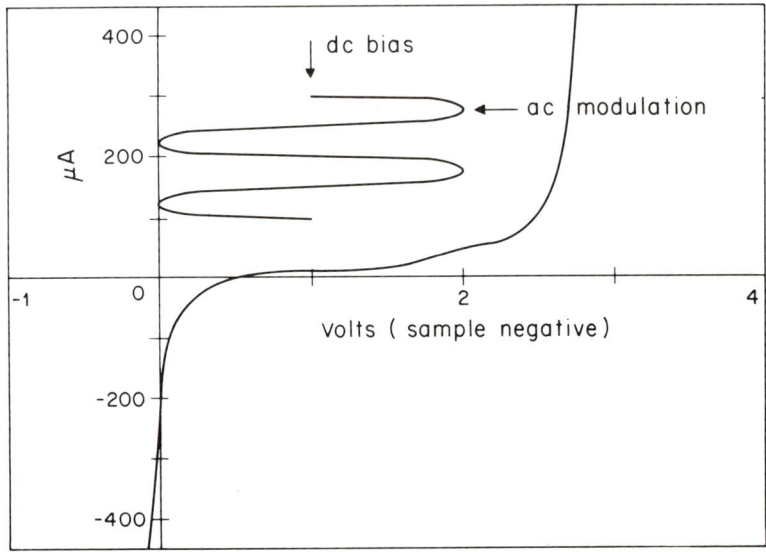

FIG. 103. Characteristic I–V curve of a p-type germanium sample with $p = 8 \times 10^{18}$ cm^{-3} showing standard operating conditions for electroreflectance measurements.

point to the center of the low-current region. A modulating ac voltage which keeps the sample in this region is superimposed to the dc bias. Electroreflectance signals are nearly always obtained with negatively biased samples. In many materials these signals disappear even for small positive biases. Since illumination of the semiconductor surface usually alters the shape of the curves of Fig. 102 (the current for a given voltage increases), it is convenient to keep the level of incident light low: the electrolytic cell should always be placed after the exit slit of the monochromator.

The temperature range of the electrolyte method is limited by the freezing and boiling points of the electrolyte. Measurements down to about $-120°C$ can be performed with either methyl ($+KCl$) or ethyl ($+H_2SO_4$) alcohol while propylene carbonate ($+KCNS$ or tetrabutyl ammonium perchlorate) permits operation up to 250°C. Organic electrolytes are also useful for measuring semiconductors which react with water.

The choice of electrolyte is also influenced by the wavelength region of interest. While water transmits quite far into the ultraviolet, its transmission region in the infrared is limited by the stretching vibrations of the OH radicals which take place at around 1.2 μ. We show in Fig. 104 the absorbance of 1 cm of water and of 2 typical

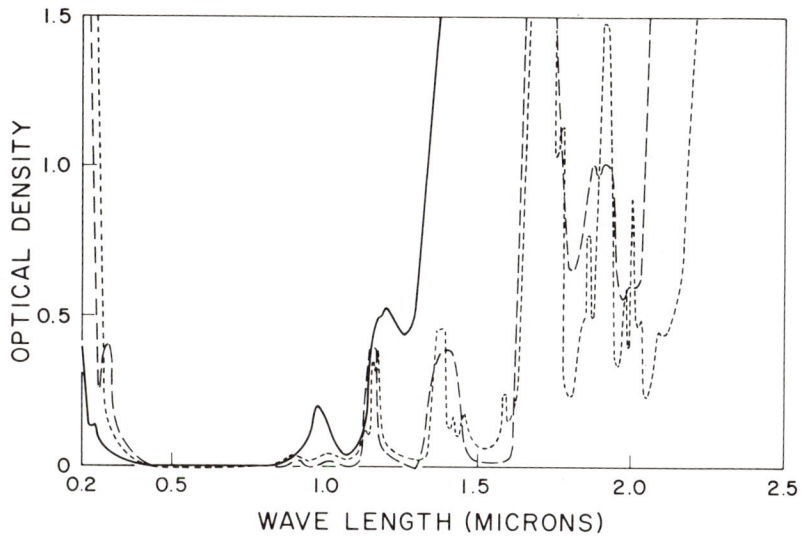

FIG. 104. Absorbance of water (solid line), acetonitrile (dotted line), and propylene carbonate (dashed line) in the region from 2000 to 25,000 Å. The liquids were in a 1 cm length cell with quartz windows.

organic electrolytes without OH radicals, acetonitrile, and propylene carbonate, in the region from 2000 to 25,000 Å. We see that water is more transparent in the ultraviolet than the organic electrolytes but in the near infrared (around 1.2 μ) the materials without OH radicals are more transparent. Acetonitrile (melting point $\sim -40°C$, boiling point 82°C, *extremely toxic*) is better than propylene carbonate in the infrared. In spite of the absorption bands of these materials in the

1 to 2.5 μ region, the absorption coefficients are relatively small ($\sim$1 cm^{-1}) and, hence, thin capillary layers ($\sim$0.01 cm) should be practically transparent in this region. This suggests the use of thin capillary layers between sample and window for electrolytic optical modulation measurements in the infrared.[70,282,283] This method can conceivably be used to extend somewhat the wavelength range of the electrolytic method in the ultraviolet. The capillary layer can be made, for aqueous electrolytes, of sufficiently high conductance to carry the modulating voltage to the sample without significant loss. If this is not possible, e.g., for organic electrolytes, it is convenient to use a conducting window (Ge, Si,[283] or SnO$_2$-coated quartz[282]) so as to decrease the loss. The window is then used as 1 of the electrodes. In order to avoid reflection from the front surface of the window one usually places window and sample at an angle with each other. This is not possible for capillary electrolyte layers but in this case the same result can be achieved by using a wedge shaped window.[70]

The electrolyte technique has, in spite of the limitations in the temperature range, a number of advantages with respect to the conventional field effect configurations of Section 25b. Optical matching of the dielectric and the transparent electrode becomes unnecessary and vibration problems are eliminated. Relatively irregular nonflat surfaces can be used. In the case of surfaces sensitive to air or humidity, e.g., AlSb, Mg$_2$Ge, the samples can be cleaved in the electrolyte thus avoiding contact with the air. The measurements are considerably faster with the electrolyte technique since the delicate preparation of the sandwich of sample, insulator, and transparent electrode is unnecessary. As we shall see in Chapter VIII, the electrolytic method is particularly suitable for measurements in the presence of a static uniaxis stress, especially when the use of polarized light is essential. Birefringence in the conventional dielectric layers usually depolarizes, at least partially, a linearly polarized incident beam. This problem is avoided with the electrolyte technique.

26. Experimental Results

a. Electroabsorption: Indirect Edge

The indirect edge of germanium and silicon has been thoroughly investigated by electroabsorption techniques.[78,257,275,284] Yacoby[257] reports having observed structure due to one- two-, and three-phonon processes while Frova et al.[78] attribute most of the many-phonon structure of Yacoby to Airy function oscillations, as predicted by Eq. (24.65) and Fig. 86. We show in Fig. 105 the electroabsorption spectrum of the indirect edge of silicon, obtained by Frova et al.[78]

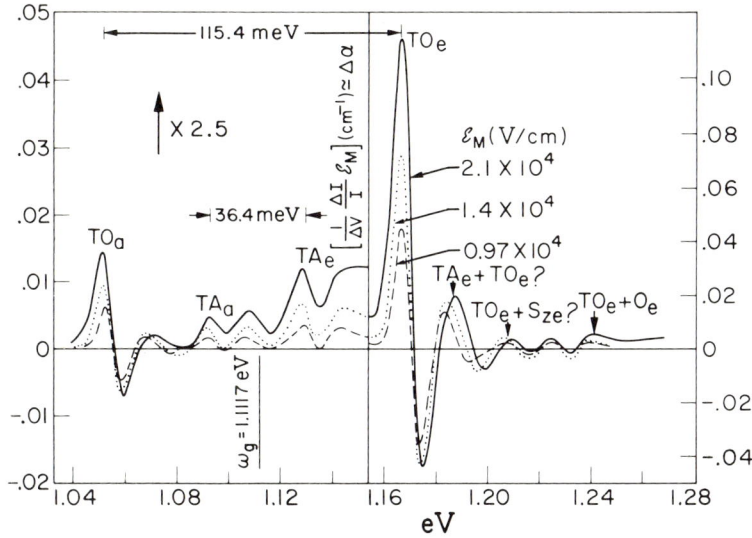

FIG. 105. Electroabsorption spectrum in the vicinity of the indirect gap of silicon at $T = 23°C$ according to A. Frova, P. Handler, F. A. Germano, and D. E. Aspnes, *Phys. Rev.* **145**, 575 (1966). Several possible phonon processes are indicated. The subindices a and e are used for absorption and emission processes, respectively. Question marks indicate the assignments according to Y. Yacoby, *Phys. Rev.* **142**, 445 (1966).

at 23°C by the p–n junction method. The energy gap ω_g is easily determined as the middle point between the strong TO_a–TO_e or TA_a–TA_e processes. The subsidiary oscillations seen in Fig. 105 have been assigned by Yacoby to many phonon processes (question marks) while Frova et al.[78] prefer the assignment to Airy function oscillations. This assignment is certainly plausible for the first peak (labeled TA + TO?) above the TO structure, since the energy of that peak shifts considerably with field. The oscillations at higher energies (TO + S?, TO + O, TA + TO?) should shift with field much more than they do if they were Airy oscillations. Also Airy oscillations should damp out more rapidly for the expected broadening ($\sim kT$). Hence we conclude that while the first oscillation above TO is mainly an Airy-function-related subsidiary oscillation associated with the TO peak, all other oscillations seem to correspond to many-phonon processes. If a collection of oscillations belongs to only 1 given phonon process, one should be able, according to Eq. (24.63), to bring the corresponding phonon absorption and emission curves to coincide with each other by shifting the horizontal scale an amount $2\omega_{phon}$ and adjusting the vertical scale by a multiplicative factor related to the phonon occupation numbers. This is illustrated in Fig. 106

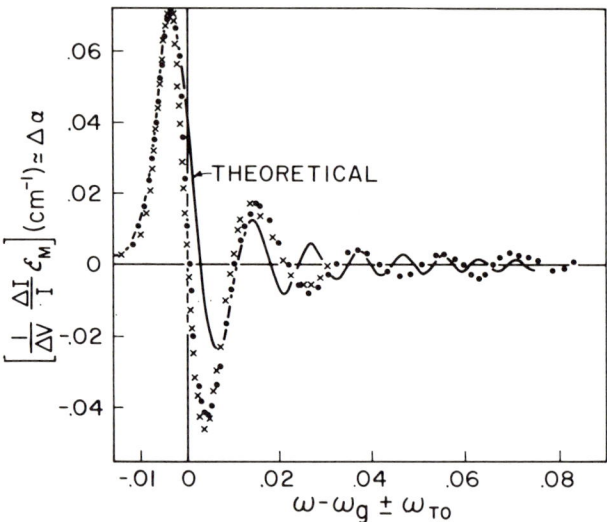

FIG. 106. TO_e (●) and TO_a (X) spectra of silicon at $T = 23°C$ brought to coincidence by shifting the energy and adjusting the vertical scale. $\mathscr{E}_M = 1.4 \times 10^4$ V cm^{-1}. Also theoretical curve calculated from Eq. (24.65). From A. Frova, P. Handler, F. A. Germano, and D. E. Aspnes, *Phys. Rev.* **145**, 575 (1966).

(from Frova *et al.*[78]). In this figure the TO_a and TO_e main peaks and the first subsidiary oscillations are brought to coincidence by the method just described. One would not be able to do so if the subsidiary oscillations were caused by different phonons. Figure 106 shows also an attempt to fit the experimental spectra with the unbroadened function of Eq. (24.65). While the calculated curve illustrates the difficulty of distinguishing between Airy oscillations and other phonon processes, it is clear that the decay in the experimental phonon emission points is too slow to be explained by the calculated curve, even having neglected broadening.

Yacoby's electroabsorption measurements[257] were performed on high-resistivity silicon, produced by fast neutron irradiation, in the configuration of Fig. 93a. His results are similar to those of Frova *et al.* except for the difference in interpretation just mentioned. This ambiguity in interpretation is quite typical of all kinds of electric field modulation spectra; it is often difficult to distinguish between Airy oscillations and extra critical points.

Figure 107 shows the dependence of the height of the main TO_e peak of Fig. 106 on applied field, e.g., voltage. If a power law $\Delta\alpha_{max} \sim V^m$ is used to represent this dependence, we obtain $m = 1.9$ for low fields and $m = 1.4$ for high fields as opposed to the uniform $m = 1.33$

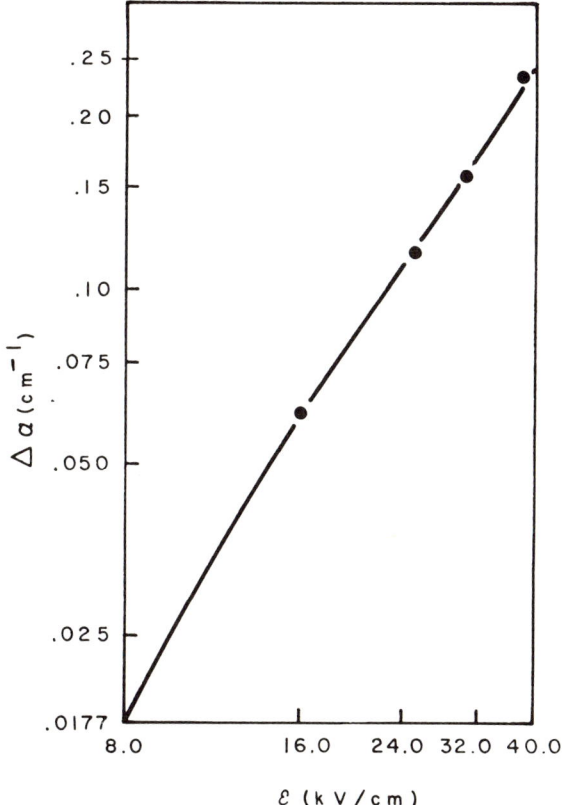

FIG. 107. Intensity of the main TO_e electroabsorption peak of Fig. 106 as a function of applied field at 77°K. From Y. Yacoby, *Phys. Rev.* **142**, 445 (1966).

dependence expected from Eq. (24.65). The higher experimental exponent for low fields can be qualitatively explained as due to broadening. For fields such that $\theta < \Gamma$ (Γ is the broadening parameter), the first oscillation must be smeared over an energy Γ. Since its unbroadened width is $\approx \theta \propto \mathscr{E}^{2/3}$, we find for the dependence of $\Delta\alpha_{\max}$ on V and exponent $m \simeq \frac{4}{3} + \frac{2}{3} = 2$. At higher fields ($\theta > \Gamma$), m must tend toward the theoretical value without broadening. This trend is clearly apparent in Fig. 107; we estimate from this figure that the field at

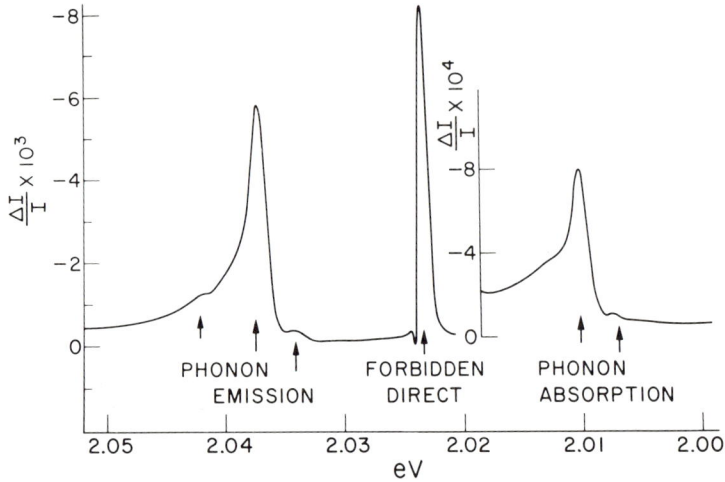

FIG. 108. Electroabsorption spectrum of Cu_2O at 77°K in the configuration of Fig. 93a (62 kV cm^{-1}). Sample thickness = 0.42 mm. This spectrum shows the zero-phonon forbidden exciton line and several associated phonon absorption and emission lines. From S. Brahms and M. Cardona, *Solid State Commun.* **6**, 733 (1968).

which $\theta = \Gamma$ is 10^4 V cm^{-1}, which yields the value of the broadening parameter $\Gamma \approx 5$ meV, quite reasonable at 82°K.

Electroabsorption at the indirect edge of silicon has also been reported by Chester and Wendland (p–n junction technique[275]). The electroabsorption spectrum of germanium (p–n junction technique) has been reported by Frova et al.[78] The various phonon energies obtained are listed in Table XI.

Figure 108 shows the electroabsorption spectrum of Cu_2O in the neighborhood of the lowest $n = 1$ (forbidden) exciton edge as observed by Brahms and Cardona.[285] Beside the zero-phonon forbidden exciton line at 2.023 eV, one observes 2 main peaks which correspond to phonon aided transitions (absorption and emission) to the $n = 1$ exciton state. These transitions have line shapes very similar to the function F of Fig. 51 and hence it is reasonable to assume that the modulation is caused by an electric-field-induced shift of the energy eigenvalue. Two considerably weaker peaks, also caused by phonon-aided exciton transitions, are seen in the phonon emission region of Fig. 108. These peaks are not well resolved in the corresponding phonon absorption spectrum.

b. Electroabsorption: Direct Edge

Because of the strong absorption coefficients in the neighborhood of direct allowed edges, electroabsorption measurement[284,286] are often confined to the region below the gap. For a perfect one-electron lowest direct edge, the absorption below the edge should be zero for $\mathscr{E} = 0$, while a finite absorption would be induced for $\mathscr{E} \neq 0$ (see Fig. 43). However, since edges are never "perfect", absorption appears below ω_g even for $\mathscr{E} = 0$. This absorption, caused in part by the broadened exciton states, extends all the way to $\omega = 0$ and, for small values of α, often follows the Urbach rule:

$$\alpha \propto \varepsilon_i \propto \exp \lambda[\omega - \omega_0], \qquad \text{for } \omega < \omega_0, \tag{26.1}$$

where λ may be a function of temperature (usually $\lambda \sim 1/T$). The electrooptic effect of an exponential absorption edge was first calculated by Franz.[72] The assumption underlying this calculation is based on the following convolution expression for $\varepsilon_i(\omega, \mathscr{E})$ for a single allowed critical point.[231,287]

$$\varepsilon_i(\omega, \mathscr{E}) = \frac{1}{\omega^2} \int_{-\infty}^{+\infty} d\omega' \, \omega'^2 \varepsilon_i(\omega', 0) \left\{ \frac{1}{|\Omega|} \text{Ai}\left(\frac{\omega' - \omega}{\Omega}\right) \right\}, \tag{26.2}$$

where

$$\Omega = 4^{-1/3}\theta = (\mathscr{E}^2/8\mu)^{1/3}.$$

The parameter θ has been defined in Eq. (24.17) and μ is the reduced mass along $\mathscr{E}$. Equation (26.2) can be derived rigorously for a parabolic one-electron edge (see Appendix II). One can assume, without justification, that a convolution expression of this type is also valid for any shape of the edge, such as that of Eq. (26.1) or an edge of excitonic character. Under this assumption one obtains for an exponential edge, such as that of Eq. (26.1), a low energy tail (see Appendix II):

$$\alpha \propto \varepsilon_i \propto \exp(\lambda^3 \Omega^3 + \lambda(\omega - \omega_0)). \quad (26.3)$$

According to Eq. (26.3) the electric field shifts the gap ω_0, i.e., the absorption edge, by an amount:

$$\Delta \omega_0 = -\lambda^2 \Omega^3 = -(\lambda^2 \mathscr{E}^2/8\mu). \quad (26.4)$$

The modulation in α produced by $\mathscr{E}$ is:

$$\Delta \alpha / \alpha = \lambda^3 \Omega^3. \quad (26.5)$$

Equations (26.4) and (26.5) suggest the use of the electrooptic effect for determining the reduced mass μ. This has been done for GaAs,[286] CdS, and CdSe,[288] and amorphous selenium.[289] Since the values of μ obtained for GaAs, CdS, and CdSe are in good agreement with those obtained by other methods, we can regard this as a confirmation of the correctness of the convolution [Eq. (26.2)] used in deriving Eqs. (26.4) and (26.5). The reduced mass obtained by Moss for GaAs is $\mu = 0.065$, a reasonable value since $m_e = 0.065$ and the average heavy hole mass $m_{hh}^* \approx 0.6$. This fact suggests that the electrooptic effect is dominated at low frequencies by transitions to the conduction band from the heavy hole band.

The measurements of Gutsche and Lange[288] were performed on the wurtzite-type materials CdS and CdSe. The electron masses of these materials at $\mathbf{k} = 0$ are quite well known[116] and hence one can use the observed reduced masses to calculate the somewhat elusive hole masses: experimental values of the hole masses are difficult to determine since one cannot make these materials p-type. Also, the hexagonal crystal field removes the degeneracy of the top valence band of zincblende and, therefore, the severe objection of having neglected interband coupling for degenerate bands in the theory of the electrooptic effect does not hold. A considerable number of parameters can be determined by this method, namely the longitudinal and transverse (to the c axis) masses of the A, B, and C valence bands (see Section 16). The results obtained for CdS and CdSe by Gutsche and Lange are listed in Table XV together with the hole masses obtained from other measurements (Zeeman effect of excitons[55,56] and those theoretically calculated.[116] A reduced mass of $m^* = 5m$ has been found by Drews by this method for amorphous selenium.[289]

The electroabsorption spectrum of germanium in the neighborhood

TABLE XV. REDUCED MASSES μ OBTAINED FROM ELECTROOPTIC MEASUREMENTS FOR TRANSITIONS INVOLVING THE A, B, AND C VALENCE BANDS OF CdS AND CdSe. ALSO, HOLE EFFECTIVE MASSES (m_h) DERIVED FROM THESE VALUES OF μ, OTHER EXPERIMENTAL VALUES OF THESE MASSES OBTAINED FROM THE ZEEMAN EFFECT OF EXCITONS, AND THEORETICAL ESTIMATES OF THESE MASSES[a]

	$\mathscr{E}$	E	Transition	μ	m_h	m_h (others) Expt	Theor[b]
CdS	$\parallel C$	$\perp C$	$\Gamma_9 \to \Gamma_7$	0.184 ± 0.01	≈ 3	5^c	2.5
	$\perp C$	$\perp C$	$\Gamma_9 \to \Gamma_7$	0.140 ± 0.01	0.42 ± 0.07	0.7^c	0.56
	$\parallel C$	$\parallel C$	$\Gamma_7 \to \Gamma_7$	0.142 ± 0.01	0.52 ± 0.1	—	0.24
	$\perp C$	$\parallel C$	$\Gamma_7 \to \Gamma_7$	0.207 ± 0.01	~ 10	—	1.2
CdSe	$\parallel C$	$\perp C$	$\Gamma_9 \to \Gamma_7$	0.123 ± 0.01	~ 2.3	—	2.5
	$\perp C$	$\perp C$	$\Gamma_9 \to \Gamma_7$	0.0885 ± 0.01	0.28 ± 0.08	0.45^d	0.42
	$\parallel C$	$\parallel C$	$\Gamma_7 \to \Gamma_7$	0.096 ± 0.006	0.37 ± 0.01	—	0.18
	$\perp C$	$\parallel C$	$\Gamma_7 \to \Gamma_7$	0.121 ± 0.01	~ 1.8	0.9^d	0.94

[a] E. Gutsche and H. Lange, *Phys. Status Solidi* **22**, 229 (1967).
[b] M. Cardona, *J. Phys. Chem. Solids* **24**, 1543 (1963).
[c] J. J. Hopfield and D. G. Thomas, *Phys. Rev.* **122**, 35 (1961).
[d] J. O. Dimmock and R. G. Wheeler, *J. Appl. Phys.* **32S**, 2271 (1961).

of the direct absorption edge has been studied by Frova et al.[78] using the p–n junction technique. The experimental results for a maximum field $\mathscr{E}_m = 3.3 \times 10^4$ V cm^{-1} are shown in Fig. 109 together with an attempt to fit them with the electrooptic function $F(\eta)$. Light- and heavy-hole contributions are given separately. Reduced masses $\mu = 0.0195$ and $\mu = 0.033$ have been assumed for the light- and heavy-hole transitions, respectively. It is quite clear in this figure that a good fit to the experimental data cannot be obtained. Interband coupling between the light- and heavy-hole bands may be partly responsible for the discrepancy, but exciton effects must have a sizeable contribution since the absorption curves at this temperature show a considerable deviation from the $(\omega - \omega_g)^{1/2}$ shape. Also, Eq. (25.21) which we have used for the calculation of $\Delta\alpha$, is only valid for fields which vary little in a distance $1/\alpha$: this condition may not be fulfilled in the experimental situation with p–n junctions in the neighborhood of ω_g. Measurements at low temperatures with high resistivity samples,[127,290] however, yield results very similar to those obtained for p–n junctions thus indicating that the rapid spatial variation of $\Delta\alpha$ is not the source of the discrepancy shown in Fig. 109. Broadening of the electrooptic function F is not expected to improve significantly the fit of theory to experiments in Fig. 109.

The question of whether the first 2 peaks in the electroabsorption spectrum (similar discussion applies to the electroreflectance) are

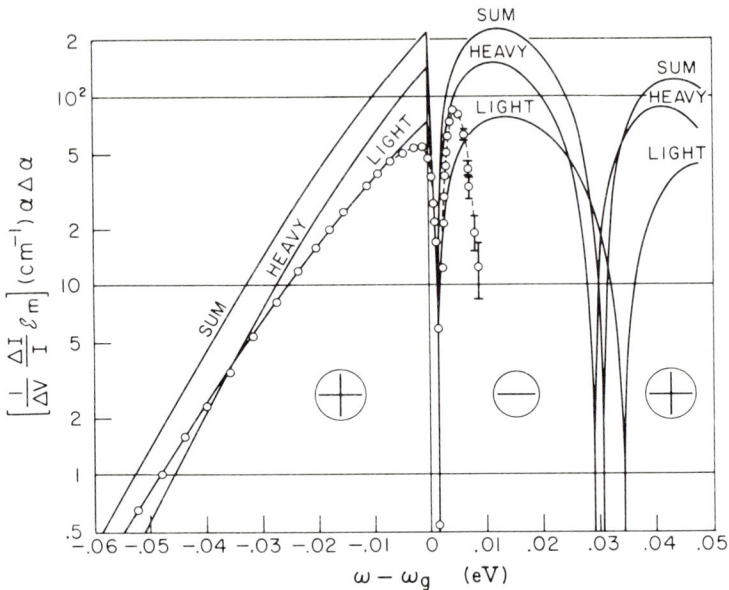

FIG. 109. Theoretical and experimental electroabsorption spectra of germanium at 7°C. $\mathscr{E}_m = 3.3 \times 10^4$ V cm^{-1}. The sign of the signal in the logarithmic vertical scale is indicated by $+$ and $-$. From A. Frova, P. Handler, F. A. Germano, and D. E. Aspnes, Phys. Rev. 145, 575 (1966).

due mainly to field shift and quenching of excitons affects considerably the determination of the energy gap from the experimental electroabsorption data. Figure 80 shows that for pure one-electron transitions the energy gap corresponds to the first positive peak position if broadening is neglected and to slightly higher energies (close to the zero in $\Delta \varepsilon_i$) for broadened transitions. If field broadening of the exciton is the main electrooptical process, as the experimental line shape (insert in Fig. 110) seems to indicate,[290] the exciton energy corresponds to the second (negative) oscillation in $\Delta \alpha$. Predominance of the field shift of the exciton would bring the exciton energy to the first zero in $\Delta\alpha$. These line shape problems cannot be conclusively solved in many cases and they often contribute an uncertainty of the order of the line width to the energy gaps determined from electric field modulation spectra.

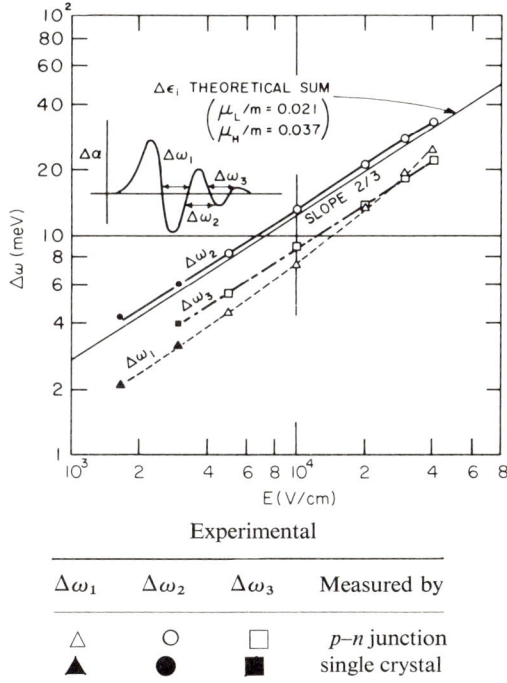

FIG. 110. Dependence of the spacing between zeros of $\Delta\varepsilon_i$ at 14°K observed by Hamakawa et al.[290] for the E_0 electroabsorption peak of germanium compared with the theoretical prediction ($\mathscr{E}^{2/3}$).

Figure 110 shows the field dependence of the energy difference between zeros of $\Delta\varepsilon_i$ observed for Ge (p–n junctions and high resistivity) by Hamakawa et al.[290] The slope of this field dependence is quite close to the theoretical $\frac{2}{3}$ slope thus indicating that the oscillations above the first one, which is most likely excitonic, are Franz–Keldysh-type oscillations whose width is, at these temperatures (14°K), determined by $\mathscr{E}$ and not by broadening.

Rees[287] has suggested the use of Eq. (26.2) to calculate the electrooptic effect with exciton interaction. If the unperturbed ($\mathscr{E}=0$) spectrum of ε_i and the reduced mass are known, the perturbed spectrum ($\mathscr{E} \neq 0$) is found by performing the convolution of $\varepsilon_i(\mathscr{E} = 0)$ with the Airy function given in Eq. (26.2). The range of validity of this equation in the presence of exciton interaction, however, is not known. Figure 111 shows the fit to experiment[291] of the ε_i calculated with Eq. (26.2) from the absorption data of Sturge[112] with a reduced mass $m^* = 0.065$ for $\mathscr{E} = 42$ kV cm^{-1}.

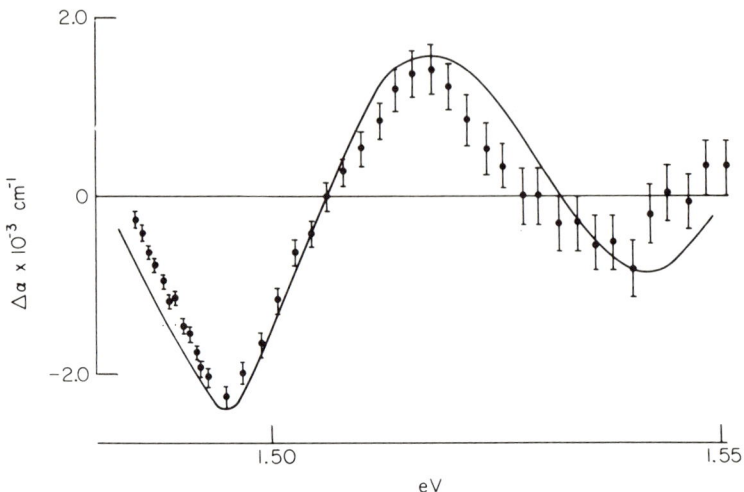

FIG. 111. Electroabsorption of GaAs at 77°K. The experimental points correspond to $\mathscr{E} = 42$ kV cm^{-1}. The theoretical curve has been calculated with Eq. (26.2) for $m^* = 0.065$ m. From H. D. Rees, *Solid State Commun.* **5**, 365 (1967).

The field induced absorption of the $n = 1$ forbidden exciton in Cu_2O, absent for $\mathscr{E} = 0$ [see Eq. (6.19)], is shown[285] in Fig. 108. Electroabsorption in KBr has been reported by Ballaro et al.[292] Measurements for CdSe have also been reported by Poehler and Abraham.[293]

c. Electroreflectance: Field Effect Configuration

Electroreflectance spectra measured with the field effect or Seraphin configuration have been published for Ge,[61] GaAs,[295] Si,[191] and CdTe.[272] We show in Fig. 112 the electroreflectance spectra in the neighborhood of the E_0 gap of germanium (n-type, 30 Ω-cm) at −65°C as measured by Seraphin.[267] for various values of the dc bias (see figure caption). It is worthwhile pointing out that different conventions are used in the literature for the sign of $\Delta R/R$. Within the convention of Fig. 112 a positive $\Delta R/R$ means an *increase* of ΔR for an increase (*positive*) of the voltage applied to the transparent field electrode. Figure 112 illustrates the vanishing of the electroreflectance

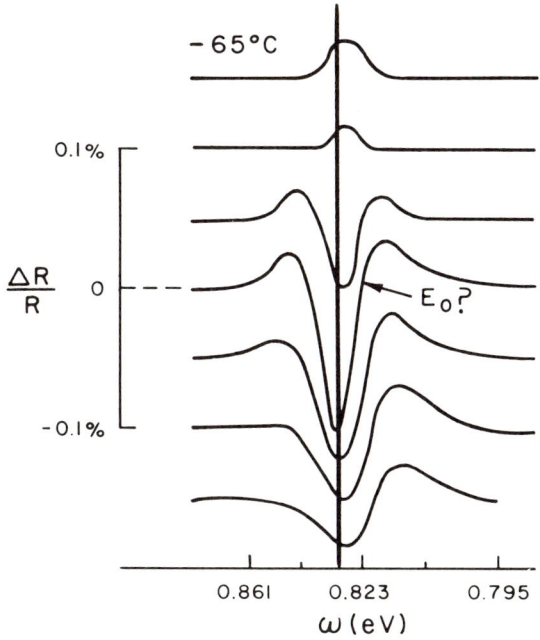

FIG. 112. Electroreflectance spectra near the direct band gap of n-type germanium (30 Ω-cm) for various bias conditions. The surface potentials as a function of bias are those of Fig. 98. A positive $\Delta R/R$ corresponds to an *increase* in R with a positive ΔV applied to the modulating electrode. dc biases equally spaced between 600 V (lower curve) and -600 V. From B. O. Seraphin, *Surface Sci.* **8**, 399 (1967).

signal at the fundamental frequency when going through the zero surface potential bias (flat band). While no signal appears for flat bands at the fundamental frequency, sizeable signals can still be seen at the second harmonic.

Figure 112 illustrates rather clearly some of the problems involved in the interpretation of electroreflectance spectra. Let us first consider the curve for zero dc bias, which corresponds to zero surface potential for the negative half-cycle of the modulating voltage. The measured

$\Delta R/R$ is simply the effect on R of the field during the positive half cycle $[R(\mathscr{E}_{max}) - R(0)]$; it should have the shape of the electrooptic function G for a one-electron transition if $\mathscr{E}$ can be assumed to be uniform. A comparison of Fig. 112 with Fig. 80 (M_0 edge, $\Delta\varepsilon_r$) shows that this is not the case. The experimental line shape, however, is quite similar to the $\Delta\varepsilon_i - M_0$ line shape of Fig. 80; this fact suggests that we may be close to the case of small field penetration depth, in which case Eq. (25.20) would reverse the roles of $\Delta\varepsilon_r$ and $\Delta\varepsilon_i$. If this situation obtains, the direct gap E_0 would be close to the first zero (labeled E_0 ? in Fig. 112) in $\Delta R/R$. The value of E_0 determined by Macfarlane et al.[294] at $-65°$ is indicated by a vertical line in Fig. 112 and is significantly higher than the energy labeled (E_0 ?). Hence we conclude, as we did in the discussion of electroabsorption, that the large negative peak in $\Delta R/R$ is due mainly to the field quenching of the exciton ground state associated with E_0. Since the binding energy of this exciton is only[294] 1 meV, this peak should practically occur at the energy of E_0.

In Fig. 113 we show, for comparison, the spectra of $\Delta\varepsilon_r$ obtained by Hamakawa et al.[290] by a Kramers–Kronig analysis of their electroabsorption data. The line shapes of Fig. 113 (14°K) agree quite well with the line shape for zero bias in Fig. 112 (208°K), especially if one takes into account that the oscillations at high energy should disappear because of broadening as the temperature is raised.

The line shape of Fig. 112 suffers a drastic change for the large positive biases. Similar effects are observed in many space–charge barrier electroreflectance measurements, including those performed with the electrolyte technique.[71] A possible interpretation of these effects is the decrease in penetration depth produced by the bias: the + sign in the bias of Fig. 112 produces an *enrichment layer* which should have a very small penetration depth. A transition from the large to the small penetration depth regime can produce a change in line shape by permuting $\Delta\varepsilon_r$ and $\Delta\varepsilon_i$ in the expression for $\Delta R/R$, according to the discussion in Section 25.

Figure 114 shows a portion of the electroreflectance spectrum of silicon obtained by Seraphin,[191] at several temperatures. The structure in Fig. 114 occurs in the immediate vicinity of the 3.3. eV reflectivity peak, usually labeled $E_0{'}$, and assigned to transitions between the $\Gamma_{25'}$ valence band and the Γ_{15} conduction band [see Fig. 29] in the vicinity of $\mathbf{k} = 0$. The complex nature of these transitions, and the fact that there may be more than 1 set of equivalent critical points involved, has been pointed out by several authors.[45] This fact is clearly evidenced by the electroreflectance data: while the II–III peaks could be due in principle to electroreflectance oscillations associated with only 1 set of critical points, peak I stands out as being due to a different process. Structure associated with I is also seen

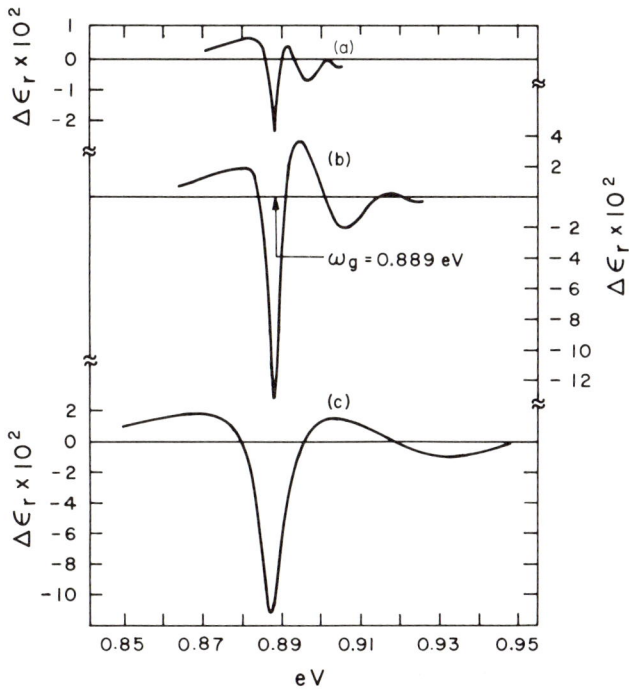

FIG. 113. $\Delta\varepsilon_r(\omega, \mathscr{E})$ obtained by Kramers–Kronig analysis of electroabsorption data for germanium at 14°K. For curves (a) $\mathscr{E} = 3 \times 10^3$ V cm^{-1}, (b) $\mathscr{E} = 10 \times 10^3$ V cm^{-1}, and (c) $\mathscr{E} = 30 \times 10^3$ V cm^{-1}. From Y. Hamakawa, F. A. Germano, and P. Handler, *Phys. Rev.* **167**, 703 (1968).

in the conventional reflectivity spectrum at 95°K, also shown in Fig. 114. It has been pointed out by Seraphin[191] that while the temperature coefficients of peaks II and III are -3.3×10^{-4} eV $\times$ (°C)$^{-1}$ (for both), the coefficient of peak I is significantly smaller (-1.5×10^{-4} eV $\times$ (°C)$^{-1}$). This fact indicates 2 different sets of critical points. It has been suggested[191] that peak I is due to the direct $\Gamma_{25'} \to \Gamma_{15}$ threshold at Γ while II–III correspond to critical points along the {[111]} directions (Λ). We shall come back to this point in Chapter VIII in connection with piezoelectroreflectance studies.

Figure 115 shows the electroreflectance spectrum of CdTe obtained by Ludeke and Paul[272] at 300°K and at 6°K. One can obtain from these spectra an accurate value for the spin orbit splitting of the E_0

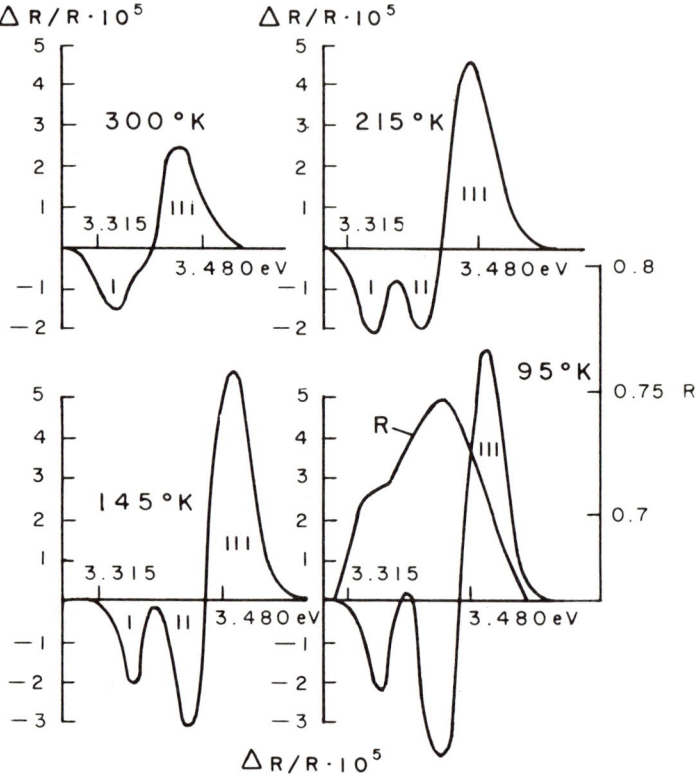

FIG. 114. Electroreflectance spectrum of *p*-type silicon at several temperatures. From B. O. Seraphin, *Phys. Rev.* **140A**, 1716 (1965). Also reflectivity spectrum of silicon at 95°K in the same spectral region.

edge of CdTe by measuring the separation between the $E_0(\Gamma_{15} \to \Gamma_1)$ and the $E_0 + \Delta_0$ peaks; because of the great similarity in the line shapes for those 2 peaks ($E_0 + \Delta_0$ is, of course, somewhat broader), it is possible to eliminate line shape uncertainties in the measurement of Δ_0 which affect E_0 and $E_0 + \Delta_0$, separately. The spin orbit splittings Δ_0 have been obtained by electroreflectance methods using the Seraphin configuration in InAs,[271] InSb,[271] and GaAs[295]: some of the obtained values of Δ_0 are listed in Table II. It is interesting to point out that the center zero in the spectrum of Fig. 115 at 6°K agrees with the exciton energy reported by Thomas[160] and hence the main structure at E_0 is thought to be excitonic. The oscillations at higher

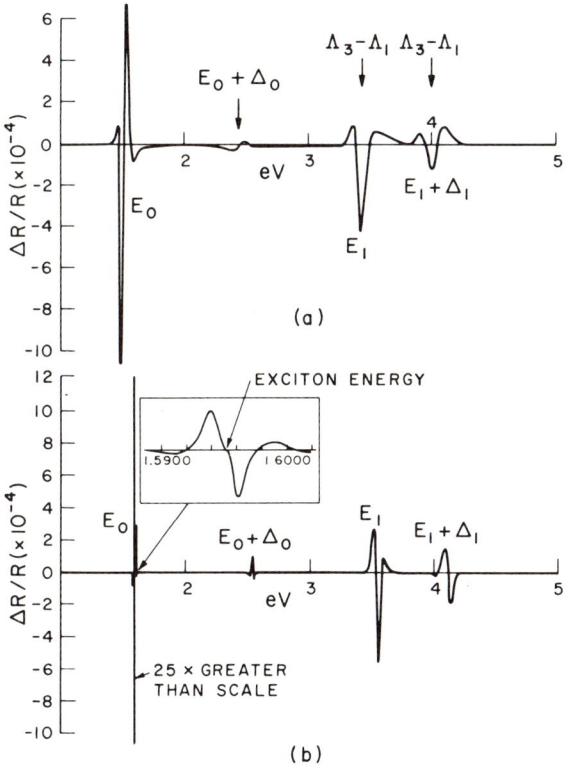

FIG. 115. Electroreflectance spectrum of CdTe at 300°K (a) and at 6°K (b), as reported by R. Ludeke and W. Paul, in "II–VI Semiconducting Compounds" (D. G. Thomas, ed.), p. 123. Benjamin, New York, 1967. The field effect package was built with an evaporated silicon layer as a dielectric and SnO_2 as a transparent electrode.

energies could be of Franz–Keldysh-type. The $E_1-(E_1+\Delta_1)$ structure seen in Fig. 115 is attributable to the transitions which cause similar peaks in Ge and InSb (Figs. 9 and 29). An interpretation in terms of Franz–Keldysh effects is not easy because of the large number of unknowns involved e.g., the fields which correspond to the positive and the negative half-cycle of the modulating voltage are not known. A Kramers–Kronig analysis of these data would be quite helpful for such interpretation. In any case, the almost identical line shapes of E_1 and $E_1 + \Delta_1$ lead again to an accurate determination of Δ_1 ($\Delta_1 = 0.58$ eV).

d. *Electrolyte Method: Germanium–Zincblende-Type Materials*

The room temperature electroreflectance spectrum of InP in the region around the $E_1 - (E_1 + \Delta_1)$ peaks is shown[296] in Fig. 116. We

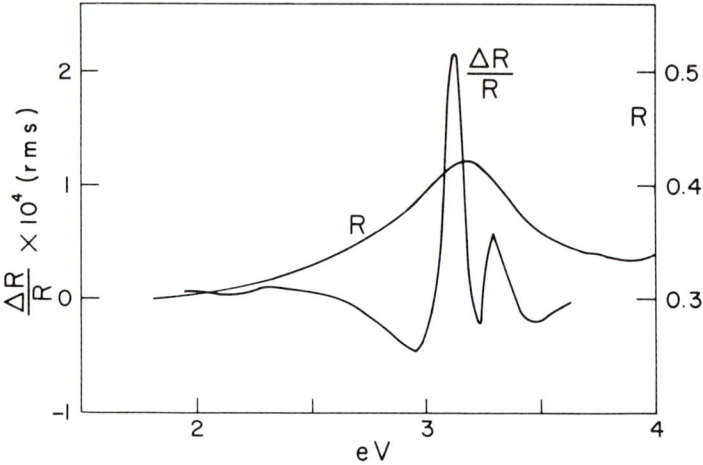

FIG. 116. Electroreflectance and reflection spectra of InP in the neighborhood of the $E_1 - (E_1 + \Delta_1)$ peaks. From K, L. Shaklee, F. H. Pollak, and M. Cardona, *Phys. Rev. Letters* **15**, 883 (1965). The sign of $\Delta R/R$ in this and all subsequent figures referring to electrolyte method is that which corresponds to a positive half-cycle applied to the sample.

also show in this figure the corresponding reflectivity spectrum as an illustration of the resolving power of electroreflectance compared to the conventional reflection method.[297] The Δ_1 spin orbit splitting, unresolved in R, is clearly resolved in $\Delta R/R$. The value obtained for the splitting ($\Delta_1 = 0.15$ eV) agrees with that obtained from conventional reflectivity measurements at low temperature.[297] We should, at this point, give the sign convention used in the display of electroreflectance data obtained by the electrolyte method. We shall plot $\Delta R/R$ as obtained when the *positive* half-cycle of the modulating voltage is applied to the sample. This convention is that used in Cardona et al.[71] in spite of an erroneous statement to the contrary in that paper.

As an illustration of the amount of structure which is obtained in the electroreflectance spectra of germanium and zincblende-type materials, we show in Fig. 117 the electroreflectance spectrum of

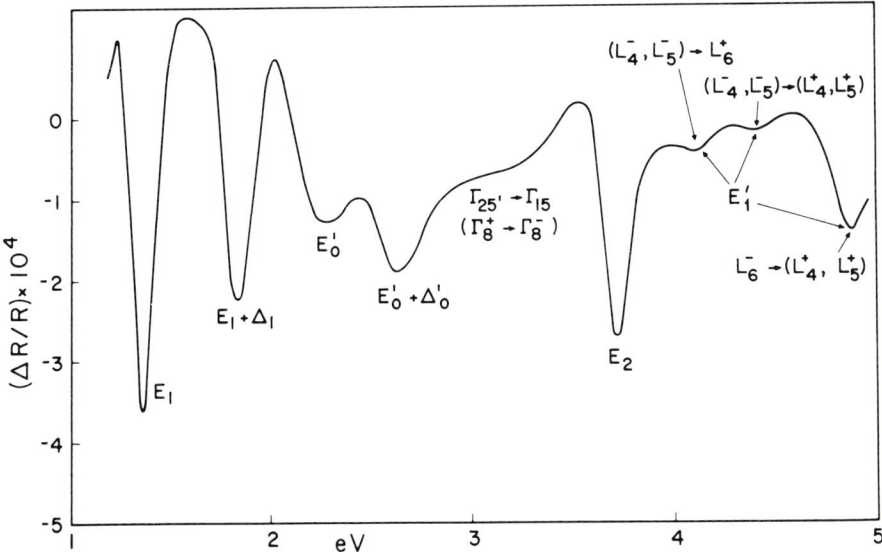

FIG. 117. Electroreflectance spectrum of n-type gray tin (α — Sn) at $-78°$C. From M. Cardona, P. McElroy, F. H. Pollak, and K. L. Shaklee, *Solid State Commun.* **4**, 319 (1966).

gray tin.[298] Owing to the fact that the E_0 gap is negative,[299a] no E_0 structure is seen; this spectrum displays otherwise practically all of the typical critical points seen in germanium–zincblende materials; the most prominent peaks are those labeled E_1, $E_1 + \Delta_1$, and E_2. The interband transitions responsible for these peaks have already been discussed a number of times, in particular in Section 8: the doublet E_1–$(E_1 + \Delta_1)$ is produced by spin orbit interaction. The peaks E_0'–$E_0' + \Delta_0'$ have been attributed[298] to critical points in the [100] (Δ) direction; these critical points can be seen in Fig. 29 but they are destroyed by the large spin orbit splitting of Γ_{15} in Fig. 9. These last situations seem to prevail in α-Sn also[298] according to $\mathbf{k} \cdot \mathbf{p}$ energy band calculations† but the critical $E_0'-E_0'+\Delta_0'$ points along Δ would reappear if the spin orbit splitting of Γ_{15} were smaller.[299b] The spin orbit origin of the $E_0'-E_0'+\Delta_0'$ is strongly suggested by the systematics of the splittings observed in all materials of the family[116] (see Table XVI). The positions of the electroreflectance peaks of

TABLE XVI. CALCULATED AND EXPERIMENTAL VALUES[a] (IN eV) OF SEVERAL SPIN ORBIT SPLITTINGS OBSERVED IN GROUP IV AND GROUP III–V MATERIALS[b]

Material	Δ_0 Expt	Δ_1 Expt	Δ_1 Calc[c]	Δ_0' Expt	Δ_0' Calc	Δ_2 Expt	Δ_2 Calc[c]
Si	0.044		0.03				0
Ge	0.29	0.19	0.20	0.19	0.17		0
α-Sn		0.48[d]	0.48	0.35	0.39		0
AlSb	0.75	0.40	0.41	0.27	0.47		0.34
GaP	0.10	0.1	0.07	0.05	0.072		0.02
GaAs	0.34	0.23	0.22	0.19	0.17		0.08
GaSb	0.80	0.46	0.46	0.29	0.48		0.32
InP	0.11	0.15	0.11	0.07	0.086	0.20	0.2
InAs	0.43	0.28	0.28		0.22		0.02
InSb	0.82	0.50	0.48	0.33	0.45		0.15

[a] The experimental values of Δ_0 have been used as adjustable parameters in the calculation of the other splittings unless otherwise indicated.
[b] From M. Cardona, K. L. Shaklee and F. H. Pollak, *Phys. Rev.* **154**, 696 (1967) unless otherwise indicated.
[c] Data taken from Table III.
[d] Used as adjustable parameter.

Fig. 117 are shown by arrows at the bottom of Fig. 33; they all correspond to peaks in the calculated ε_i spectrum. In particular, the E_1' peaks correlate quite well with critical points at L between the spin orbit split L_3 valence bands (Fig. 9) and the L_3 (also spin orbit split) conduction bands.

The electrolyte method has yielded the spin orbit splitting Δ_0 of the top of the valence band of a number of materials of this family. As an example, and for comparison with Fig. 115, we show in Fig. 118 the electroreflectance spectrum of CdTe[71] as obtained with the electrolyte

† Pseudopotential calculations which include spin orbit interaction give smaller values of the Δ_{15} splitting of the Γ_{15} conduction band.[299b]

method. It is clear from this figure that the technique gives a rather accurate value of the spin orbit splittings Δ_0 and Δ_1 irrespective of line shape considerations. The values obtained for Δ_0 are tabulated in Table II. The values of Δ_1 for group IV and group III–V compounds are given in Table XVI. The Δ_1 splittings of several II–VI compounds

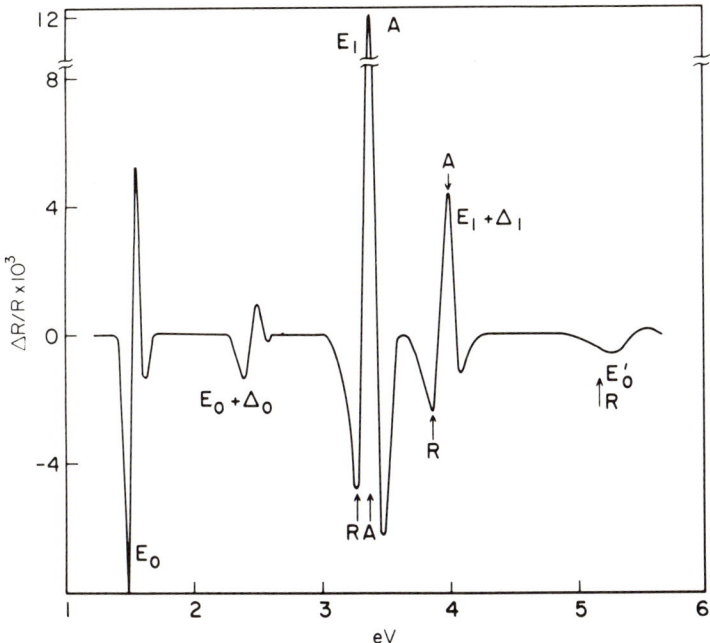

FIG. 118. Electroreflectance spectrum of CdTe as obtained by the electrolyte method. From M. Cardona, K. L. Shaklee, and F. H. Pollak, *Phys. Rev.* **154**, 696 (1967).

have also been determined by this method. They can be obtained by subtracting the E_1 energy from the $E_1 + \Delta_1$ energy in Table XVII.

Shallow impurity levels have been observed as optical transitions below the energy gap E_0 in the electroreflectance spectra of GaAs and InP.[71,295] We show in Fig. 119 the electroreflectance spectra of 3 typical GaAs samples near E_0. We see that E_0 is preceeded in all 3 cases by a peak E_1 (often stronger than E_0) which has been attributed to impurity levels.[71,295] It has also been suggested[300] that these peaks (E_1) are caused by electroabsorption of the signal reflected at the back surface of the sample. We believe that these E_1 peaks are indeed

TABLE XVII. ENERGIES (IN eV) OF THE PEAKS OBSERVED IN THE ELECTROREFLECTANCE SPECTRA OF GROUP IV, III–V, AND II–VI SEMICONDUCTORS WITH GERMANIUM–ZINCBLENDE STRUCTURE[a]

Material	E_0	$E_0 + \Delta_0$	$E_1(1)$	$E_1(2)$	$E_1(1) + \Delta_1$	$E_1(2) + \Delta_1$	E_0'	$E_0' + \Delta_0'$	E_2	$E_2 + \delta$	E_1'
Si	4.06 ±0.1	4.13 ±0.1					3.32 3.38		4.31 4.49		
Ge	0.798	1.09	2.12		2.34		3.13	3.32	4.42		4.11, 4.39
α-Sn			1.36		1.84		2.28	2.63	3.72		4.89
AlSb	2.22	~3	2.81	2.88	3.21	3.30	3.72	3.99	4.25	4.6	
GaP	2.74	2.84	3.66	3.80			4.78	4.83	5.27	5.74	
GaAs	1.43	1.77	2.89	2.96	3.12	3.19	4.44	4.63	4.99	5.33	
GaSb		1.52	2.03		2.49		3.27	3.56	4.20	4.57	5.50
InP	1.34	1.44	3.12	3.19	3.27	3.34	4.72	4.79	5.04 5.24	5.6	
InAs			2.50		2.78		4.44		4.70	5.18	
InSb			1.88		2.38		3.16	3.49	4.08	4.66	5.25
CdTe	1.49	2.41	3.28	3.38	3.87	3.99	5.30				
ZnTe	2.25	3.18	3.61	3.67	4.18	4.27	5.40				
CdS	2.42										
HgSe			2.85		3.15		5.08				
HgTe			2.12		2.78		4.14		4.79	5.24	

[a] All data taken at room temperature except for α-Sn which was measured at −78°C.

due to impurity levels since they also appear in wedge shaped samples; the back-reflected signal can actually be seen in a number of samples at energies below E_1. Also, a splitting of the $E_0 + \Delta_0$ peak, attributed to the E_1 level,[71] can be often seen at an energy at which the back-reflected signal is completely absorbed. We do not

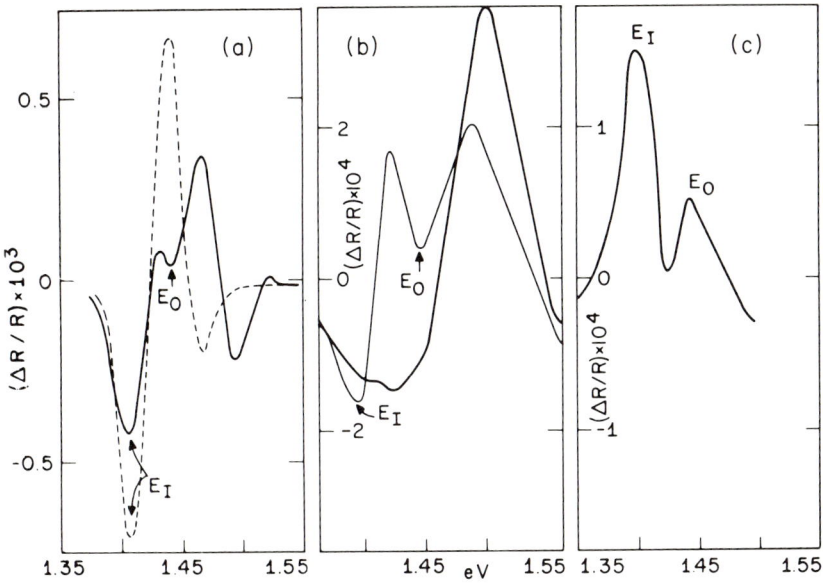

FIG. 119. Electroreflectance spectra of several GaAs samples showing impurity effects (E_1) near the fundamental edge (E_0). For (a), $N = 6 \times 10^{15}$ cm^{-3}, $V_{dc} = 0.35$ V, and $V_{ac} = 0.25$ V. For (b), $N = 10^{17}$ cm^{-3}, $V_{dc} = 0.75$ V, and $V_{ac} = 2$ V. For (c), $P = 2 \times 10^{17}$ cm^{-3}, $V_{dc} = 1.5$ V, and $V_{ac} = 0.5$ V. From M. Cardona, K. L. Shaklee, and F. H. Pollak, *Phys. Rev.* **154**, 696 (1967).

know the mechanism responsible for the reflectivity modulation of the $E_1 - (E_1 + \Delta_0)$ peaks; modulation of the population of these levels by the electric field in the penetration depth may have a large contribution to the effect.

Figure 120 shows the results of the Kramers–Kronig analysis of the data in Fig. 118. The $\Delta\varepsilon_r$ and $\Delta\varepsilon_i$ shown in Fig. 120 correspond to the *difference* between two spectra of the type shown in Fig. 80, for two values ($\mathscr{E}_{max}$ and $\mathscr{E}_{min}$) of the electric field. Since we do not know $\mathscr{E}_{max}$ and $\mathscr{E}_{min}$ (simultaneous field effect measurements were not performed), we must make the reasonable assumption that $\mathscr{E}_{min} \approx 0$ if we want to analyze the experimental line shapes. Figure 98 suggests that this situation should hold, except for very small amplitudes of the modulating voltage. If one makes this assumption, a comparison of the $E_0 - (E_0 + \Delta_0)$ peaks with Fig. 80 suggests that the line shape is that of an M_0 edge. Unfortunately, a similar line shape is expected

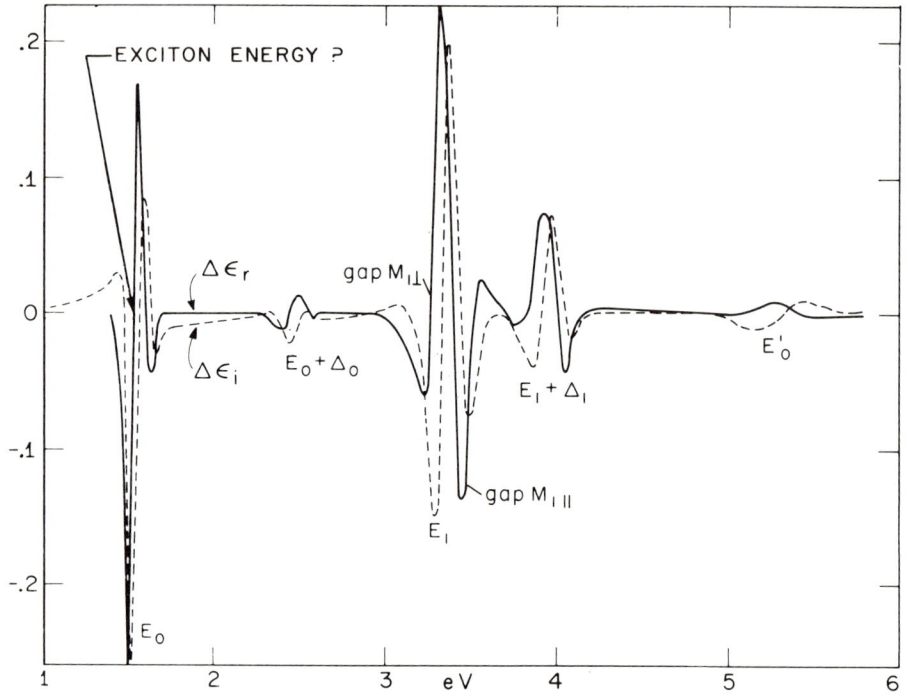

FIG. 120. $\Delta\varepsilon_r$ and $\Delta\varepsilon_i$ as obtained from the Kramers–Kronig analysis of the data in Fig. 118 for CdTe. From M. Cardona, K. L. Shaklee, and F. H. Pollak, *Phys. Rev.* **154**, 696 (1967).

(except for the oscillation at the highest energy, which must be of Franz–Keldysh origin) for the field quenching of the exciton. According to the discussion in Sections 26b–c, this field quenching should contribute significantly to the line shape in Fig. 120 and lower the energy of the first zero in $\Delta\varepsilon_r$ from the energy gap by the exciton binding energy.

The E_1, $E_1 + \Delta_1$ peaks in Fig. 120 have a shape which suggests (Fig. 80) M_1 critical points with the field in the positive mass direction

($M_{1\perp}$). As discussed in Section 24a, when a cubic material has $M_{1\perp}$ critical points, one also expects $M_{1\parallel}$ critical points, at least for certain orientations of $\mathscr{E}$. It is generally admitted that the critical points responsible for the $E_1 - (E_1 + \Delta_1)$ peaks are along the {[111]} directions (see Figs. 9 and 29). These critical points have a small transverse mass (~ 0.1 for CdTe) and a rather large longitudinal mass (~ 1); hence, one expects the longitudinal M_1 contribution to be small and probably lost in the broadening. It is curious to note that the curves of Fig. 120 near E_1 and $E_1 + \Delta_1$ can be fitted almost as well with $M_{1\parallel}$ critical points but such a fit would imply a higher energy gap: in the $M_{1\parallel}$ configuration the gap occurs near the high-energy oscillation (see Fig. 80). Because of the calculated longitudinal and transverse masses we feel that, if these peaks are of Franz–Keldysh origin, the $M_{1\perp}$ critical points must produce the main contributions to the observed electroreflectance. The possibility of excitonic contributions to these electroreflectance peaks has been suggested[301] and is reasonable since deviations from the one-electron shapes, attributable to excitons, have been observed in the conventional optical spectra.[46] It is curious to note that the L transitions (e_1, $e_1 + \Delta_1$), which seem to have been observed below E_1, $E_1 + \Delta_1$ in a number of zincblende materials[117], are not resolved in the electroreflectance spectra of Cardona et al.[71] Ghosh[208] has reported $e_1 - (e_1 + \Delta_1)$ structure for germanium (see Fig. 121). Since the position of these peaks does not agree with that reported by Potter,[154] we feel further work is necessary to resolve the controversy.

We show in Figs. 122–123 the unpublished $\Delta\varepsilon_r$ and $\Delta\varepsilon_i$ obtained from the $\Delta R/R$ data of Cardona et al.[71] for InP and InAs. The spectra look quite similar to those of Fig. 120, except for a sign reversal of the InAs spectra of Fig. 123. This sign reversal is to be attributed to the type of the InAs measured (p). A reversal in the sign of the spectra is usually encountered in going from n- to p-type for this family of materials. This reversal indicates that the Fermi energy is clamped by surface states at the surface of the material somewhere near the center of the gap (in germanium, however, the sign reversal occurs for n-type material with $N \sim 10^{17}$ cm^{-3}, hence the Fermi energy at the surface is close to the bottom of the conduction band).

The dc bias applied to the sample was negative for all the electrolyte measurements described. As discussed in Section 25, usually the semiconductor electrolyte interface is most blocking for this sign of the bias and hence, the signals are strongest for this dc polarity; it is not

248   VII. ELECTRIC FIELD MODULATION

even possible, in many cases, to obtain signals with a positive bias. In the case of silicon, however, signals are obtained for a wide range of biases of either sign. This may be caused by the presence of oxide on the silicon surfaces which prevents direct contact of the silicon with the electrolyte. It is possible, even at room temperature, to reverse

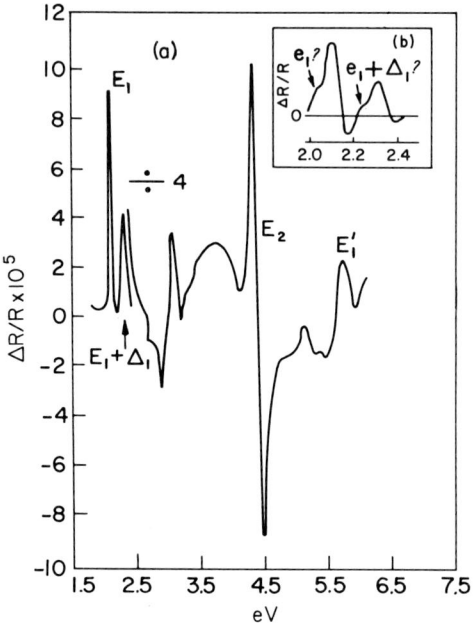

FIG. 121. Electroreflectance spectra of germanium at room temperature reported by Ghosh,[208] (a) for 3 V ac; (b) for 1 V ac.

the sign of the electroreflectance peaks by reversing the dc bias. This is illustrated in Figs. 10–11 of Cardona *et al.*[71] for *n*-type silicon. We have plotted in Figs. 124–125 the Kramers–Kronig analyses of these data. It is seen that considerable change in the line shapes results from the change in bias. Such changes in shape can be understood as due to a variation in field penetration depth. For a positive dc bias, a depletion layer results at the surface of *n*-type silicon, with a relatively large field penetration depth. The electroreflectance signal can be calculated with Eq. (8.9a). If the field penetration depth is very small, as may be the case for a negative bias (enrichment layer), one must use Eq. (25.19) and the Kramers–Kronig analysis does not yield $\Delta\varepsilon_r$ and $\Delta\varepsilon_i$ but the quantities in Eq. (25.20) instead. In this manner one can qualitatively account for the observed changes in shape.

The theory of Section 24a predicts normal incidence electroreflectance spectra which depend on the direction of the electric field of the

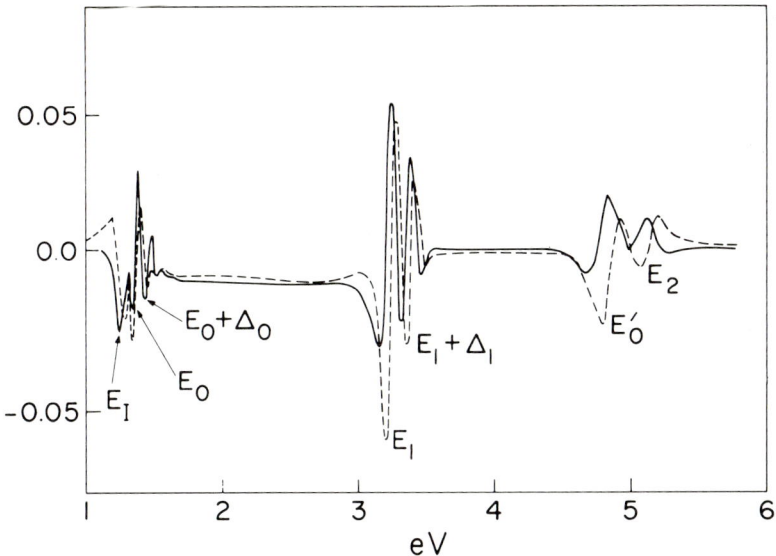

FIG. 122. $\Delta\varepsilon_r$ (solid line) and $\Delta\varepsilon_i$ (dashed line) obtained by the Kramers–Kronig analysis of the data of Fig. 21 of M. Cardona, K. L. Shaklee, and F. H. Pollak, *Phys. Rev.* **154**, 696 (1967) for n-type InP.

light **E** even for cubic semiconductors, when the reflecting surface is other than a (111) or a (001) surface. As an illustration,[274] we show in Fig. 126 the electroreflectance of a (110) and a (001) silicon surface for the 2 normal modes of **E**. While anisotropy is practically absent for the (001) surface, it is quite considerable for the (110) surface. The I–II doublet is well resolved for $\mathbf{E}\|[1\bar{1}0]$. This doublet is the same as that shown in Fig. 114.

The plasma edge of a heavily doped semiconductor should also appear as a peak in the electroreflectance spectrum in a manner similar to that shown in Fig. 42: the electric field modulates the plasma frequency at the surface. The modulation spectrum of the plasma edge of n-type GaAs ($N = 2 \times 10^{18}$ cm^{-3}), as observed by Axe and

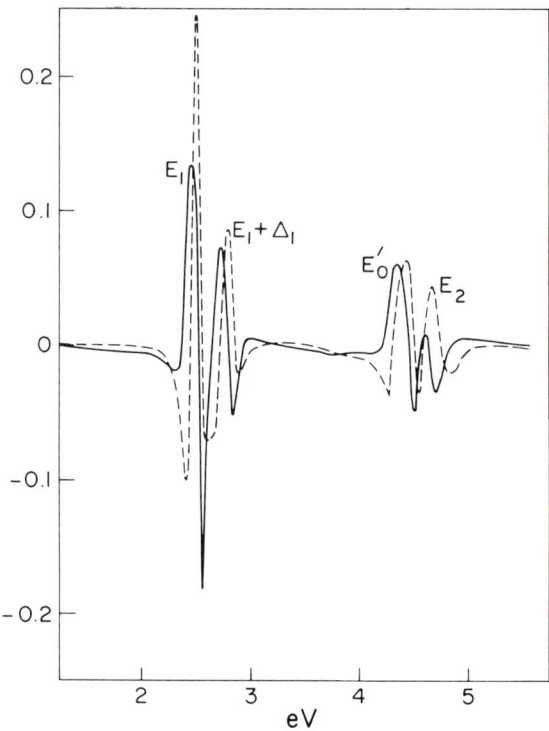

FIG. 123. $\Delta\varepsilon_r$ (solid line) and $\Delta\varepsilon_i$ (dashed line) obtained by the Kramers–Kronig analysis of the data in Fig. 22 of M. Cardona, K. L, Shaklee, and F. H. Pollak, *Phys. Rev.* **154**, 696 (1967) for p-type InP.

Hammer,[302] is displayed in Fig. 127. This spectrum shows also structure at the transverse optical phonon frequency (Reststrahlen). For comparison, the conventional reflection spectrum is included. These measurements were performed by the electrolyte method (methanol + NaCl) keeping the layer of electrolyte about 0.0002 inch thick in order to prevent absorption by the liquid in the infrared. The solid lines were fitted by assuming a depletion layer of semiconductor near the surface without any free carriers. The thickness of this layer is modulated by the electric field. The dielectric constant of the bulk was taken to be:

$$\varepsilon = \varepsilon_\infty + \frac{\Delta\varepsilon\omega_0^2}{(\omega_0^2 - \omega^2) - i\omega\gamma} - \frac{\varepsilon_\infty \omega_p^2}{\omega^2 + i\omega/\tau}, \qquad (26.6)$$

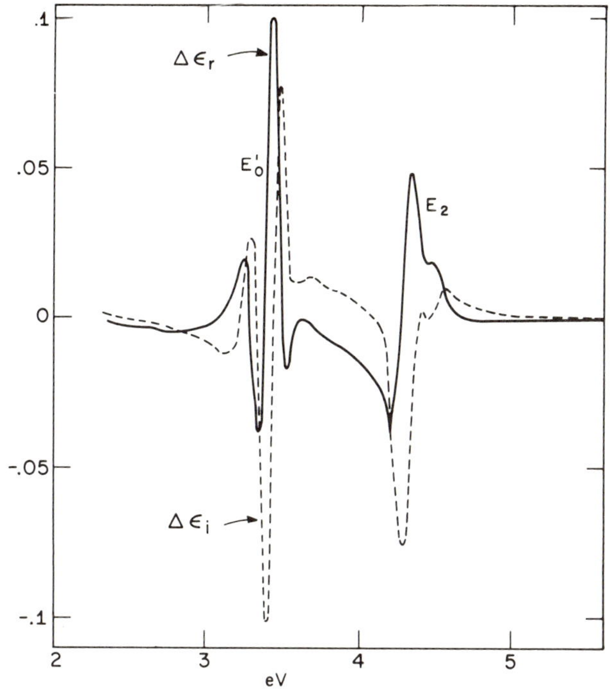

FIG. 124. $\Delta\varepsilon_r$ and $\Delta\varepsilon_i$ for a (110) n-type silicon surface with the electric field of the light along [001] ($E\|[001]$) and a *positive* dc bias applied to sample. From M. Cardona, K. L. Shaklee, and F. H. Pollak, *Phys. Rev.* **154**, 696 (1967).

where $\omega_0 = 3.39 \times 10^{-2}$ eV, $\varepsilon_\infty = 10.9$, $\Delta\varepsilon = 2.0$, $\omega_p = 6.08 \times 10^{-2}$ eV, $\omega_p \tau = 8.3$, and $\gamma = 6.8 \times 10^{-4}$ eV. It is interesting to note that the peak caused by the plasma edge occurs almost exactly at the plasma frequency (see Section 17).

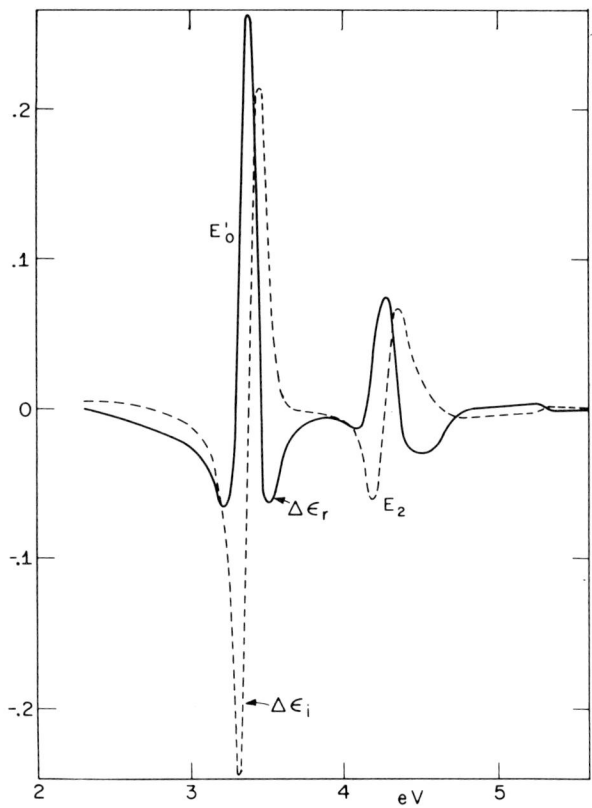

Fig. 125. $\Delta\varepsilon_r$ and $\Delta\varepsilon_i$ for a (110) n-type silicon surface with the electric field of the light along [001] ($E\|[001]$) and a *negative* dc bias applied to the sample.

26. EXPERIMENTAL RESULTS

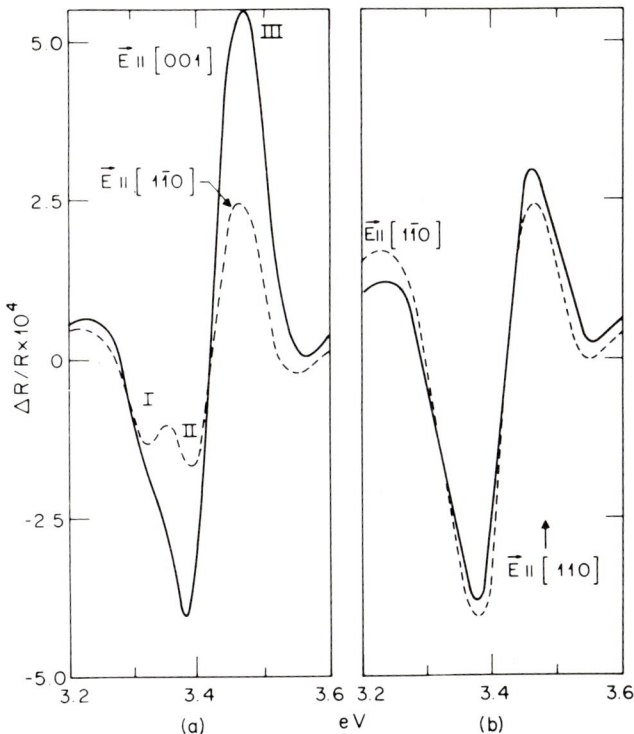

Fig. 126. Electroreflectance spectrum of (a) (110) and (b) (001) face of n-type silicon ($N = 10^{16}$ cm^{-3}) at room temperature showing polarization effects. From M. Cardona, K. L. Shaklee, and F. H. Pollak, *Proc. Phys. Soc. Japan Suppl.* **21**, 20 (1966). (Arrows over letter are represented by boldface in the text.)

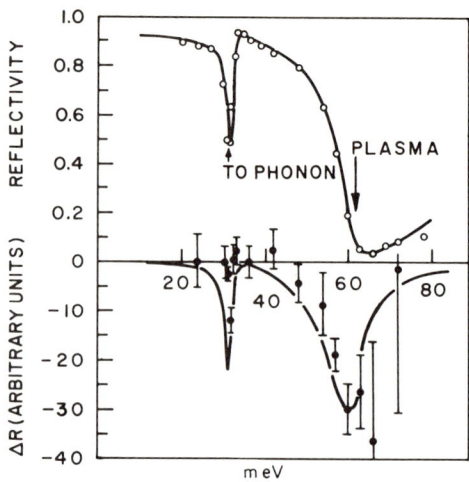

FIG. 127. Infrared electroreflectance and reflectance spectra of n-type GaAs ($N = 2 \times 10^{18}$ cm^{-3}). The solid curves were calculated from the known parameters of this semiconductor. From J. D. Axe and R. Hammer, *Phys. Rev.* **162**, 700 (1967).

e. *Electrolyte Method: Wurtzite-Type Materials*

The electroreflectance spectrum of a ZnO crystal surface which contains the hexagonal c axis is shown in Fig. 128 for the 2 possible normal modes of polarization. This spectrum illustrates very clearly the polarization selection rules expected for the E_0 edge of wurtzite-type materials: since the spin orbit splitting Δ_0 of ZnO is negligible at room temperature ($\Delta_0 \ll kT$) one would expect simply a shift between the spectra for $\mathbf{E} \parallel c$ and $\mathbf{E} \perp c$ owing to the hexagonal crystal field.[158] It is interesting to note that 2 additional oscillations are seen above the main negative peak for each direction of polarization ($L_1 - L_2$). These oscillations are not Franz–Keldysh oscillations since the period increases slightly instead of decreasing with increasing energy, as would be required by the theory of Section 24a. From the discussion in Section 26d, one would expect the peaks at the gap to be excitonic in origin. The higher energy oscillations L_1 and L_2 correspond to structure in the optical constants which has been attributed to phonon-exciton complexes.[303] Modulation techniques, and in particular electroreflectance, may be very useful for studying such complexes.

The electroreflectance spectra of CdS and CdSe near the lowest absorption edge at room temperature (E_0) have been measured by

Gutsche and Lange[300] by the electrolyte method. These authors concluded that the observed peaks are due to the quenching of excitons by the field.

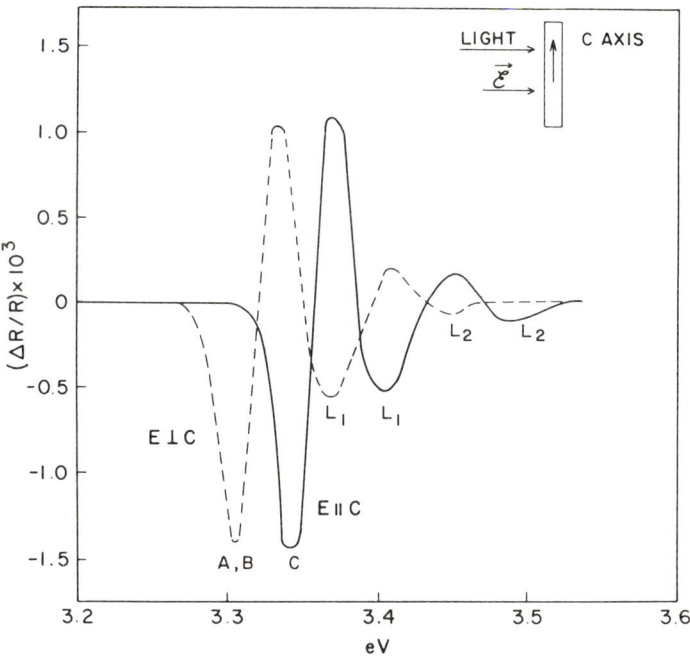

FIG. 128. Electroreflectance spectrum of ZnO. From M. Cardona, K. L. Shaklee, and F. H. Pollak, *Phys. Rev.* **154**, 696 (1967).

Figure 129 shows the electroreflectance spectrum[71] of hexagonal CdSe for several orientations of $\mathscr{E}$ and E. Two groups of peaks are seen in this figure: 1 of them around 2 eV, which correspond to the direct $\mathbf{k} = 0$ edge and its spin orbit splitting, and another between 4 and 5 eV, in the region where the E_1, $E_1 + \Delta_1$ peaks of cubic CdSe occur.[272] The spin orbit splitting of the direct edge is clearly resolved $[AB - C]$. The crystal field splitting $[A - B]$, which should produce two separate peaks for $E \perp c$, is not resolved, but it can be estimated from the shift between the peak for $E \parallel c$ (due *completely* to B) and the peak for $\mathbf{E} \perp c$ (due *mostly* to A). The value obtained by this method for the $A - B$ crystal field splitting is 0.029 ± 0.005.

The structure in Fig. 129 between 4 and 5 eV shows a striking selection rule: the E_1 and E_2 peaks appear only for $\mathbf{E} \perp c$, while the B peak appears for either direction of polarization. The similarity between the crystal structures of wurtzite and zincblende leads to the conclusion that the $A_1 - A_2$ peaks are the analog of the contribution to the $E_1 - (E_1 + \Delta_1)$ peaks of zincblende caused by transitions along the $\pm[111]$ set of $\{[111]\}$ valleys. The degeneracy of these valleys is

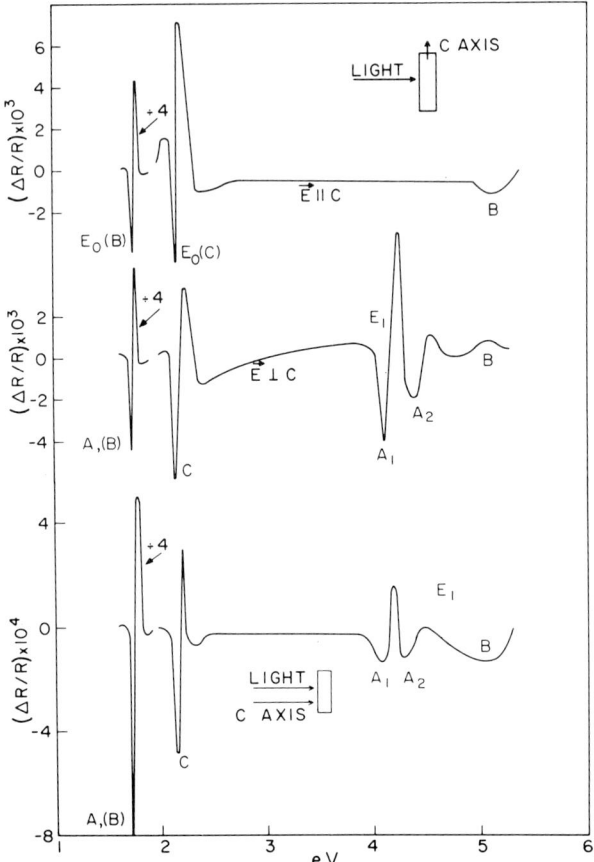

FIG. 129. Electroreflectance spectrum of n-type hexagonal CdSe at room temperature. From M. Cardona, K. L. Shaklee, and F. H. Pollak, *Phys. Rev.* **154**, 696 (1967).

broken in wurtzite into 1 set along the c axis and 3 other sets along lower-symmetry directions. These other 3 sets are presumably responsible for the B peak.

In order to illustrate this qualitative consideration, we show in Fig. 130 the Brillouin zone and the band structure of hexagonal CdSe as obtained by Bergstresser and Cohen.[304] The $E_1(A_1 - A_2)$ peaks have been assigned by these authors to $\Gamma_6 \to \Gamma_3$ transitions: these bands are actually nearly parallel along the c axis (Δ) and thus the main critical point need not be exactly at $\mathbf{k} = 0$. These transitions are indeed the analog of the {[111]} transition of zincblende. The B transitions have been assigned[304] to a U critical point along the L–M lines. These points have the postulated analogy with the $[\bar{1}11]$, $[1\bar{1}1]$, $[11\bar{1}]$ directions of zincblende.

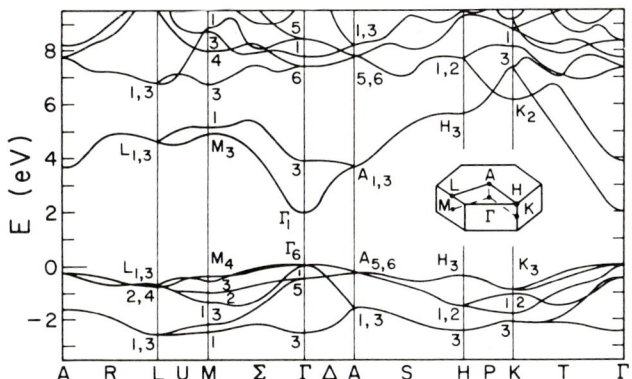

FIG. 130. Band structure of hexagonal CdSe according to T. K. Bergstresser and M. L. Cohen, *Phys. Rev.* **164**, 1069 (1967).

f. Electrolyte Method: IV–VI Compounds

The optical constants (Fig. 34) and the band structure (Fig. 35) of these materials (PbS, PbSe, PbTe, SnTe, and GeTe) have been briefly discussed in Section 9b. Figure 131 shows the electroreflectance and electroabsorption spectra of epitaxial films of PbS: the ease of preparation and high crystalline perfection of those films[27] makes their use very convenient for optical measurements. Electroreflectance and electroabsorption data are obtained for this material with either positive or negative sample biases. The spectra change considerably with bias; the sign of ΔR can be changed without changing the shape, by inverting the bias. Since the epitaxial layers are usually deposited on alkali halides, it is necessary to use for these measurements a nonaqueous electrolyte. Propylene carbonate proved to be the most appropriate electrolyte for these materials. Measurements with aqueous electrolytes are possible for bulk crystals.

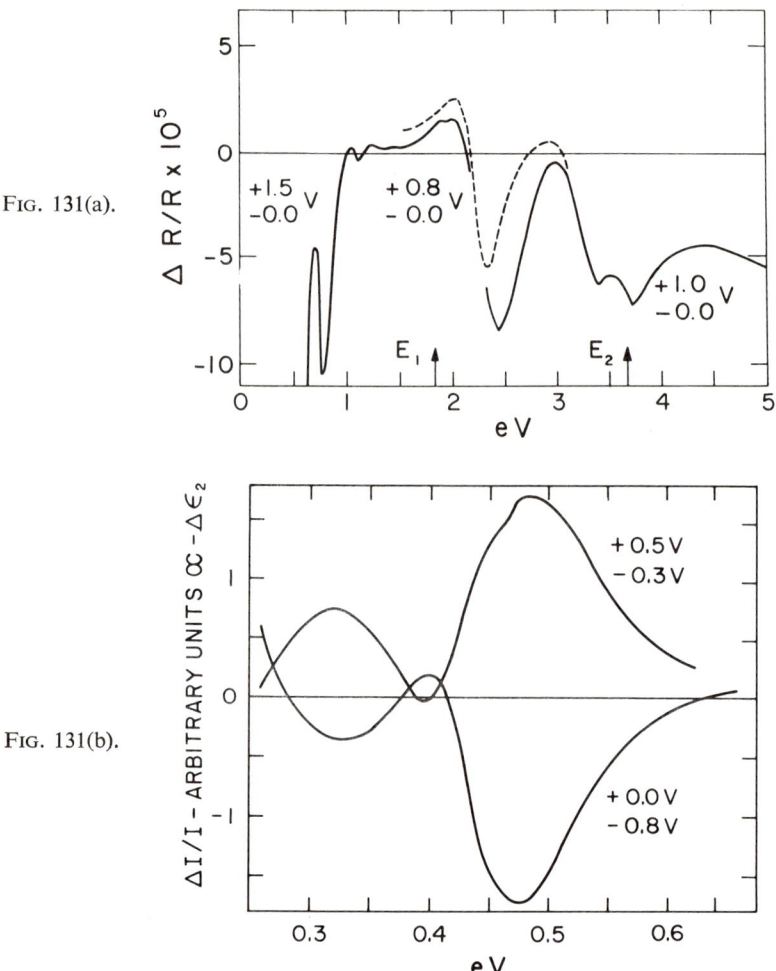

FIG. 131(a).

FIG. 131(b).

FIG. 131. Electroreflectance (a) and electroabsorption (b) spectra (near the lowest gap) of PbS obtained by the electrolyte method. [D. E. Aspnes and M. Cardona, *Phys. Rev.* **173**, 714 (1968).]

Figure 131b reproduces once more the exciton Franz–Keldysh controversy which has already been discussed a number of times: it is not easy to distinguish between a broadened $\Delta \varepsilon_i$ line in Fig. 80 and a quenched exciton, although the narrow central peak and broad wings of Fig. 131 are somewhat suggestive of exciton quenching. The position of the E_1 and E_2 peaks of Fig. 34 is indicated by arrows in Fig. 131; electroreflectance structure appears also at these energies.

The E_1 peak becomes a doublet in the electroreflectance spectrum while another peak, not seen in the reflectivity, appears below E_2. The variation of the E_1 splitting from PbS to PbSe suggests that it is produced by spin orbit interaction. An examination of Fig. 35 indicates that this is probably the splitting of the $L_3(1)$ valence band. The splitting of the E_2 peak into a doublet, clearly shown in Fig. 131a, can be just barely seen in conventional reflectivity measurements by Belle.[305] Table XVIII lists the energies of the electroreflectance and electro-

TABLE XVIII. ENERGIES (IN eV) OF THE PEAKS OBSERVED IN THE ELECTROREFLECTANCE SPECTRA OF PbS, PbSe, AND PbTe COMPARED WITH THOSE OF THE ASSIGNED CRITICAL POINTS OBTAINED FROM LIN AND KLEINMAN[a]

Transition	Peak	Type	PbS (eV)	PbSe (eV)	PbTe (eV)
$L_1^6(2) \to L_2^{6'}(2)$	E_0^*, E_1^{**}	M_0	0.25	0.11	1.30
$\Sigma_1^5(3) \to \Sigma_1^5(4)$		M_1	1.48		1.62
$L_1^6(2) \to L_3^{6'}(1)$	E_1^*, E_0^{**}	M_0	1.72	1.48	0.21
$L_3^{45}(1) \to L_2^{6'}(2)$		$M_0 + M_1$	1.92	1.85	2.06
$L_3^6(1) \to L_2^{6'}(2)$	E_2^{**}	$M_0 + M_1$	2.00	2.00	2.50
$L_1^6(2) \to L_3^{45'}(1)$		M_0	2.25	2.00	1.46
$\Sigma_1^5(3) \to \Sigma_4^5(2)$		M_1	2.52		1.15
$\Sigma_4^5(1) \to \Sigma_1^5(4)$		M_1	2.85		2.60
$\Delta^6(4) \to \Delta^6(5)$		M_2	3.02	2.81	1.80
$L_3^{45}(1) \to L_3^{6'}(1)$		$M_0 + M_1$	3.37	3.19	0.98
$L_3^6(1) \to L_3^{6'}(1)$	E_2^*	$M_0 + M_1$	3.45	3.34	1.41
$\Sigma_3^5(1) \to \Sigma_1^5(4)$		M_1	3.84		3.90
$\Sigma_4^5(1) \to \Sigma_4^5(2)$		M_1	3.89		1.95
$L_3^{45}(1) \to L_3^{45'}(1)$		$M_0 + M_1$	3.94	3.68	2.21

[a] P. J. Lin and L. Kleinman, Phys. Rev. **142**, 478 (1966).
* PbS, PbSe.
** PbTe.

absorption peaks observed for PbS, PbSe, and PbTe compared with the energies of the gaps to which the structure has been assigned.

Figure 132 shows the electroreflectance spectra of SnTe and GeTe obtained by Aspnes.[306] The main peak around 5 eV has been interpreted[306] as caused by the Burstein–Moss shift of transitions involving the highest valence band maximum.[307] The electric field modulates the depth of the space–charge layer which, in turn, modulates ε_i via a Burstein–Moss shift. The transitions involved are likely to be the $\Sigma_1 - \Sigma_4$ transitions of Fig. 133.

260   VII. ELECTRIC FIELD MODULATION

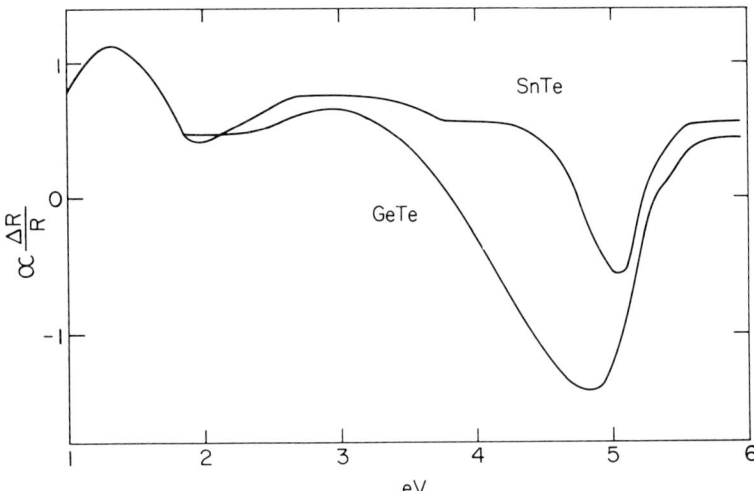

FIG. 132. Electroreflectance spectra of SnTe and GeTe (D. E. Aspnes, unpublished data).

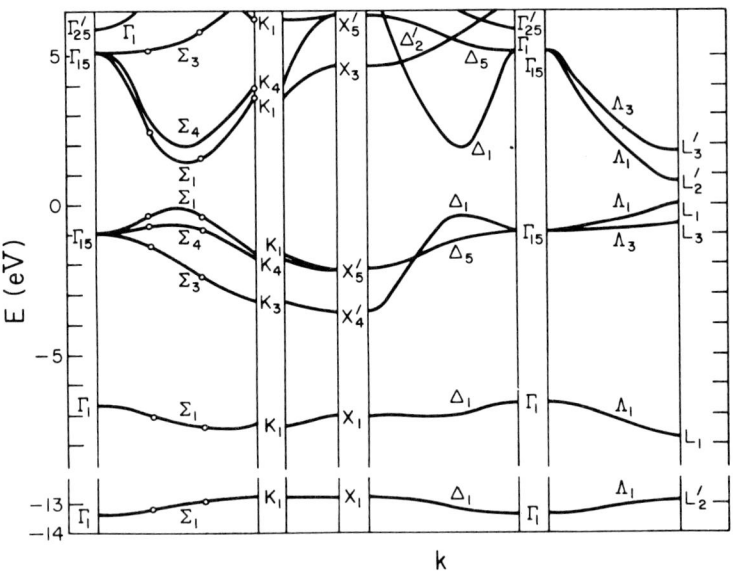

FIG. 133. Band structure of SnTe. From P. J. Lin, W. Saslow, and M. L. Cohen, *Solid State Commun.* **5**, 973 (1967).

g. *Electrolyte Method:* Mg_2Si, Mg_2Ge, *and* Mg_2Sn

These are cubic materials with antifluorite structure. All of them react with water to a greater or lesser extent; propylene carbonate was found to be a convenient electrolyte also for these materials. Figure 134 shows the electroreflectance spectrum[308] of Mg_2Ge.

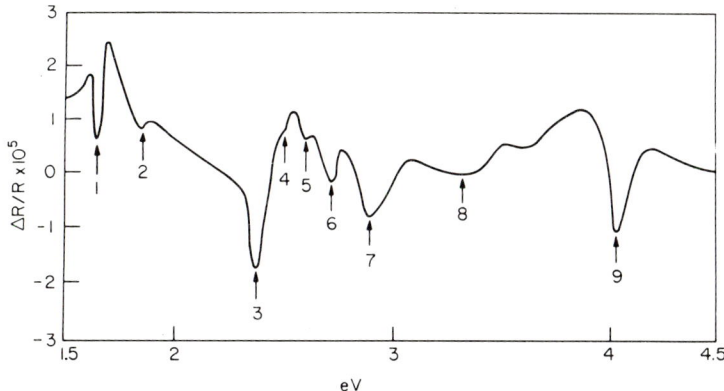

FIG. 134. Electroreflectance spectrum of *p*-type Mg_2Ge. From F. Vázquez, R. A. Forman, and M. Cardona *Phys. Rev.* **176**, 905 (1968).

The observed peaks have been labeled 1–9 in order of increasing energies. Similar peaks and shapes are observed for the other 2 materials of the family.[308]

The Bravais lattice of these materials is, as that of germanium and zincblende, face centered cubic. Moreover, the unit cells of the 2 families are isoelectronic. Hence, since the free electron bands of Mg_2X and germanium are the same, one would expect similarities between the band structure of both families of materials. However, quantitatively reliable band structures of Mg_2X are not yet available: considerable discrepancies exist between the 2 available sets of calculations.[309,310] It is hoped that the structure derived from the electroreflectance measurements, once properly identified, will be used to determine accurate band structures with any of the usual parametrization methods.

A comparison of the spectrum of Fig. 134 with that of germanium–zincblende suggests some simple analogies. The 1–2 structure seems to correspond to the $E_0 - (E_0 + \Delta_0)$ structure of germanium (transitions at $\mathbf{k} = 0$). The 1–2 splitting (0.2 eV) is reasonable in comparison with that of Ge (0.29 eV): a small admixture of Mg wave functions to the valence band wave functions could easily lower the spin orbit splitting of germanium. A spin orbit splitting of 0.2 eV has been found by Lott and Lynch[311] by the intravalence band transitions method (in p-type material).

Without any strong justification, one is tempted to assign the 3–4 peaks to L or Λ transitions [$E_1 - (E_1 + \Delta_1)$ peaks of germanium] and the peak 9 either to X transitions (E_2 of germanium) or to $L_{3'} \to L_3$ transitions (E_1'). We shall not, at present, attempt any identification of the remaining peaks. The energies of all peaks observed by electroreflectance in the 3 materials of this family which have been measured (no measurements have been reported for Mg$_2$Pb) are listed in Table XIX together with the energies of corresponding peaks observed in

TABLE XIX. ENERGIES OBSERVED IN THE ELECTROREFLECTANCE SPECTRA[a] OF Mg$_2$Si, Mg$_2$Ge, AND Mg$_2$Sn, COMPARED WITH THE ENERGIES OF CORRESPONDING STRUCTURE IN THE R AND ε_i SPECTRA[b]

Peak	Mg$_2$Si			Mg$_2$Ge			Mg$_2$Sn		
	R	ε_2	E.R.	R	ε_2	E.R.	R	ε_2	E.R.
1	2.17	2.18	2.27	1.67	1.7	1.64			
2						1.84			
3	2.75	2.70	2.51	2.57	2.57	2.37	2.12	2.12	1.96
—			2.61						2.11
4						2.50	2.30	2.26	2.24
5						2.60			
6	2.85		2.78			2.71	2.50	2.45	2.48
7			3.05	3.00	2.90	2.88	2.73	2.63	2.70
8	3.33		3.28	3.47		3.31	2.96		2.96
9	4.04	3.80	3.80	4.19	4.07	4.03	3.61	3.50	3.60
10							4.06		3.89

[a] F. Vázquez, R. A. Forman, and M. Cardona *Phys. Rev.* **176**, 3 (1968).
[b] W. Scouler (to be published).

the reflectivity spectra by Scouler[312] and the energies of the peaks in ε_i obtained by the Kramers–Kronig analysis[312] of the reflectivity.

h. Ferro- and Paraelectric Materials

As discussed in Section 24d, the electrooptic effect of these materials seems related almost exclusively to field-induced ionic displacements.

26. EXPERIMENTAL RESULTS

The main reasons for this conclusion are[249]:

(i) The apparent energy gaps usually increase with increasing fields.
(ii) The effects are several orders of magnitude larger than those expected from Franz–Keldysh theories.
(iii) The gap shifts are a function only of the total polarization **P** (spontaneous plus induced) regardless of temperature and field needed to produce this polarization.
(iv) The gap shifts are proportional to **P**² for small values of *P*.

Hence, in the paraelectric phase and at a given temperature, they are proportional to E^2 and not to the fractional powers of E required by the Franz–Keldysh theory.

The electroabsorption of $KTaO_3$ (paraelectric) below the lowest edge has been studied by Frova and Boddy[247]; that of $BaTiO_3$ above and below the Curie temperature ($T_c \simeq 120°C$) has been reported by Gähwiller.[249] The absorption edge of these materials follows the Urbach rule [Eq. (26.1)] below ω_0. From Eq. (26.1) we obtain:

$$\int_{\alpha(\mathscr{E})}^{\alpha(0)} d\alpha/\alpha = \lambda[\omega_0(\mathscr{E}) - \omega_0(0)] = \lambda \, \Delta\omega_0(\mathscr{E}). \tag{26.7}$$

Hence, the gap shift $\Delta\omega_0$ can be obtained from the change in absorption coefficient measured in a transmission modulation experiment. Figure 135 shows the shift $\Delta\omega_0$ produced by a field $\mathscr{E} = 2 \times 10^3$ V cm^{-1} applied to $BaTiO_3$ with the transparent electrode configuration of Fig. 93a. For the small fields used in these measurements, $\Delta\omega_0$ is proportional to $\mathscr{E}^2$, e.g., to P^2 above T_c and to $\mathscr{E}$ below T_c, due to the existence of a spontaneous polarization P_{sp} beside the induced polarization P_i:

$$\Delta\omega_0 \propto (P_{sp} + P_i)^2 \propto \begin{cases} \mathscr{E}, & \text{for } T < T_c \\ \mathscr{E}^2, & \text{for } T > T_c. \end{cases}$$

Gähwiller[249] has suggested that for any field in the experimental range the gap shift $\Delta\omega_0$ is well represented by the equation:

$$\Delta\omega_0 = aP^2 + bP^4, \tag{26.8}$$

with $a = 0.24$ eV $\times$ C^{-2} $\times$ m^4 and $b \simeq 3.4$ eV $\times$ C^{-4} $\times$ m^8 for $BaTiO_3$. Equation (26.8) also accounts for the increase in ω_0 for $\mathscr{E} = 0$ at temperatures below T_c due to the spontaneous polarization.

The materials of the SrTiO$_3$ family under discussion can be measured by the electrolyte technique: they can be made highly conducting, either by doping or by reduction (heating in a vacuum). No surface states seem to be present in these materials and hence the dc bias changes the quiescent surface potential. Figure 136 shows the electroreflectance

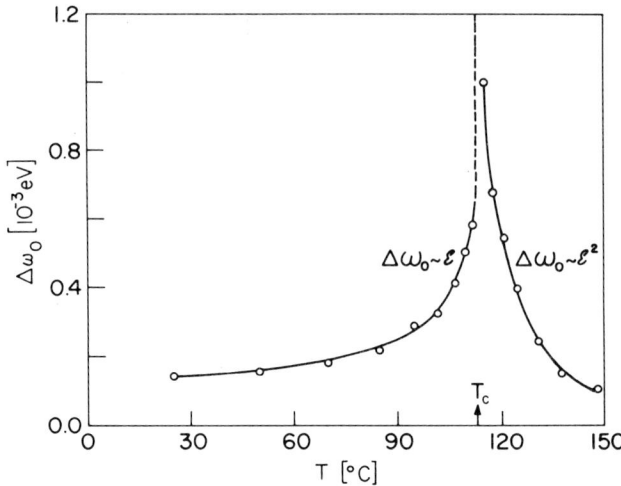

FIG. 135. Shift of the absorption induced by a field of 2×10^3 V cm^{-1} in BaTiO$_3$ versus temperature ($T_c = 113°$C), for the $E \perp c$ polarization. From C. Gahwiller, *Solid State Commun.* **5**, 65 (1967); *Phys. Kond. Mat.* **6**, 269 (1967).

spectrum of KTaO$_3$ for 2 dc biases and a small modulating voltage.[247] It is interesting to note that with these materials the blocking direction corresponds to positive bias on the sample. The quiescent surface field $\mathscr{E}_s$ was determined by means of capacitance measurements [see Eq. (25.7)]. In order to check the dependence of the position of critical points on polarization, the surface polarization P_s was determined with the equation:

$$P_s = \tfrac{1}{4}\pi^{-1} \int_0^{\mathscr{E}_s} (\varepsilon - 1)\, d\mathscr{E}_s, \qquad (26.9)$$

where the dielectric constant of the surface layer for the field $\mathscr{E}_s$ is found from the surface capacitance C through

$$\varepsilon = \tfrac{1}{2}\pi^{-1}[eN_s\, d(C^{-2})/dV]^{-1} \qquad (26.10)$$

The surface doping N_s can be determined from Eq. (26.10) for small V for which ε has the known low-field value.

Frova and Boddy have indicated that the curves in Fig. 136 are similar in shape to the derivatives of the corresponding reflection spectrum: the peaks in the reflectivity correspond to inflection points

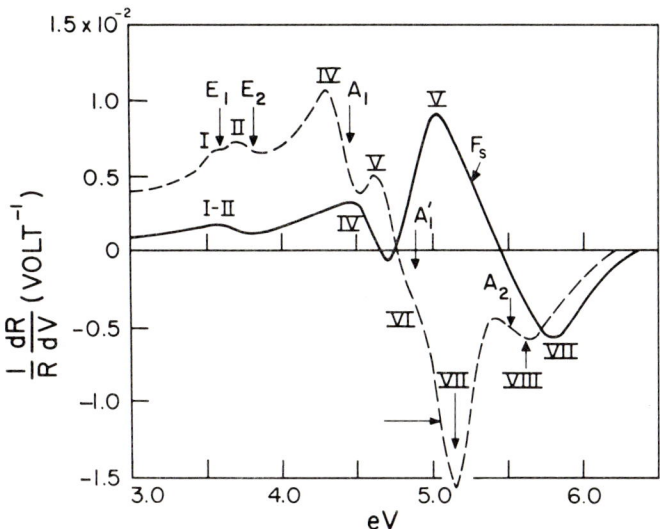

FIG. 136. Electroreflectance of a cleaved (100) surface of $KTaO_3$ with $N = 3.8 \times 10^{18}$ cm^{-3} (calcium doped). Sign convention: ΔR positive means increasing reflectance with *decreasing* surface field. $\mathscr{E}_s = 2 \times 10^6$ V cm^{-1} (solid line) and $\mathscr{E}_s = 3 \times 10^5$ V cm^{-1} (dashed line). From A. Frova and P. J. Boddy, *Phys. Rev.* **153**, 606 (1967).

in $\Delta R/R$. The intensity of peak V in this figure increases with increasing $\mathscr{E}_s$ at the expense of that of peak IV while the IV–V splitting increases. This increase in the IV–V splitting is linear in P_s; the linearity must, however, break down by symmetry for $P_s = 0$. No increase in the IV–V splitting with P_s occurs for either a (111) or a (110) surface. This fact suggests the assignment of the IV–V peaks to transitions along [100]. A look at the band structure of $SrTiO_3$ [Fig. (90)], which is expected to be similar to that of $KTaO_3$, suggests a large number of possibilities, among them: from $X_{5'}$ (upper or lower) to either X_3 or X_5. A closer identification is not possible at present.

The electroreflectance spectrum of rutile (TiO_2) has been studied by Frova et al.[248] for several surface orientations perpendicular and parallel to the tetrogonal (c) axis. The most salient feature of this work is the absence of the large critical point shifts with field found for the KTaO₃ materials. This fact lead Frova et al.[248] to postulate that the observed effects are actually Franz–Keldysh effects (see Section 24a) at critical points with large broadening. The fields required to explain the extremely large modulation amplitudes observed would, however, be much larger than those applied, especially for

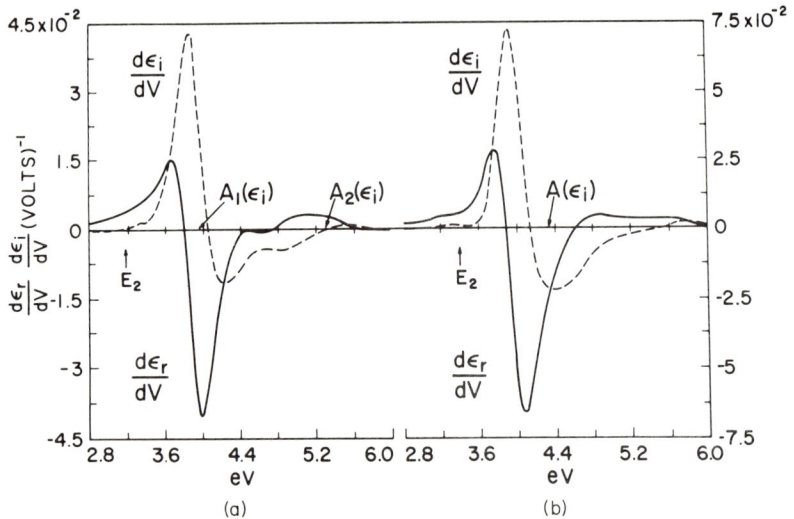

FIG. 137. $d\varepsilon_r/dV$ and $d\varepsilon_i/dV$ obtained for rutile with $\mathscr{E} = 10^6$ V cm⁻¹ by Kramers–Kronig analysis of electroreflectance data. The surface measured was (100) oriented (containing the c axis). Data given for (a) $E \perp c$ and for (b) $E \parallel c$. The arrows indicate the position of the corresponding structure in ε_i, from A. Frova, P. J. Boddy, and Y. S. Chen, *Phys. Rev.* **157**, 700 (1967).

the large broadening parameters necessary to explain the observed line shapes. Internal field corrections, which could be large in the case of materials with a large dielectric constant, could account for the high fields required; such corrections, however, lie outside the scope of the theory developed in Section 24a.

We show in Fig. 137 the spectral dependence of $d\varepsilon_r/dV$ and $d\varepsilon_i/dV$ (V = applied voltage) as obtained through Kramers–Kronig analysis of the electroreflectance data of rutile.[248] The measured surface was

(100) oriented and thus contained the tetragonal axis c. Anisotropy is expected depending on whether the light is polarized parallel or perpendicular to c. Such anisotropy has also been observed in the ordinary reflection spectrum; we show for comparison in Fig. 137 the positions of the peaks (A_1, A_2, A) in ε_i (derived from R by Kramers–Kronig analysis). As indicated earlier for $KTaO_3$, the positions of the peaks in the optical constants should be those of the inflection points in the corresponding modulation spectra.

Gähwiller[249] has reported electroreflectance measurements on insulating $BaTiO_3$ with an electrode configuration similar to that of Fig. 93a, but with the light incident on the electrode-free side of the sample.

i. Metals

The Debye screening length of metals is of the order of 1 Å and, therefore, the electric fields produced in the field effect configuration are expected to penetrate into a metal less than an interatomic distance: possible optical field modulation effects are not expected to be describable in terms of the band structure of the bulk metal.

In spite of the fact that the high-field region is very small, measureable modulation effects may appear because of the high fields present. Feinleib[313] reported electroreflectance spectra of silver and gold measured with the electrolyte technique. His results for silver are shown in Fig. 138. The sharp peak near 4 eV corresponds to the minimum in the reflectivity, which is caused by a combination of the plasma edge and the $L_{32} \to L_{2'}$ interband edge (see Fig. 40): Feinleib interpreted these electroreflectance spectra as related to a modulation in the real refractive index of the electrolyte. This modulation would presumably be produced in the Gouy layer by the ionic displacements caused by the electric field. A model calculation of this effect by Prostak and Hansen[314] indicates that it is not possible to interpret the peak of Fig. 139 as due to a thin layer of electrolyte whose dielectric constant varies with applied field. This is illustrated in Fig. 139 for gold. The peak seen by Feinleib in the electroreflectance spectrum cannot be fitted under the assumption of modulation of n_r in a layer of electrolyte (dashed curve). A peak of the right shape is obtained if the optical constants of a very thin metal layer at the interface are modulated. The model used for calculating the dotted-dashed curve of Fig. 139 assumes[314,315] that the metal modulation consists of a shift in energy of the spectral dependence of the real and imaginary parts of the dielectric constant of the metal by 0.1 eV. The thickness of the modulated surface layer in the metal is assumed to be 1.5 Å. While the experimental peak around 2.5 eV is fitted well by field modulation of the optical properties of the metal, the rise in $\Delta R/R$ above 4 eV indicates that there may, nevertheless, be a contribution of electrolyte modulation (dashed curve) to the observed spectrum.

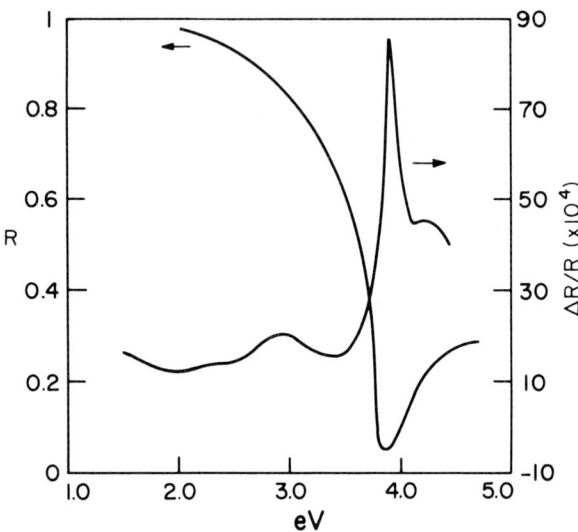

FIG. 138. Electroreflectance of silver as reported by J. Feinleib, *Phys. Rev. Letters* **16**, 1200 (1966). Also reflectivity spectrum of the same sample in air.

A somewhat different method of measuring electroreflectance in metals has been suggested by Stadler.[316] A condenser is made by evaporating metal electrodes on the surfaces of a thin ferroelectric slab. Application of a voltage to the condenser results in the orientation of the ferroelectric domains; the ferroelectric acquires the polarization **P**. As a result, a charge $\sigma = \mathscr{E}/4\pi + \mathbf{P}$ appears per unit surface in the metal electrodes; this charge can be as high as 10^8 V cm^{-1} and produces fields of the same order at the surface of the metal in contact with the ferroelectric. These fields are comparable to those obtained with the electrolyte technique for an applied voltage of 1 V. The optical properties of the surface of the metal thin film in contact with the ferroelectric can be studied by reflection on the free surface of the metal; since the metal electrode is very thin, the optical constants of the surface in contact with the ferroelectric affect the reflection at the free surface.

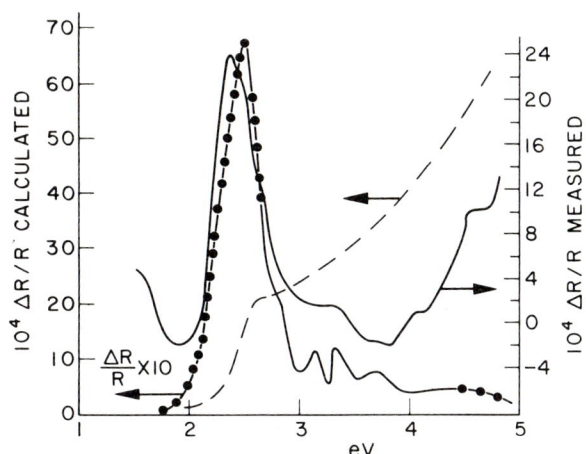

FIG. 139. Electroreflectance of gold as measured by J. Feinleib, *Phys. Rev. Letters* **16**, 1200 (1966) (solid curve) compared with 2 model calculations by A. Prostak and W. N. Hansen, *Phys. Rev.* **160**, 600 (1967). The dashed curve corresponds to modulation of a 100 Å layer of electrolyte, the dotted curve to modulation of a 1.5 Å layer of metal.

j. Electroreflectance in Insulators

Electroreflectance measurements with sample configurations of the type shown in Figs. 93c–d have been reported for GaAs[258,317] ZnS, ZnSe, and CdS.[256] Measurements have also been performed on semi-insulating (gold doped) silicon at 77°K.[317] We show, in Fig. 140, electroreflectance spectra of 2 GaAs surfaces: (100) (Forman *et al.*[317] and (111) (Rehn and Kyser[258]). The spectrum of the (100) surface does not show any polarization effects. That of the (111) surface shows strong polarization dependence of the E_1, $E_1 + \Delta_1$ structure. The E_0 peaks, not shown in Fig. 140, exhibit no significant polarization effects. The absence of polarization effects at E_1 and $E_1 + \Delta_1$ for (100) surfaces, as opposed to the (111) surface, is strong confirmation of the assignment of these peaks to [111] critical points. According to the Franz–Keldysh theory of direct allowed interband transitions described in Section 23a, no anisotropy is expected for [111] transitions with $\mathscr{E}$ along [001]; anisotropy should appear, however, if the field is along

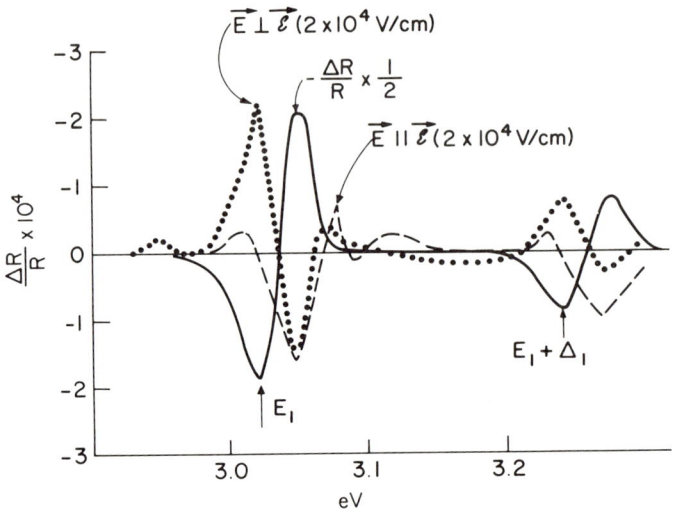

FIG. 140. Electroreflectance spectrum of the E_1 and $E_1 + \Delta_1$ transitions for a (100) surface of GaAs at 77°K (solid line) [R. A. Forman, D. E. Aspnes, and M. Cardona (to be published)] and a (111) surface (dashed line $\mathbf{E} \parallel [1\bar{1}0]$, dotted line $\mathbf{E} \perp [1\bar{1}0]$; $\mathscr{E}$ field along [1$\bar{1}$0]) [V. Rehn and D. S. Kyser, *Phys. Rev. Letters* **18**, 848 (1967).]

either [1$\bar{1}$0] or [111]. We have tried to fit the data of Rehn and Kyser to the appropriate sums of electrooptic functions F and G without success. A Kramers–Kronig analysis of these data, not available yet, would be very useful for this purpose.

The E_0 spectrum of the [100] face[258,317] of GaAs shows at least 4 oscillations. An attempt to fit them with electrooptic functions (G in this case) encounters the same problems as those discussed in Section 26b and d. Hence it seems that the inclusion of exciton effects is necessary in order to interpret quantitatively the E_0 line shape.

Figure 141 shows 2 examples of electroreflectance spectra of excitons in the neighborhood of the lowest absorption edge of II–VI compounds. The selection rules for the *A*, *B*, *C* peaks of wurtzite appear quite clearly in Fig. 141a; peak *A* is only allowed for $\mathbf{E} \perp c$.

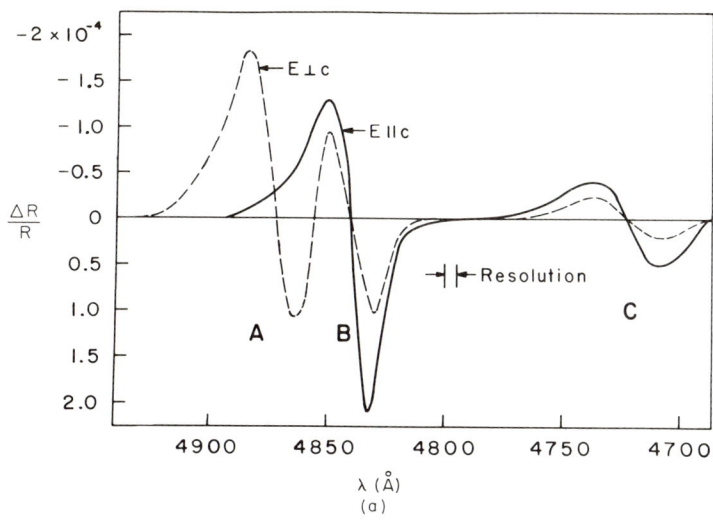

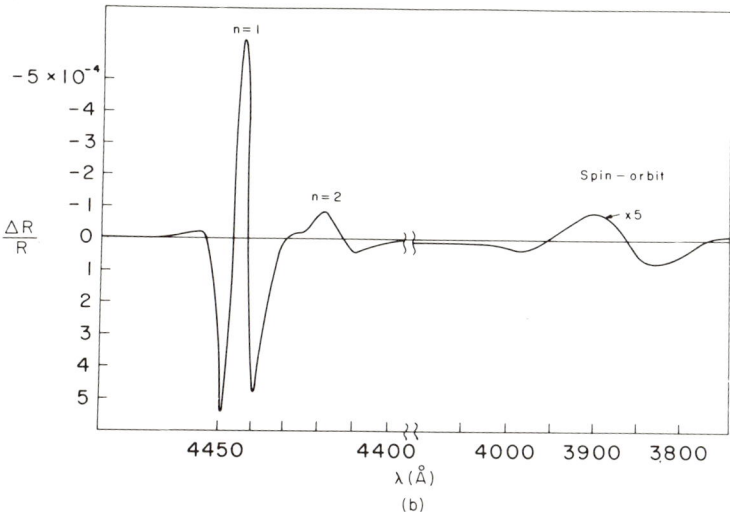

Fig. 141. (a) Electroreflectance spectrum of the exciton peaks in the neighborhood of the lowest edge of hexagonal CdS at 77°K, for an *ac* face with $\mathscr{E} \parallel a$ and $\mathscr{E} \sim 1.8 \times 10^4$ V cm^{-1}. (b) Electroreflectance spectrum of the exciton peaks in the neighborhood of the lowest edge of cubic ZnSe at 77°K for a cleaved face with $\mathscr{E} = 2 \times 10^4$ V cm^{-1}. From R. Forman and M. Cardona, *in* "II–IV Semiconducting Compounds" (D. G. Thomas, ed.), p. 100. Benjamin, New York, 1967.

The A–B peaks, not resolved at room temperature for $\mathbf{E} \perp c$, in Fig. 129 (for CdSe) are well resolved at liquid nitrogen temperature for CdS.

It is interesting to comment on the difference in shape between the peaks of Figs. 141a and b. The peaks of Fig. 141a have the full shape of ε_r due to an exciton [$F_1(W)$, Fig. 50] and hence one may think that this spectrum corresponds to the complete destruction of the excitons by the electric field. The applied fields ($\sim 2 \times 10^5$ V cm^{-1}) are indeed high enough to ionize completely these excitons, according to the discussion of Section 24c. The magnitude of the observed $\Delta R/R$ signal ($\sim 10^{-4}$), however, advises us against this conclusion: changes in R of the order of 30% are produced by these excitons. We must, therefore, conclude that the field seen by the sample surface is considerably smaller than the nominal 1.8×10^4 V cm^{-1}. This could be due to surface conductance or photoconductivity, but the exact mechanism is not known. Therefore, the spectrum in Fig. 141a must be a differential spectrum. Equations (17.3) and Fig. 50 suggest that we are dealing with a differential broadening spectrum of these excitons (see Section 24c).

Figure 141b, however, exhibits line shapes qualitatively similar to F_1' in Fig. 50. Therefore, we conclude that these spectra are related mainly to a field shift of the exciton lines. It is interesting to note, however, that the wings on both sides of the $n = 1$ peak are considerably stronger than those of Fig. 50. Stronger wings would appear in the derivative spectrum of ε_r if the exciton line shape were Gaussian instead of Lorentzian.[318]

k. Other Optical Modulation Measurements

The possibility of reflection modulation by an electric field in a semiconductor under total reflection conditions has been suggested by Harrick.[319] As is well known, the reflection coefficient for total reflection is smaller than 1 if some absorption is present ($n_i \neq 0$). The reduction in reflection coefficient caused by absorption is largest at incidences near the critical angle. One can cumulate the effect by having the light undergo a large number of total reflections at incidence near the critical angle with the scheme shown in Fig. 142.

The amount of absorption (n_i) at the reflecting surface can then be modulated in the field effect configuration, either with a dry capacitor (as shown in Fig. 142) or with an electrolyte. A considerable modulation in the light transmitted through the package of Fig. 142 can be produced by an ac electric field. An *increase* in the absorption index n_i produces a *decrease* in the transmitted intensity. This modulation method is only usable in the region of transparency of the sample

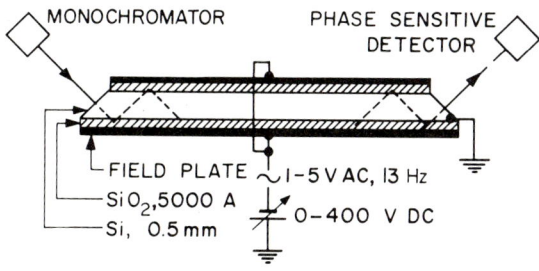

Fig. 142. Schematic diagram of the system used for electromodulation at total reflection. From N. J. Harrick, *J. Phys. Chem. Solids* **8**, 106 (1958); *Phys. Rev.* **125**, 116 (1962).

(n_i small) i.e., below the lowest absorption edge of the semiconductor. It should, therefore, be useful for the study of intraband transitions (free carrier effects) in semiconductors, impurity absorption (impurity level to impurity level and impurity level to energy band), and surface states.[319,320] The various mechanisms which can contribute to total reflection modulation are:

(1) transitions from valence band to empty surface states,
(2) transitions from filled surface states to conduction band,
(3) free electron absorption at the surface layer,
(4) free hole absorption at the surface layer,
(5) transitions between various valence bands (*p*-type surfaces),
(6) transitions between various conduction band (*n*-type surfaces).

All effects mentioned above can produce electrooptical modulation either by a field modulation of the carrier populations or by a field mixing of the wave functions similar to the Franz–Keldysh effect. Since the carrier populations at the surface change strongly with applied field, it is believed that this effect is dominant in the observed spectra.

Figure 143 shows the spectra obtained by Samoggia *et al.*[320] for germanium (20 Ω-cm, *n*-type) by the total reflection electromodulation technique. The various curves correspond to several values of the surface potential ($u_s = e\phi_s/kT$ is the reduced surface potential) obtained by varying the ambient gases (O_3—O_2, O_2, and O_2—H_2O mixtures). For large negative values of u_s the surface ($u_s = -2.75$) becomes p-type and intravalence band transitions (5) are observed in the modulation spectrum. For large positive values of u_s ($u_s = +5$),

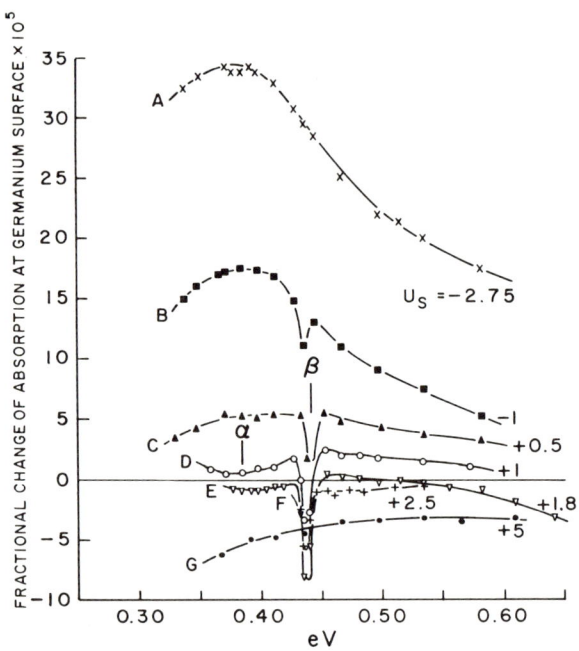

FIG. 143. Field effect modulated optical absorption through the package of Fig. 142 (total reflection measurements). The various curves refer to different surface potentials obtained by changing the atmosphere surrounding the sample. From G. Samoggia, A. Nuccioti, and G. Chiarotti, *Phys. Rev.* **144**, 749 (1966).

the surface is n-type and the modulation spectrum reflects free carrier ($\sim \omega^{-2}$) absorption (3). Between these 2 extreme cases, 2 peaks (α and β) appear at 3.8 eV (broad) and at 4.4 eV (sharp). These peaks were assigned by Samoggia et al.[320] to transitions from occupied surface states to the conduction band.

An interesting method of obtaining electroreflectance spectra of semiconductors has been reported by Wang et al.[273] The field in the space–charge layer is modulated by shining chopped white light from an intense source near grazing incidence. The monochromatic light from the spectrometer is reflected at normal incidence and the signal in the reflected light beam synchronous with the white light chopper is detected in the usual manner. This method removes the temperature and wavelength limitations of more conventional electroreflectance techniques. The sensitivity achieved in practice, which is a function of the sample photoconductivity, seems to be smaller than that obtained with either the electrolyte method or the Seraphin technique. Measurements by this method have been reported[273] for Ge, CdS, and ZnTe. This method is known as *photoreflectance*.

Another technique which eliminates wavelength limitations of more conventional methods is the sample rotation technique referred to as rotoreflectance.[274] This method is usable for cubic materials with surface states and reflecting surfaces other than (111) and (100): the surface field at the space–charge layer, always present if the material has surface states (except for accidental flat bands) produces an optical anisotropy of the surface. This anisotropy can be measured under normal incidence reflection with linearly polarized light, by rotating the sample surface about its normal. A component at twice the rotation frequency, caused by the difference in reflectivity between the 2 normal modes of polarization, appears in the reflected beam. Measurements by this technique have been reported only for (110) surfaces of silicon[274]· structure connected with the E_0' and E_2 peaks of this material was observed.

VIII. Modulation Techniques and Dependence of Band Structure on Static Parameters

27. Introduction: Effects of Temperature and Doping

Several of the modulation techniques described in previous chapters can be used to find the derivative of energy gaps with respect to the modulation parameter. For example, we have discussed in Chapter VI the use of stress modulation data for determining the ratio of the shear to the hydrostatic deformation potentials. However, because of the difficulties involved in the accurate determination of the amplitude of the modulation, it is often more convenient to use a modulation technique to sharpen the features of the optical spectra and then measure the dependence of these modulation spectra on the desired static external parameter. This method has also the advantage of permitting the study of large external parameters which produce nonlinear effects on the band structure. As an example, we mention the case of a strong magnetic field which completely alters the nature of the optical spectra by introducing Landau quantization. Modulation techniques can be used in this case to enhance the magneto-optical structure produced by Landau quantization. Measurements in a strong magnetic field with stress modulation have been performed by reflection[69] (magnetopiezoreflectance) and by transmission[194] (magnetopiezoabsorption). Electroreflectance measurements in a strong magnetic field have also been reported.[271]

Several modulation techniques have been used in conjunction with static uniaxial stress. Wavelength modulation transmission[67] experiments under static uniaxial compression and stress modulation under static uniaxial compression (piezopiezoreflection) have been reported by Balslev.[67,189,321] Electroreflectance measurements under static uniaxial stress have been reported by Pollak et al.[322,323]

The sharp optical structure obtained by modulation methods is usually not washed out by alloying in binary or pseudobinary solid solutions. Hence, many of the modulation methods described in previous chapters can be used to study the dependence of energy gaps on composition in binary or pseudobinary solid solutions. Electroreflectance data have been reported for the GaAs–GaP,[190] the GaAs–InAs,[324] and Ge–Si[325] systems. The determination of temperature coefficients of gaps by nonthermal modulation techniques, e.g., using the temperature as a static parameter, has been mentioned several times in the preceeding chapters, e.g., the temperature coefficients of the E_0' peaks of silicon in electroreflectance, (see Section 26d). A study of the effect of impurities on the electroreflectance peaks of GaAs by the electrolyte method has also been reported[71]; strong Burstein–Moss shifts are observed for the E_0 and $E_0 + \Delta_0$ peaks of heavily doped n-type GaAs and for the E_0 peak of p-type GaAs. These shifts, shown in Fig. 144, are caused by the filling of the botton/top of the conduction/valence band in n/p-type material. The

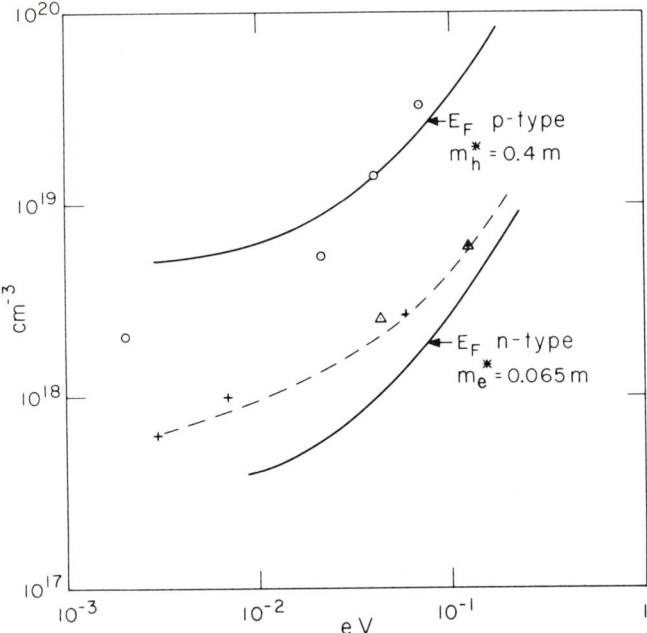

FIG. 144. Burstein–Moss shifts of the E_0 (p-type (○), n-type (+)) and $E_0 + \Delta_0$ (n-type (△)) edges of GaAs at 297°K observed by electroreflectance techniques. The solid curves are calculated including nonparabolicity corrections for n-type but not for p-type materials (see text). From M. Cardona, K. L. Shaklee, and F. H. Pollak, *Phys. Rev.* **154**, 696 (1967).

calculated solid curves shown in Fig. 144 include nonparabolicity effects[206] in *n*-type material. For *p*-type material, the calculated curve represents the shift of the Fermi level in the valence band: no allowance is made for the corresponding shift in the conduction band which one would have for direct transitions. The agreement between this calculated shift and the observed one suggests that in *p*-type material of high-doping levels, the E_0 peak is mainly caused by indirect transitions from the Fermi level of the valence band to the bottom of the conduction band.

We show in Fig. 145 the dependence on doping of the various

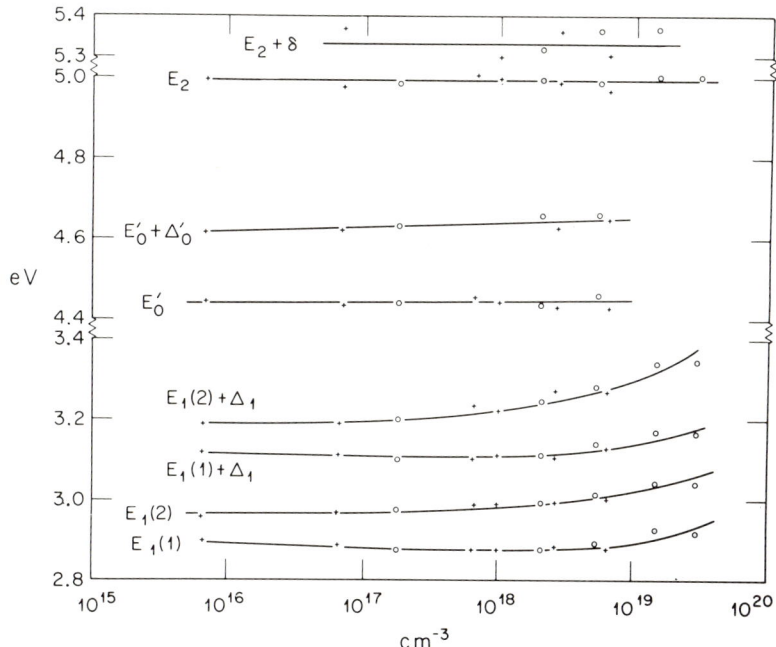

FIG. 145. Dependence on doping of the energy of the peaks observed in the electroreflectance spectrum of GaAs at 297°K (*p*-type (○) and *n*-type (+)). From M. Cardona, K. L. Shaklee, and F. H. Pollak, *Phys. Rev.* **154**, 696 (1967).

peaks observed above E_0 and $E_0 + \Delta_0$ in the electroreflectance spectrum[71] of GaAs. Except for the $E_1 - (E_1 + \Delta_1)$ peaks [the energies of the 2 oscillations (1) and (2) of each one of the $E_1 - (E_1 + \Delta_1)$ peaks is given], little dependence on doping is observed. The high-energy (2) oscillations of the $E_1 - (E_1 + \Delta_1)$, peaks shift more with doping than their low energy (1) mates; this can be interpreted[71] as loose evidence of the fact that the corresponding gaps occur closer to to the (1) than to the (2) components. The E_1, $E_1 + \Delta_1$, E_0' and E_2 electroreflectance peaks have been observed[71] for p-type germanium with hole concentrations as high as 2.4×10^{20} cm^{-3}.

28. Electroreflectance in Binary and Pseudobinary Alloys

Figure 146 shows the electroreflectance spectra of 2 germanium–silicon alloys at room temperature.[325] The spectrum of the germanium rich sample (10.9 at. % Si) is nearly as sharp as that of germanium and shows all the peaks observed for the pure material. As indicated in Fig. 147, the $E_0 - (E_0 + \Delta_0)$ peaks can be tracked to about 46 at. % Si. At higher silicon concentrations these peaks become enmeshed in the superposition of the E_1, $E_1 + \Delta_1$, and E_0' peaks. It is possible to extrapolate linearly the energy of the E_1 peak, and, thus, to obtain the experimentally elusive $\Gamma_{25'} - \Gamma_{2'}$ gap of silicon. We find from a least square fit of a straight line to E_0 below

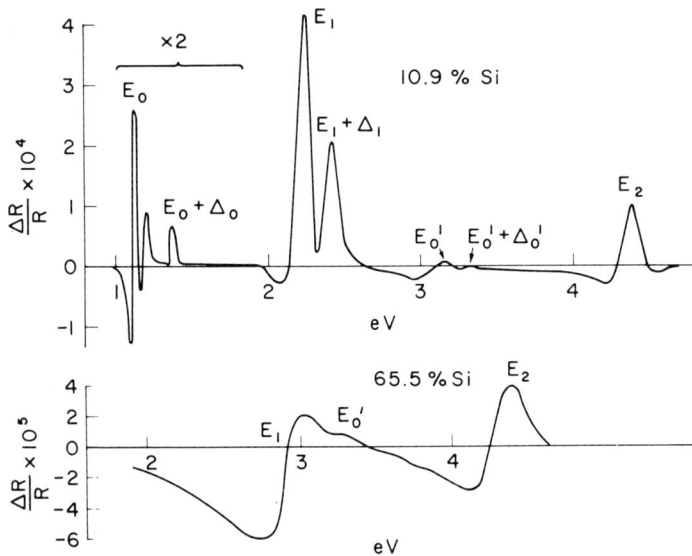

Fig. 146. Electroreflectance spectra of 2 Ge–Si alloy samples. From J. S. Kline, F. H. Pollak, and M. Cardona, *Helv. Phys. Acta* **172**, 816 (1968).

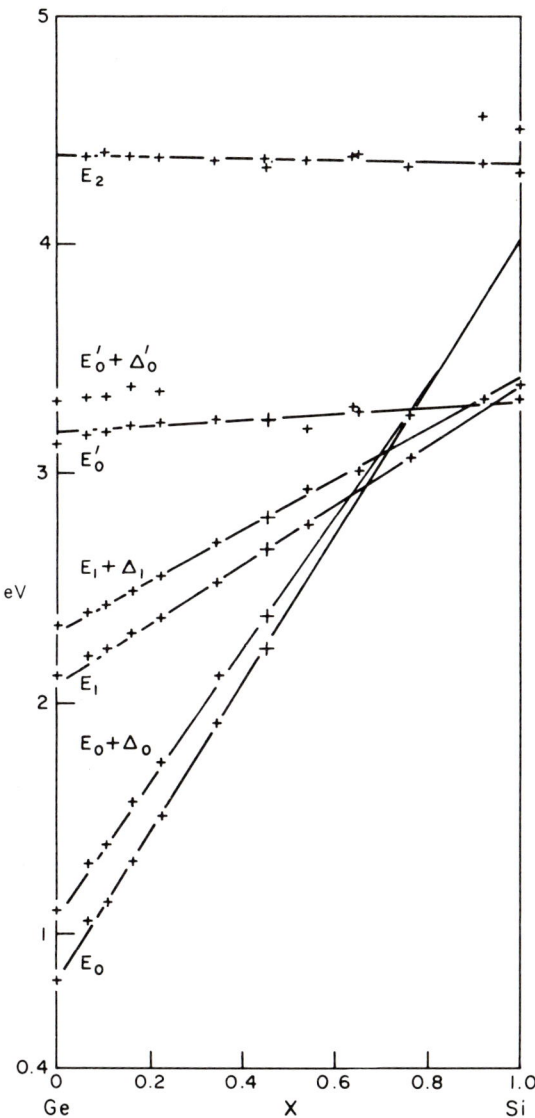

FIG. 147. Variation with composition of the room temperature energies of the electroreflectance peaks for the Ge–Si alloy system. From J. S. Kline, F. H. Pollak, and M. Cardona, *Helv. Phys. Acta* **172**, 816 (1968).

46 at. % Si, $E_0(\text{Si}) = 3.99 \pm 0.02$ eV. This value would be slightly larger if the variation of E_0 with concentration had a small quadratic term (see below).

The $E_1 - (E_1 + \Delta_1)$ peaks appear clearly split in Fig. 147 up to 55 at. % silicon concentrations. Near 100 at. % Si the E_1, $E_1 + \Delta_1$, E_0'

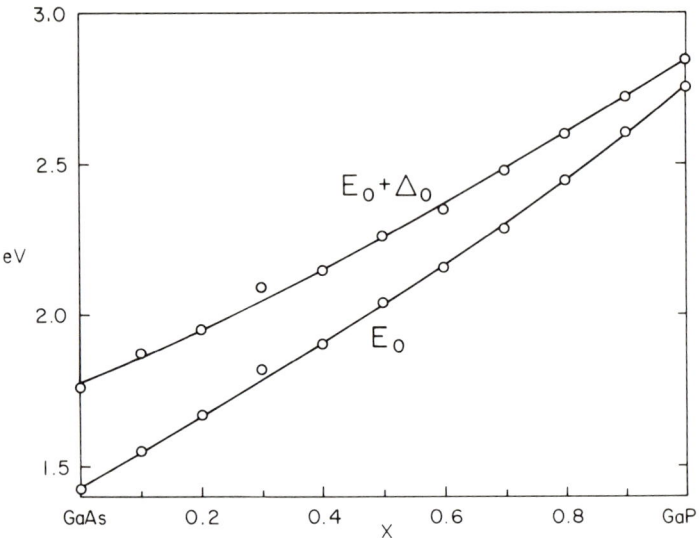

FIG. 148. Variation of the lowest direct energy gap E_0 and its spin orbit split component $E_0 + \Delta_0$ with composition for the $\text{GaAs}_{1-x}\text{P}_x$ system. The points are experimental. The solid curves represent the least square fit expressions $E_0 = 0.210x^2 + 1.091x + 1.441$ eV and $E_0 + \Delta_0 = 0.182x^2 + 0.884x + 1.776$ eV. From A. G. Thompson, M. Cardona, K. L. Shaklee, and J. C. Woolley, *Phys. Rev.* **146**, 601 (1966).

and $E_0' + \Delta_0'$ peaks are all superimposed: the bands must be parallel over large regions of **k**-space, as already suggested by band calculations.[162,166] The Δ_0' splitting can be followed up to 25% silicon concentrations and the splitting of E_2 seen in pure silicon can also be seen for 92% silicon.

The electroreflectance spectra of 2 pseudobinary alloy systems of III–V compounds (GaAs–GaP[190] and GaAs–InAs[324]) have also been reported. The E_0 and $E_0 + \Delta_0$ peaks of the GaAs–GaP systems (those of the GaAs–InAs systems have not been reported) vary with composition x in a slightly nonlinear manner (see Fig. 148). This variation, which can be described by the approximate quadratic forms:

$$E_0 = 0.210x^2 + 1.091x + 1.441 \quad \text{eV}$$
$$E_0 + \Delta_0 = 0.182x^2 + 0.884x + 1.776 \quad \text{eV,} \quad (28.1)$$

has been attributed[137] to the fluctuation potential which must be added to the average periodic virtual crystal potential. Figure 149 shows the variation of Δ_1 with concentration for the GaAs–InAs alloy

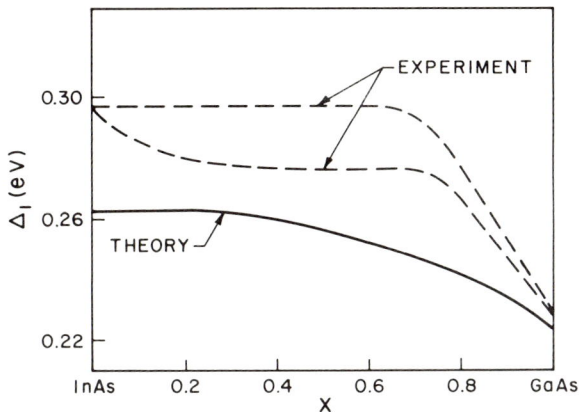

FIG. 149. Variation of Δ_1 with composition for the Ga_x–In_{1-x}As alloy system. The dashed curves are obtained from experiment and correspond to 2 different interpretations of the experimental line shapes [see A. G. Thompson and I. C. Woolley, Canad. J. Phys. **45**, 2597 (1967).] The solid curve has been calculated by the $\mathbf{k} \cdot \mathbf{p}$ method [J. Barber (unpublished)].

system as obtained from electroreflectance measurements. The experimental (dashed) curves correspond to 2 different interpretations of the line shapes. In either case Δ_1 shows a surprisingly nonlinear variation with concentration. Similar results have also been reported[190] for the Δ_1 splitting in the GaAs–GaP system. These nonlinearities have been tentatively[190] attributed to a sliding in **k**-space of the Λ critical points which are responsible for the $E_1 - (E_1 + \Delta_1)$ transitions: $\mathbf{k} \cdot \mathbf{p}$ calculations[326] with parameters obtained by admixture of the parameters of GaAs and InAs proportional to concentration yield the solid curve of Fig. 149. A nonlinear variation of Δ_1, which corresponds to the sliding of critical points discussed above, is observed in this theoretical curve.

284   VIII. MODULATION TECHNIQUES

29. Uniaxial Stress Measurements

a. Stressing Techniques

All measurements which have been reported using optical modulation techniques with the sample under static uniaxial stress have been performed with a stressing method similar to that of Cuevas and Fritzsche.[327] A detail of a typical stressing apparatus is shown in Fig. 150. The stress is applied with a spring and a lever and measured by measuring the spring elongation with a linear variable differential transformer (LVDT). This method is preferable to simply hanging weights on the lever arm: the stress can be applied smoothly with the screw–spring arrangement and the LVDT method gives continuous

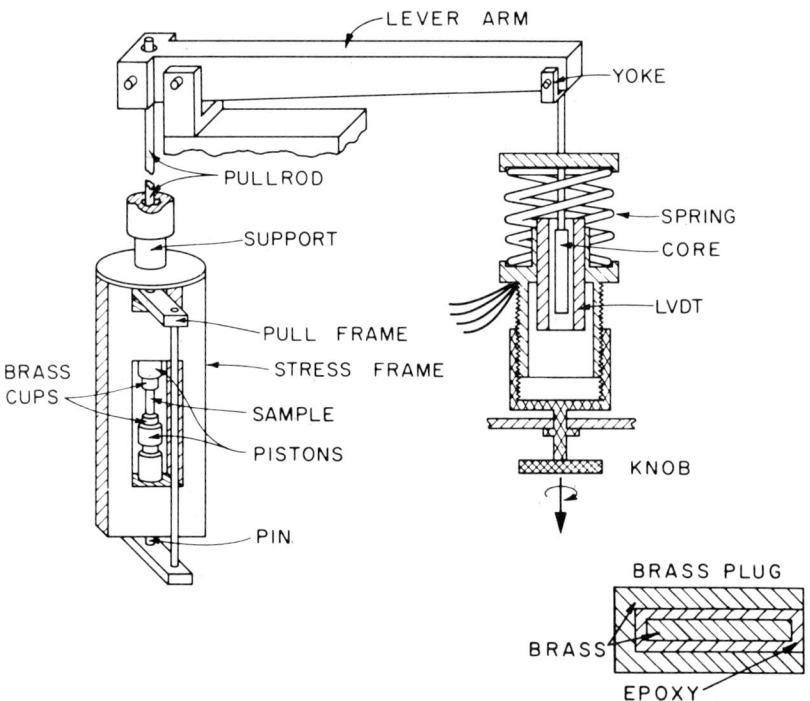

Fig. 150. Stress apparatus used for piezoelectroreflectance measurements. From F. H. Pollak and M. Cardona, *Phys. Rev.* **172**, 816 (1968); M. Cuevas and H. Fritzsche, *Ibid.* **137**, A1847 (1965).

indication of possible sample fracture and slippage into the epoxy brass cups. Typical dimensions of the exposed parts of the sample are 1.5 × 1.5 × 10 mm. The apparatus can be used at low temperatures if the stress frame and the matching pistons are made out of invar.

Reflection measurements under uniaxial stress can also be performed with the bending method of Philipp et al.[328] No modulation measurements under static bending have been reported to this date.

b. Indirect Transitions under Static Uniaxial Stress: Wavelength Modulation

Balslev[67] has reported the effect of uniaxial stress on the wavelength modulation spectrum of indirect excitons in germanium and silicon. The corresponding spectra of the unstressed materials have already been discussed in Section 16 and shown in Figs. 52–53. The effect of stress on these indirect excitons reflects mainly the splittings and shifts of the corresponding band edges: the change in exciton binding energies with stress has been discussed by Balslev[67] and found to be nearly negligible within experimental error. The shifts and splittings of the indirect band edges of germanium and silicon under uniaxial stress have been discussed in Section 20d. The indirect edge of silicon splits into 2 edges (because of the splitting of the $\Gamma_{25'}$ valence band) under the action of uniaxial stress along [111] and into 4 edges (both the $\Gamma_{25'}$ valence band and the Δ_1 conduction bands split) under [100] stress. The opposite situation obtains for germanium since the lowest conduction band minima lie along [111]. This is illustrated in Fig. 151. For identifying the various transition α_i, we have used in this figure the angular momentum notation for the split valence band; the $\Gamma_{25'}$ valence band splits under [100] or [111] stress into states with wave functions which have the same symmetry as the $(\frac{3}{2}, \pm \frac{3}{2})$ and the $(\frac{3}{2}, \pm \frac{1}{2})$ angular momentum wave functions (see Section 20b). The sign of the valence band deformation potentials b and d, which determines which one of the angular momentum states shifts upwards in energy (see Section 10b), is not conclusively obtained from these experiments. As we shall see in the next section, this sign can be determined by observing nonlinearities in the stress dependence of the split lines: the $(\frac{3}{2}, \pm \frac{1}{2})$ states are coupled by the stress to the spin orbit $(\frac{1}{2}, \pm \frac{1}{2})$ states and hence, a nonlinear stress dependence of their energies results. The $(\frac{3}{2}, \pm \frac{3}{2})$ states are decoupled from other valence band states for either [100] or [111] stress and, hence, their energies should vary linearly with stress. As will be shown in the next section, the quadratic shift of the $(\frac{3}{2}, \pm \frac{1}{2})$ states is:

$$(\delta\omega)^2/2\Delta_0, \tag{29.1}$$

where $\delta\omega$ is the corresponding linear splitting of the $(\frac{3}{2}, \pm \frac{3}{2})$, $(\frac{3}{2}, \pm \frac{1}{2})$ valence bands and Δ_0 the spin orbit splitting. From the known value of Δ_0 (see Table I) and the value of $\delta\omega$ for silicon (Fig. 151), we obtain for [111] stress a quadratic shift of 10 MeV for a stress of 8×10^9 dyn cm^{-2}. The nonlinear shift of the α_1 peak ([111] stress) has the sign required by this mechanism if α_1 is associated with the $(\frac{3}{2}, \pm \frac{1}{2})$ states, although its magnitude is somewhat small. The α_2 peak, however does not shift linearly with stress, as it would be expected for transitions involving the $(\frac{3}{2}, \pm \frac{3}{2})$ states, and, therefore, the assignment of these peaks to $(\frac{3}{2}, \pm \frac{3}{2})$ and $(\frac{3}{2}, \pm \frac{1}{2})$ transitions cannot be made solely on the basis of the nonlinearities in the stress dependence of

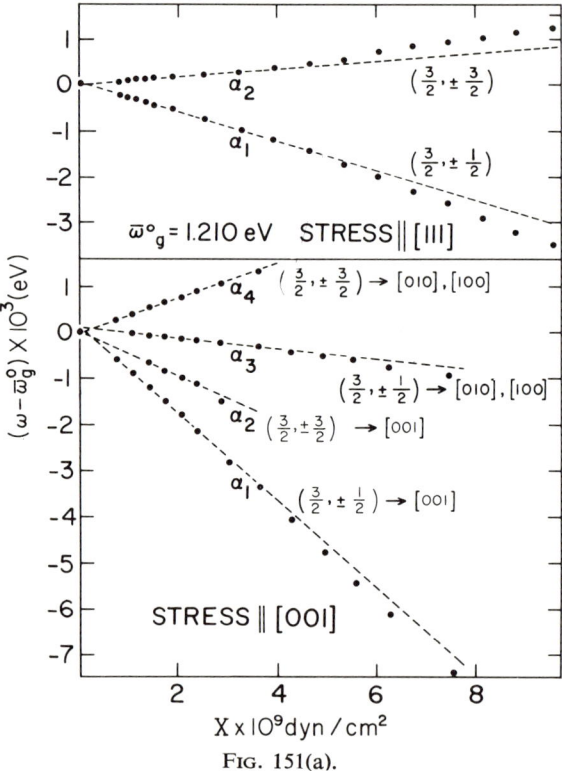

FIG. 151(a).

FIG. 151(b).

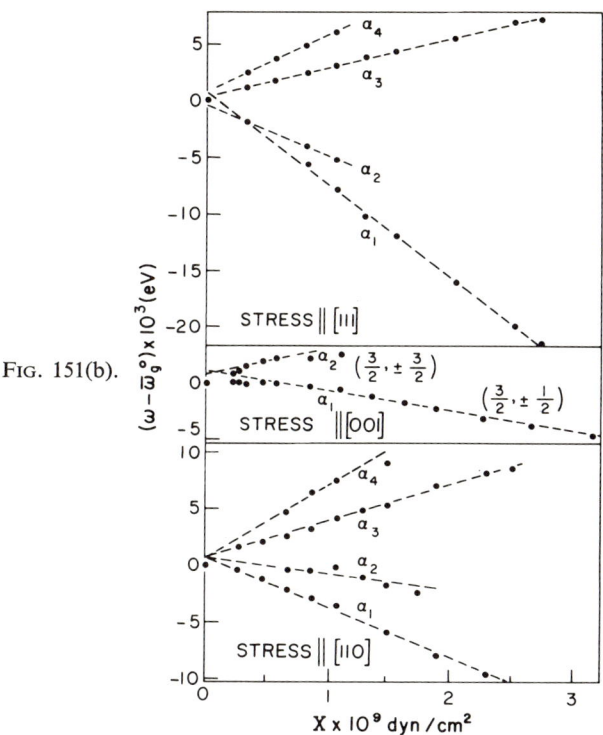

FIG. 151. Shifts and splittings of the indirect edges of silicon and germanium at 80°K under uniaxial stress, as measured by I. Balslev *Phys. Rev.* **143**, 636 (1966) by means of wavelength modulated transmission. For (a) $\bar{\omega}_g{}^0 = 1.210$ eV. For (b) $\omega_g{}^0 = 0.761$ eV.

their energy. Nor can it be easily made, in general, by considering the polarization selection rules for the optical transition (Section 20d): the existence of a large number of intermediate states makes these selection rules rather complicated.[221] Balslev[67] has suggested a simple way of obtaining information about the sign of the deformation potentials from the intercept of the straight lines of Fig. 151. He has attributed this intercept to a change of the exciton binding energy with stress. Under this assumption, the intercepts of Fig. 151 can be interpreted as caused by the fact that the excitons tied to the upper split valence bands have a smaller binding energy than those tied to the lower ones. A calculation of the binding energies of the excitons tied to both bands can be used to discern the sign of the deformation potentials a and b. It is found[67] that the exciton tied to the $(\frac{3}{2}, \pm\frac{3}{2})$ valence bands have a higher binding energy than those tied to the $(\frac{3}{2}, \pm\frac{1}{2})$ bands: hence the $(\frac{3}{2}, \pm\frac{1}{2})$ states split toward higher energies under compressive stress.

It is possible to give a heuristic argument showing why the binding energy of the excitons tied to the $(\frac{3}{2}, \pm\frac{3}{2})$ states is higher than that of those tied to $(\frac{3}{2}, \pm\frac{1}{2})$. The argument is based on the fact that for a given reduced mass:

$$\frac{1}{m^*} = \frac{1}{3}\left(\frac{2}{m_\perp} + \frac{1}{m_\parallel}\right) \tag{29.2}$$

($m_\perp$ and $m_\parallel$) are the longitudinal and transverse masses of the valence bands decoupled by the [100] or [111] stress) the exciton binding energy is higher the higher the mass anisotropy.[67,329] This result can be guessed from the fact that the binding energy of the two-dimensional hydrogenic exciton is 4 times that of its three-dimensional counterpart, as shown in Section 6. The $m_\perp$ and $m_\parallel$ valence band masses are given by (for [100] stress)[215b]:

$$\left.\begin{aligned}-\frac{1}{m_\parallel} &= A + B \\ -\frac{1}{m_\perp} &= A - \tfrac{1}{2}B\end{aligned}\right\}(\tfrac{3}{2}, \pm\tfrac{1}{2}) \qquad \left.\begin{aligned}-\frac{1}{m_\parallel} &= A - B \\ -\frac{1}{m_\perp} &= A + \tfrac{1}{2}B\end{aligned}\right\}(\tfrac{3}{2}, \pm\tfrac{3}{2}), \tag{29.3}$$

In Eqs. (29.3) A and B are valence band parameters which are negative for both Ge and Si[116] ($|A|$ is larger than $|B|$). Equations (29.3) show that for the known values of A and B[116] the mass anisotropy is larger for the $(\frac{3}{2}, \pm\frac{3}{2})$ than for the $(\frac{3}{2}, \pm\frac{1}{2})$ bands; the reduced masses obtained with Eq. (29.2) are the same for both bands. Therefore, the binding energy of excitons tied to the $(\frac{3}{2}, \pm\frac{3}{2})$ bands is expected to be larger than that of those tied to $(\frac{3}{2}\pm\frac{1}{2})$.

From the $\alpha_1 - \alpha_3$ splittings of Fig. 151, one can also obtain the $\mathscr{E}_2$ deformation potentials [Eq. (20.13)]. The shift with stress of the average of all peaks observed yields the hydrostatic deformation potential $\mathscr{E}_1$ [Eq. (20.13)] of the energy difference between the $\Gamma_{25'}$ valence and the lowest conduction band (Δ_1 for silicon, L_1 for germanium). All deformation potentials obtained by this method are given in Table XX together with those obtained by other methods including, for germanium, piezoelectroreflectance, piezopiezoreflectance, and wavelength modulated reflectance of the direct gap (see next Section).

TABLE XX. DEFORMATION POTENTIAL CONSTANTS OF THE VALENCE BAND AND THE LOWEST CONDUCTION BANDS OF GERMANIUM AND SILICON OBTAINED BY MODULATION TECHNIQUES AND BY OTHER OPTICAL METHODS

	$|b|$	$|d|$	$\mathscr{E}_2$	$d\omega_g/dp \times 10^{12}$
Germanium				
Wavelength modulation:				
Indirect transitions[a]	1.8 ± 0.4 eV	3.7 ± 0.4 eV	16.2 ± 0.2 eV	3 ± 1 eV cm^2/dyn
Indirect transitions[b]	2.4 ± 0.4	3.5 ± 0.4		
Piezoreflectance[c]	2.4 ± 0.2	4.1 ± 0.4		
Piezoelectroreflectance[d]	2.6 ± 0.2	4.7 ± 0.3		
Piezoabsorption	2.7 ± 0.3[e]	4.7 ± 0.5[e]		4[f]
Piezobirefringence[g]			18.9 ± 1.7	
Silicon				
Wavelength modulation:				
Indirect transitions[a]	2.4 ± 0.2	5.3 ± 0.4	8.6 ± 0.2	-3.8 ± 0.5
Piezoabsorption[h]				-1

[a] I. Balslev, *Phys. Rev.* **143**, 636 (1966).
[b] I. Balslev, *Phys. Letters* **24A**, 113 (1967).
[c] I. Balslev, *Solid State Commun.* **5**, 315 (1967).
[d] F. H. Pollak and M. Cardona, *Phys. Rev.* **172**, 816 (1968).
[e] A. M. Glass, *Can. J. Phys.* **43**, 12 (1965).
[f] W. Paul and D. M. Warschauer, *J. Phys. Chem. Solids* **5**, 89 (1958).
[g] K. J. Schmidt-Tiedemann, *Proc. Intern. Conf. Phys. Semiconductors*, 6th Exeter, p. 191 (1962).
[h] W. Paul and D. M. Warschauer, *J. Phys. Chem. Solids* **5**, 102 (1958).

c. Lowest Direct Gap: Piezoelectroreflectance, Piezopiezoreflectance, and Wavelength Modulated Reflectance

The electrolyte technique of electroreflectance, discussed in Section 25, is particularly suitable for measurements under a large static uniaxial stress. The sample, with a suitably polished and etched face, can be mounted in the stressing apparatus of Fig. 150 and immersed, together with the surrounding frame, into the electrolyte. Care must be taken to insure the good electrical insulation of the sample from the metal frame; even a small leakage short-circuits the voltage applied to the sample through the metal frame electrolyte interface. Insulation is obtained by means of the epoxy in the brass cups of Fig. 150. For measurements in the infrared with the thin (capillary) electrolyte layer method, one can use a cell with a window which protrudes inward so as to make contact with the sample surface without interfering with the stressing frame.

The effect of stress on the $\Gamma_{25'} - \Gamma_{2'}$ direct edge of germanium (or the $\Gamma_{15} - \Gamma_1$ edge of GaAs) is described by the Hamiltonian of Eq. (20.6): the $J = \frac{3}{2}$ component of this edge splits under uniaxial stress because of the splitting of the valence band, and the "center of mass" of the 2 split components shifts because of the hydrostatic component of the stress. The $J = \frac{1}{2}$ spin orbit split edge suffers only a hydrostatic shift. The uniaxial splitting of the valence band is linear in stress provided it is much smaller than the spin orbit splitting. At higher stresses the coupling between the $(\frac{3}{2}, \pm\frac{1}{2})$ and the $(\frac{1}{2}, \pm\frac{1}{2})$ states produces nonlinear effects. These nonlinear effects are usually negligible in a stress modulated experiment and they have not been considered in Chapter VI. However, they become appreciable when a large static stress is applied. Such nonlinear effects can be described by referring the Hamiltonian of Eq. (20.6) to the 3 degenerate $\Gamma_{25'}$ (or Γ_{15}) states. We obtain[323]:

$$\begin{Vmatrix} |\tfrac{3}{2},\tfrac{3}{2}\rangle & |\tfrac{3}{2},\tfrac{1}{2}\rangle & |\tfrac{1}{2},\tfrac{1}{2}\rangle \\ \tfrac{1}{3}\Delta_0 - \delta\omega_H - \tfrac{1}{2}\delta\omega_{001} & 0 & 0 \\ 0 & \tfrac{1}{3}\Delta_0 - \delta\omega_H + \tfrac{1}{2}\delta\omega_{001} & 2^{-1/2}\delta\omega_{001} \\ 0 & 2^{-1/2}\delta\omega_{001} & -\tfrac{2}{3}\Delta_0 - \delta\omega_H \end{Vmatrix}. \quad (29.4)$$

By diagonalizing the above Hamiltonian we find the simultaneous effect of spin orbit splitting and stress. The change in the energy gap between the 3 valence bands and the conduction band produced by spin and stress becomes, for a compression along [100]:

$$\Delta[\Gamma_1 - (\tfrac{3}{2}, \tfrac{3}{2})] = -\tfrac{1}{3}\Delta_0 + \delta\omega_H + \tfrac{1}{2}\delta\omega_{001}$$
$$\Delta[\Gamma_1 - (\tfrac{3}{2}, \tfrac{1}{2})] = \tfrac{1}{6}\Delta_0 + \delta\omega_H - \tfrac{1}{4}\delta\omega_{001}$$
$$\qquad - \tfrac{1}{2}[\Delta_0^2 + \Delta_0\,\delta\omega_{001} + \tfrac{9}{4}(\delta\omega_{001})^2]^{1/2} \quad (29.5)$$
$$\Delta[\Gamma_1 - (\tfrac{1}{2}, \tfrac{1}{2})] = \tfrac{1}{6}\Delta_0 + \delta\omega_H - \tfrac{1}{4}\delta\omega_{001}$$
$$\qquad + \tfrac{1}{2}[\Delta_0^2 + \Delta_0\,\delta\omega_{001} + \tfrac{9}{4}(\delta\omega_{001})^2]^{1/2},$$

where $\delta\omega_H$ is the hydrostatic shift and δ_{001} the uniaxial linear splitting of the $(\tfrac{3}{2}, \tfrac{3}{2}) - (\tfrac{3}{2}, \tfrac{1}{2})$ states [Eq. (20.7)]. By expanding Eq. (29.5) to second order in $\delta\omega_{001}/\Delta_0$ we obtain:

$$\Delta[\Gamma_1 - (\tfrac{3}{2}, \tfrac{3}{2})] = -\tfrac{1}{3}\Delta_0 + \delta\omega_H + \tfrac{1}{2}\delta\omega_{001}$$
$$\Delta[\Gamma_1 - (\tfrac{3}{2}, \tfrac{1}{2})] = -\tfrac{1}{3}\Delta_0 + \delta\omega_H - \tfrac{1}{2}\delta\omega_{001} + [(\delta\omega_{001})^2/2\Delta_0] \quad (29.6)$$
$$\Delta[\Gamma_1 - (\tfrac{1}{2}, \tfrac{1}{2})] = \tfrac{2}{3}\Delta_0 + \delta\omega_H + [(\delta\omega_{001})^2/2\Delta_0].$$

Equations (29.6) are also valid for [111] stress provided one replaces $\delta\omega_{001}$ by the corresponding linear splitting under [111] stress. For stress directions of lower symmetry, the situation is considerably more complicated: the 3 components of the split valence band should show nonlinear splittings. A discussion of the case of [110] stress is given by Pollak and Cardona.[323] It can be shown that even in this case one of the valence bands is decoupled (shifts linearly) when the isotropy condition $\delta\omega_{001} = \delta\omega_{111}$ is fulfilled.

The effect of [001] stress on the E_0 and $E_0 + \Delta_0$ electroreflectance peaks of germanium is shown in Fig. 152. The $E_0 + \Delta_0$ peak exhibits

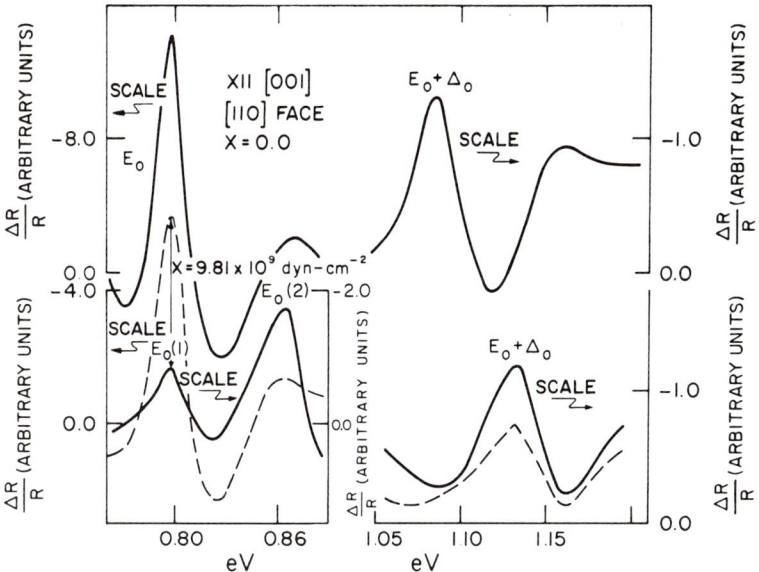

FIG. 152. Electroreflectance spectra of the $E_0 - (E_0 + \Delta_0)$ peaks of germanium for zero stress and for a [001] stress of 9.81×10^9 dyn cm^{-2}, with light polarized parallel (solid line) and perpendicular (dashed line) to the stress axis [F. H. Pollak and M. Cardona, *Phys. Rev.* **172**, 816 (1968)].

only the hydrostatic shift discussed above, but no splitting. This shift equals the shift in the average position of the split E_0 peaks. It should be made clear that the high-energy component of the E_0 peaks under stress (at around 0.96 eV) is indeed the split component of E_0 and not a Franz–Keldysh oscillation, since it can be tracked as a split peak from small stresses to the highest stresses available. The motion of these peaks with stress is shown in Fig. 153. The nonlinear shift in one of the E_0 peak is conclusively observed and agrees with the value (see flag in Fig. 153) estimated with Eq. (29.6).† The nonlinear shift

† This agreement is only qualitative. See recent work by J. C. Hensel and K. Suzuki, *Bull. Am. Phys. Soc.* **14**, 113 (1969).

of the $E_0 + \Delta_0$ peak is difficult to observe because of the lower accuracy in the determination of the energy of this broader peak. The observed nonlinear shift of the low-energy E_0 peak proves that this peak is related to the $(\frac{3}{2}, \pm\frac{1}{2})$ valence band states and determines the sign of the corresponding deformation potential: b must be negative. The identification of the split E_0 peaks can also be made from the intensities of the E_0 peaks under stress seen in Fig. 153: the high-

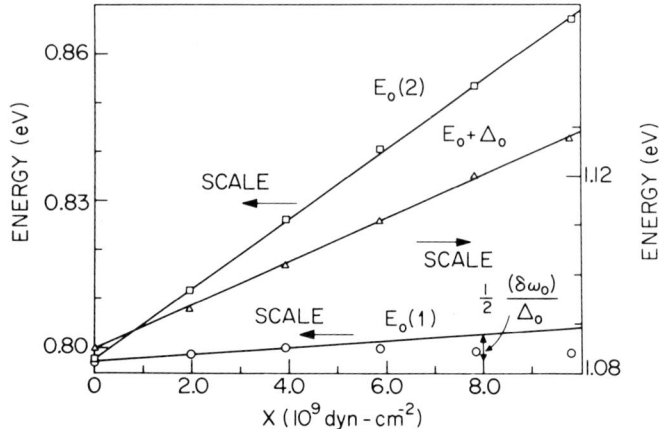

FIG. 153. Energies of the $E_0(1)$, $E_0(2)$, and $E_0 + \Delta_0$ peaks of the electroreflectance spectrum of germanium as a function of a [001] ($X \parallel$ [001]) uniaxial compression with a [110] face. The flag at 8×10^9 dyn cm^{-2} is the nonlinear term calculated with Eq. (29.6). [F. H. Pollak and M. Cardona, Phys. Rev. **172**, 816 (1968).]

energy component is almost nonexistent for $\mathbf{E} \parallel X$. In view of the selection rules discussed in Section 20b, this component of E_0 must be associated with the $(\frac{3}{2}, \pm\frac{3}{2})$ valence bands. The intensities of the split E_0 peaks agree, at least qualitatively, with the matrix element derived from the wave functions of Eq. (9.3). The intensities expected from these matrix elements are:

$$(\tfrac{3}{2}, \tfrac{3}{2}) \to \Gamma_2' \begin{cases} \propto \tfrac{1}{2} & \mathbf{E} \perp X \\ = 0 & \mathbf{E} \parallel X \end{cases}$$

$$(\tfrac{3}{2}, \tfrac{1}{2}) \to \Gamma_2' \begin{cases} \propto \tfrac{1}{6} & \mathbf{E} \perp X \\ \propto \tfrac{2}{3} & \mathbf{E} \parallel X. \end{cases} \quad (29.7)$$

The electrooptic effect could also influence the line intensities discussed above since it introduces factors dependent on effective masses.

However, these masses are dominated by the electron mass, which is independent of stress, and, hence, most of the contribution to the difference in line intensities comes from the matrix elements. The coupling between $(\frac{3}{2}, \pm \frac{1}{2})$ and $(\frac{1}{2}, \pm \frac{1}{2})$ states introduces a mixing of these wave functions which produces a change in the intensities proportional to stress. These effects, which are observed experimentally,[323] can be calculated with the eigenvectors of Eq. (29.4).

Piezoelectroreflectance measurements similar to those discussed above have been also performed for stresses along [111] and [110] in germanium and for GaAs.[323] The deformation potentials obtained from these measurements are listed in Table XXI.

TABLE XXI. DEFORMATION POTENTIALS OF Ge AND GaAs OBTAINED BY MEANS OF REFLECTANCE MODULATION EXPERIMENTS IN COMPARISON WITH VALUES DETERMINED BY OTHER TECHNIQUES *

Potentials	Ge		GaAs	
	Piezoelectro-reflectance	Previous results	Piezoelectro-reflectance	Previous results
$d\omega_g/dp$ ** $(10^{-6}$ eV-bar$^{-1})$	$+12.0 \pm 0.5^a$ $+12.5 \pm 0.5^b$ $+12.5 \pm 0.7^c$	$+13 \pm 1^d$ $+13.7 \pm 1.5^e$ $+10.6^f$	$+11.5 \pm 0.5^a$ $+12.0 \pm 0.5^b$ $+11.5 \pm 0.5^c$	$+11.5 \pm 1^d$ $+11.8 \pm 0.6^k$ $+10.5^f$
b (eV)	-2.6 ± 0.2^a	-2.7 ± 0.3^e -2.4 ± 0.2^f -1.8 ± 0.2^g -2.1 ± 0.2^h	-2.0 ± 0.2^a	-1.7 ± 0.2^f -1.9 ± 0.1^k
d (eV)	-4.7 ± 0.3^b	-4.7 ± 0.5^e -4.1 ± 0.4^f -3.7 ± 0.4^g -7.0 ± 1.5^h -6.0 ± 0.6^i	-6.0 ± 0.4^b	-4.4 ± 0.6^f -5.4 ± 0.3^k
$\mathscr{E}_1$ (eV)	-4.5 ± 0.4^a	-5.7 ± 0.4^d -5.7 ± 0.3^j	-4.0 ± 0.4^a	
$\mathscr{E}_2$ (eV)	$+5.1 \pm 1.0^{a,b}$	$+5.1 \pm 1.0^j$	$+7.4 \pm 0.7^b$	

* The $\mathscr{E}_1$ and $\mathscr{E}_2$ potentials correspond to the direct E_1 edge; they are not to be mistaken with those of the indirect transitions given in Table XX.
** Hydrostatic coefficient of the direct gap.
[a] [100] stress measurements.
[b] [111] stress measurements.
[c] [110] stress measurements.
[d] J. Feinleib, S. Groves, W. Paul, and R. Zallen, *Phys. Rev.* **131**, 2070 (1963); R. Zallen and W. Paul, *Phys. Rev.* **134**, A1628 (1964).
[e] A. M. Glass, *Can. J. Phys.* **43**, 12 (1965).
[f] I. Balslev, *Solid State Commun.* **5**, 315 (1967).
[g] I. Balslev, *Phys. Rev.* **143**, 636 (1966).
[h] J. J. Hall, *Phys. Rev.* **128**, 68 (1962).
[i] J. C. Hensel, *Solid State Commun.* **4**, 231 (1966).
[j] U. Gerhardt, *Phys. Status Solidi* **11**, 801 (1965).
[k] R. M. Bhagarva and M. Nathan, *Phys. Rev.* **161**, 695 (1967).

Measurements of the lowest direct gap under static uniaxial stress have been performed by Balslev for germanium and for GaAs with a combination of a small modulated stress and a large superimposed static stress[189]; nonlinear effects were not reported in this work. The differential piezoreflectance measurements with a superimposed static uniaxial stress were performed with the apparatus of Fig. 69. The results of these measurements are listed in Table XXI.

d. Higher Direct Gaps

Modulation measurements under static uniaxial stress have been reported for the $E_1 - (E_1 + \Delta_1)$ peaks of germanium and GaAs and for the E_0' peak of silicon.[323] We shall first discuss the measurement of the $E_1 - (E_1 + \Delta_1)$ peaks: in contrast to the E_0' peak of silicon, the symmetry of the transitions which cause the E_1, $E_1 + \Delta_1$ peaks of germanium and GaAs is quite well understood (see Section 19a). These peaks should suffer intervalley splitting for [111] stress according to Eq. (20.13). This effect is illustrated in Fig. 154 for GaAs: 2 components of the E_1 peak, and somewhat less clearly, of the $E_1 + \Delta_1$ peak, are seen for $\mathbf{E} \perp X$ while only the high-energy component appears, for $\mathbf{E} \parallel X$. This is to be expected since the E_1 tran-

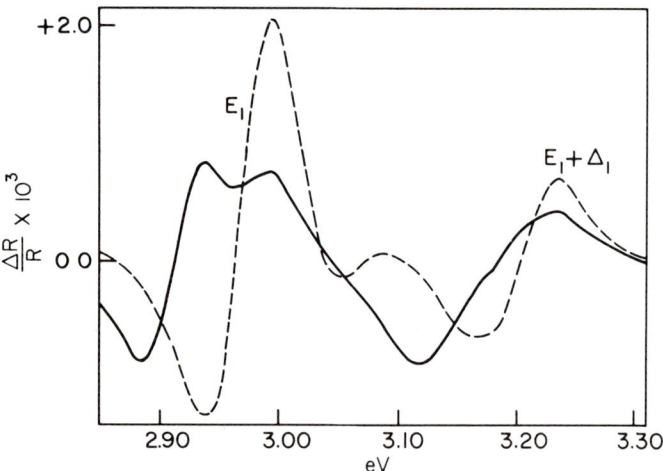

FIG. 154. E_1 and $E_1 + \Delta_1$ peaks of GaAs with a [11$\bar{2}$] face under a [111] stress of 8.05×10^9 dyn cm^{-2}, for $E \parallel X$ (dashed line) and $E \perp X$ (solid line). From F. H. Pollak and M. Cardona *Phys. Rev.* **172**, 816 (1968).

sitions ($\Lambda_3 \to \Lambda_1$) are only allowed for **E** perpendicular to the direction of the [111] valley under consideration. Hence, for $\mathbf{E} \perp [111]$, only the $\pm [\bar{1}11]$, $\pm [1\bar{1}1]$, and $\pm [11\bar{1}]$ transitions are allowed, while for $\mathbf{E} \perp [111]$, the $\pm [111]$ transitions are also allowed. The deformation potentials $\mathscr{E}_1$ and $\mathscr{E}_2$ of Eq. (20.13) derived from these measurements are given in Table XXI.

For a [100] compression there is no intervalley splitting of the [111] transitions E_1 and $E_1 + \Delta_1$. In the absence of spin orbit splitting, however, the [100] stress should produce a splitting of the valence bands (intraband splitting). The orbital valence band wave functions have symmetries $\bar{X}$ and $\bar{Y}$ where $\bar{x}$ and $\bar{y}$ are the coordinates with respect to axes such that $\bar{z}$ is along the direction of the critical point, i.e., [111]. For the [111] critical point we can take $\bar{X} = 2^{-1/2}(X - Y)$ and $\bar{Y} = 6^{-1/2}(X + Y - 2Z)$, and, therefore, a stress along [100] (X symmetry) makes $\bar{X}$ and $\bar{Y}$ inequivalent: intravalley splittings result. In the presence of spin-orbit coupling, the intravalley orbital "splitting" produces only a repulsion between the spin orbit states and thus an apparent increase in the spin orbit splitting. It also produces a mixture of the valence band wave functions which has, as a result, a linear change in the intensities of the E_1 and $E_1 + \Delta_1$ peaks. The stress induced quadratic shift in the E_1, $E_1 + \Delta_1$ splitting and the linear change in peak intensities can be obtained from the Hamiltonian:

$$\begin{array}{cc} |2^{-1/2}(X+iY)\uparrow\rangle & |2^{-1/2}(X-iY)\uparrow\rangle \end{array}$$

$$\left\| \begin{array}{cc} \tfrac{1}{2}\Delta_1 + \delta_H & -\tfrac{1}{2}\delta_1 \\ -\tfrac{1}{2}\delta_1 & -\tfrac{1}{2}\Delta_1 + \delta_H \end{array} \right\| \quad (29.8)$$

where δ_H is the hydrostatic shift and δ_1 the intraband stress splitting which one would obtain in the absence of spin orbit interaction. The stress induced change in the energies of E_1 and $E_1 + \Delta_1$ is:

$$\Delta E_1 = \tfrac{1}{2}\Delta_1 + \delta_H - \tfrac{1}{2}[\Delta_1^2 + \delta_1^2]^{1/2}$$
$$\simeq \delta_H - (\delta_1^2/4\Delta_1)$$
$$\Delta(E_1 + \Delta_1) = -\tfrac{1}{2}\Delta_1 + \delta_H + \tfrac{1}{2}[\Delta_1^2 + \delta_1^2]^{1/2} \quad (29.9)$$
$$\simeq \delta_H + (\delta_1^2/4\Delta_1).$$

An increase with [100] stress in the splitting of the $E_1 - (E_1 + \Delta_1)$ peaks has been observed by the piezoelectroreflectance method. Figure 155 shows the results obtained for germanium compared with the predictions of Eq. (29.9). The agreement between the calculated solid line and the experimental points is not bad in view of the large experimental scatter: the total increase in the $E_1 - (E_1 + \Delta_1)$ splitting for X between 0 and 10^{10} dyn cm^{-2} is only 0.01 eV. Changes in line

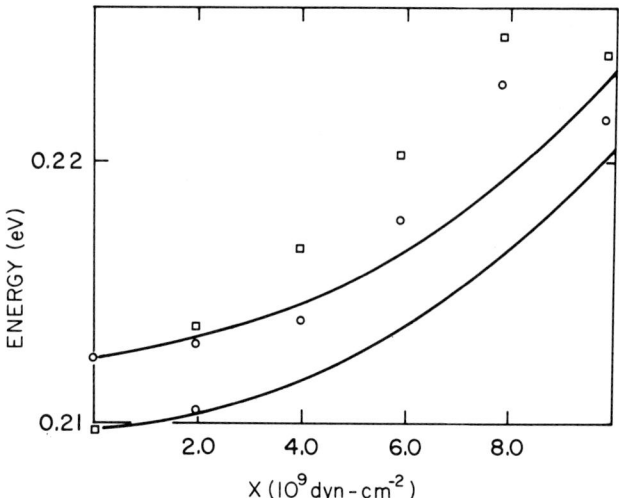

FIG. 155. Splitting of the $E_1 - (E_1 + \Delta_1)$ peaks of germanium with [001] stress for a [110] face. From F. H. Pollak and M. Cardona, *Phys. Rev.* **172**, 916, (1968). The solid lines were calculated with Eq. (29.9). The parallel polarizations are represented by circles and the perpendicular ones by squares.

shape produced by the changes in intensity with stress to be discussed next, coupled with the overlap in the peaks, could also contribute to apparent changes in this splitting.

In the absence of stress the wave functions of the valence band responsible for the $E_1 - (E_1 + \Delta_1)$ peaks have the symmetry of $(X + iY)\uparrow$ and $(X - iY)\uparrow$ and their time reversed partners. Therefore, the optical matrix elements for these 2 transition are the same at the critical point; experimentally $E_1 + \Delta_1$ is usually weaker than E_1 but this is caused at least in part, by a larger broadening of $E_1 + \Delta_1$. The application of the stress mixes $(X + iY)$ and $(X - iY)$ according to Eq. (29.8). This fact produces a stress and polarization dependence of the intensities of E_1 and $E_1 + \Delta_1$. From the eigenvectors of Eq. (29.8) one obtains for the intensities of the E_1 and $E_1 + \Delta_1$ peaks for light polarized with E parallel ($I^{\parallel}$) and perpendicular ($I^{\perp}$) to the stress direction the expressions[323] (to first order in $\alpha_1 = \delta_1/\Delta_1$):

$$I_{E_1}^{\parallel}(X) = I_{E_1}^{\parallel}(0)(1 + \alpha_1),$$

$$I_{E_1}^{\perp}(X) = I_{E_1}^{\perp}(0)(1 - \tfrac{1}{2}\alpha_1);$$

$$I_{E_1 + \Delta_1}^{\parallel}(X) = I_{E_1 + \Delta_1}^{\parallel}(0)(1 - \alpha_1),$$

$$I_{E_1 + \Delta_1}^{\perp}(X) = I_{E_1 + \Delta_1}(0)(1 + \tfrac{1}{2}\alpha_1). \tag{29.10}$$

The observed dependence of the intensities of the $E_1 - (E_1 + \Delta_1)$ electroreflectance peaks on [100] stress is given in Fig. 156. The

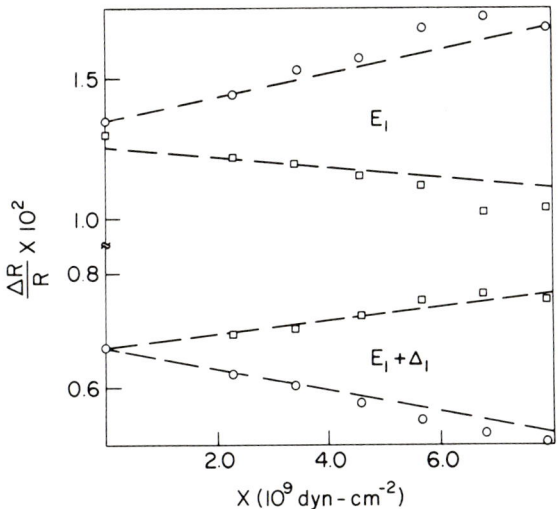

FIG. 156. Intensities of the E_1 and $E_1 + \Delta_1$ electroreflectance peaks of GaAs with a [100] as a function of [001] stress face for light polarized with E parallel (circles) ($I^{\parallel}$) and perpendicular (squares) ($I^{\perp}$) to the stress axis. The dashed curves have been calculated with Eq. (29.10) assuming that δ_1 is the same as δ_{100}, the splitting of the top valence bands at $\mathbf{k} = 0$. [F. H. Pollak and M. Cardona, *Phys. Rev.* **172**, 816 (1968).]

dashed lines were calculated with Eq. (29.10) assuming that $\delta_1 = \delta\omega_{001}$ [see Eq. (20.9)]. This assumption is reasonable[323] in view of the fact that the X and Y wave functions of the valence bands at the Λ critical point are almost the same as those at Γ.[103] The agreement between theoretically predicted and experimentally observed intensities is excellent. Similar changes of peak heights with stress have been observed by Gerhardt[331] in the conventional reflection spectrum of germanium, but a quantitative comparison of these experiments with theory is not possible because of the large background onto which the $E_1 - (E_1 + \Delta_1)$ peaks are superimposed. Electroreflectance measurements in the region of the E_1, $E_1 + \Delta_1$ peaks under static [100] stress have only been performed for Ge and GaAs.[323]

A similar change in the intensity of the $E_1 - (E_1 + \Delta_1)$ peaks due to $\pm[\bar{1}11]$, $\pm[1\bar{1}1]$, $\pm[11\bar{1}]$ transitions is also expected for [111] stress. A theoretical calculation of this effect and a comparison with the experimental result for Ge and GaAs is given by Cardona and Pollak.[323]

The E_0' peaks of silicon shown in the electroreflectance spectrum of Fig. 126 are, as already discussed, the result of transitions between bands parallel over a large region of the Brillouin zone; a number of nonequivalent critical points may be present. Hence, the effect of uniaxial stress on these peaks is expected to be quite complex; intervalley splittings of several nonequivalent critical points and intravalley splittings due to the small spin orbit interaction may result. We show in Fig. 157 the electroreflectance spectra of a [110] silicon face with zero stress (a) and with large stress along [001] (b), [111] (c), and [110] (d). The complicated multiple splitting under [011] stress confirms the expectation discussed above. The $E_0'(1)$ peak seems to be the component which already existed for $\mathbf{E} \parallel [1\bar{1}0]$ with zero stress.

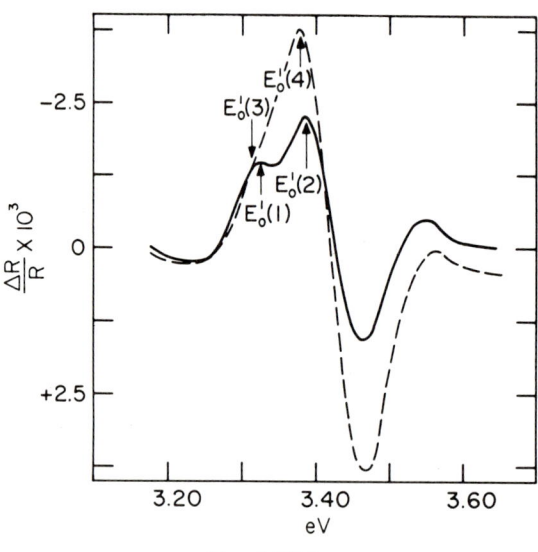

Fig. 157(a).

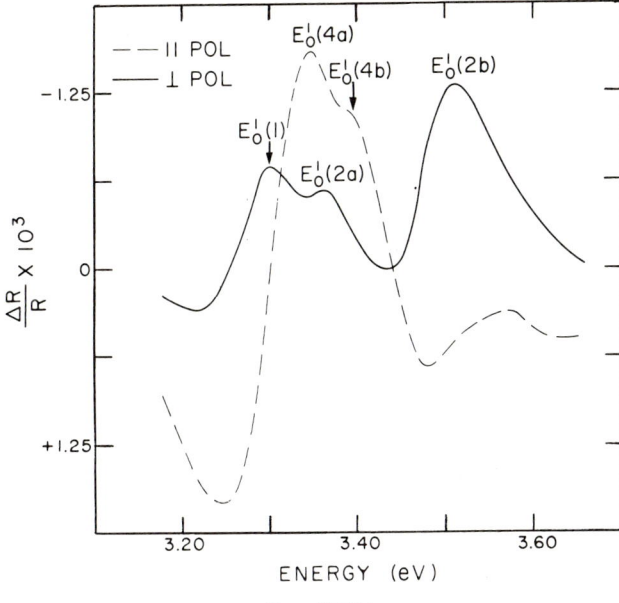

Fig. 157(b).

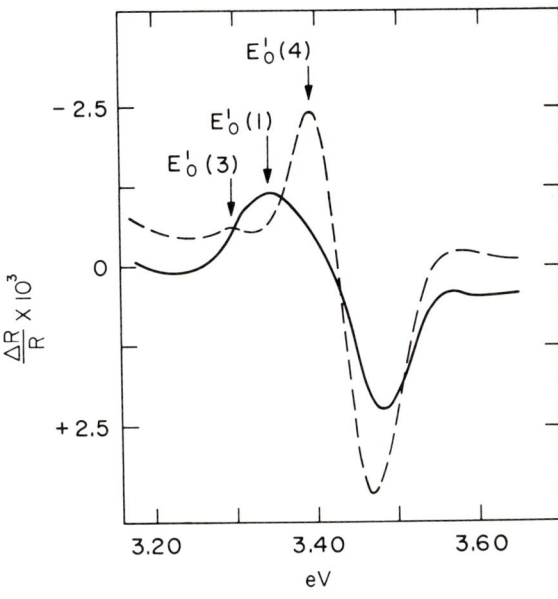

Fig. 157(c).

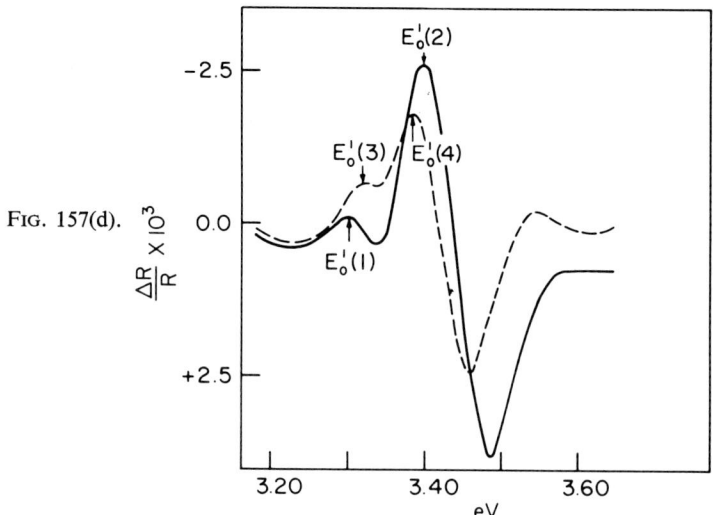

FIG. 157. Electroreflectance spectra of the E_0' peaks of a $(1\bar{1}0)$ surface of silicon with zero stress $X = 0.0$ (a), under [001] stress $X = 14.22 \times 10^9$ dyn cm^{-2} (b), under [111] stress $X = 9.48 \times 10^9$ dyn cm^{-2} (c), and under [110] stress $X = 10.96 \times 10^9$ dyn cm^{-2} (d). From F. H. Pollak and M. Cardona, *Phys. Rev.* **172**, 816 (1968). (Dashed line $\parallel$, solid line $\perp$.)

All other peaks seem to have originated from $E_0'(2)$: the splitting of $E_0'(2)$ is therefore of the order of 0.2 eV. The fact that for other stress directions the splittings of $E_0'(2)$ are smaller suggests transitions along [100] (Δ lines). The peaks E_0' (4b), $E_0'(2a)$, and $E_0'(2b)$ can be attributed to inter- and intravalley splittings of Δ critical points, but a complete explanation of the whole spectrum in terms of only these transitions is not possible. Intraband splitting can also be invoked to explain the $E_0'(1) - E_0'(4)$ splitting for [111] stress and its intensity dependence on polarization. A detailed theoretical analysis and interpretation of these meaurements should yield a great wealth of information about deformation potentials.

30. MODULATION TECHNIQUES IN THE PRESENCE OF A STEADY MAGNETIC FIELD

a. Theory

We shall first review the theory of the effect of a magnetic field on a nondegenerate band extremum. Let us assume, for the sake of simplicity, a conduction band minimum with spherical symmetry at $\mathbf{k} = 0$.

The wave function in the presence of a uniform magnetic field along z is[329, 332]:

$$\psi(r) = u_0(r)F(r), \quad (30.1)$$

where $u_0(r)$ is the periodic part of the Bloch function at the extremum and $F(r)$ fulfills the effective mass equation (see Section 6b):

$$\frac{1}{2m^*}\left(\mathbf{p} + \frac{1}{c}\mathbf{A}\right)^2 F(r) = \omega_H F(r). \quad (30.2)$$

In Eq. (30.2) m^* is the effective mass of the extremum, $\mathbf{A}$ the vector potential, and ω_H the energy of the corresponding electron state measured from the bottom of the unperturbed band. We chose the gauge so that $\mathbf{A} = (0, Hx, 0)$, where H is the magnitude of the static field. Equation (30.2) becomes:

$$-\frac{1}{2m^*}\left\{\frac{\partial^2}{\partial x^2} + \left(\frac{\partial}{\partial y} + \frac{iHx}{c}\right)^2 + \frac{\partial^2}{\partial z^2}\right\}F(r) = \omega_H F(r). \quad (30.3)$$

The translational invariance of Eq. (30.3) along y and z indicates that the solution is of the form:

$$F(r) = \exp[i(k_y y + k_z z)]f(x). \quad (30.4)$$

By replacing Eq. (30.4) into Eq. (30.3) we find:

$$-\frac{1}{2m^*}\left\{\frac{\partial^2}{\partial x^2} - \left(k_y + \frac{Hx}{c}\right)^2\right\}f(x) = \left(\omega_H - \frac{k_z^2}{2m^*}\right)f(x). \quad (30.5)$$

As is well known, Eq. (30.5) is analogous to the equation of a one-dimensional harmonic oscillator with the energy shifted by k_z^2/m^* and the minimum of the potential at $x_0 = -(c/H)k_y$. Hence, the eigenvalues of the energy are:

$$\omega_H = (k_z^2/2m^*) + \omega_c(n + \tfrac{1}{2}), \quad n = 0, 1, 2, \ldots, \quad (30.6)$$

with the cyclotron resonance energy $\omega_c = H/m^*c$. The energy ω_H is independent of k_y. The eigenfunctions of Eq. (30.5) are the well-known wave functions of the harmonic oscillator centered at x_0, $\Phi_n(x - x_0)$. It is worthwhile noting that the energy dependence on k_z is not affected by the magnetic field, as expected: a magnetic field has no effect on motion along its direction. Hence, we find for the energy, a collection of one-dimensional subbands separated by the energy ω_c (Landau levels or subbands). The density of states within each subband has the one-dimensional shape of Eq. (4.6).

An analogous reasoning can be applied to a valence band maximum. Optical transitions between Landau subbands must occur with k_y and k_z conservation. If no phonons or imperfections are included (direct transitions), the combined density of states for the transitions is similar to that of Eq. (4.6): direct transitions are expected to become considerably sharper in the presence of a magnetic field than in its absence, because of the one-dimensional nature of the Landau subbands. A conventional absorption spectrum in the presence of a large magnetic field H should have line shapes of the $(\omega - \omega_g)^{-1/2}$ type, similar to those obtained in the derivative spectra for $H = 0$, for each allowed transition between Landau subbands. The line shapes of the derivative spectra for $H \neq 0$ are expected to be of the $(\omega - \omega_g)^{-3/2}$ type, even sharper than those of ordinary magneto absorption or magneto reflection. The effect of Lorentzian broadening on the $(\omega - \omega_g)^{-1/2}$ line shapes of the conventional optical spectra for $H \neq 0$ is given by the function $F(W)$ of Eq. (15.4) and Fig. 51:

$$\varepsilon_r(H \neq 0) \propto \eta^{-1/2} F(-W)$$
$$\varepsilon_i(H \neq 0) \propto \eta^{-1/2} F(W). \tag{30.7}$$

The shapes of the derivative spectra for $H \neq 0$ are obtained from the function $F'(W)$. This function[127] is shown in Fig. 158 together with $F(W)$; the sharpening characteristic of all derivative spectra is clearly evident in this figure.

The probability for direct optical transitions between 2 Landau subbands n_i and n_j belonging to bands i and j is proportional to the square of the matrix element:

$$\int f_{n_i}^*(x) u_i(\mathbf{r}) \mathbf{p} \cdot \mathbf{\hat{n}} u_j(\mathbf{r}) f_{n_j}(x) \, dV_\mathbf{r}, \tag{30.8}$$

where $\mathbf{\hat{n}}$ gives the direction of polarization of $\mathbf{E}$.

The matrix element of Eq. (30.8) can be evaluated if one assumes that the wave functions f_n vary slowly compared with the Bloch functions u_j [this is a good approximation for small magnetic fields; f varies little in a distance smaller than the extent of the ground state harmonic oscillator wave function $L_H = (c/H)^{-1/2}$]. The matrix element of (30.8) can, in this case, be written as[333]:

$$\langle u_i | \mathbf{p} \cdot \mathbf{\hat{n}} | u_j \rangle \int f_{n_i}^*(x) f_{n_j}(x) \, dx$$
$$+ \langle u_i | u_j \rangle \int f_{n_i}^*(x) \mathbf{p} \cdot \mathbf{\hat{n}} \, f_{n_j}(x) \, dx. \tag{30.9}$$

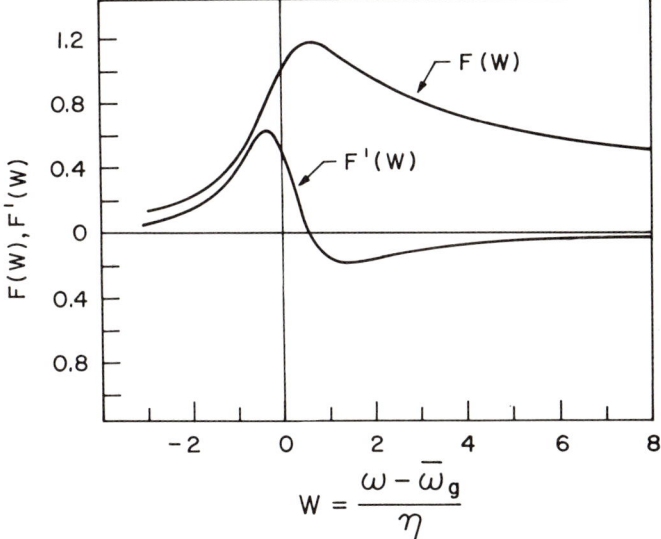

FIG. 158. Functions $F(W)$ (see Fig. 51) and $F'(W)$. These functions determine the line shape of broadened direct transitions between Landau levels (see text).

The second term in Eqs. (30.9) vanishes for interband transitions because of the orthogonality of the Bloch functions, while the first term usually can be neglected for intraband transitions. We shall confine our discussion to interband transitions: eqs. (30.9) gives for this case the selection rule $n_i = n_j$. The transition probability vanishes for $n_i \neq n_j$ because of the orthogonality of harmonic oscillator wave functions which belong to different quantum number n. The absorption spectrum for $\langle u_i | \mathbf{p} \cdot \hat{\mathbf{n}} | u_j \rangle \neq 0$ (allowed transitions for $H \neq 0$) is composed of a series of asymmetric peaks at the energies:

$$\omega_n = \omega_g + (n + \tfrac{1}{2})\left(\frac{1}{m_c^*} + \frac{1}{m_v^*}\right)\frac{H}{c}; \quad \text{for } n = 0, 1, 2, \ldots. \quad (30.10)$$

The shape of these peaks has already been discussed [Fig. (158)]. A plot of the energy of the nth peak as a function of H gives ω_g as the intercept. The slope yields the reduced mass $1/\mu = 1/m_c^* + 1/m_v^*$.

We shall now discuss the shape of magneto optical lines for indirect phonon-aided transitions.[333] As mentioned in Section 5, these transitions take place via an intermediate state. The photon produces vertical transitions to or from the intermediate state with n and k_z

conservation while the phonon transition conserves neither k_z nor n: transitions between any initial subband n_0 and any final one n_f are possible. Assuming that the probability for the phonon transition does not depend strongly on k_z, n_0, or n_f, we find that the total probability for transitions between subband n_0 and n_f for a frequency ω is proportional to:

$$\int_0^{\omega - \omega_g - H(n_0/m_v^* + n_f/m_c^*)/c} \omega_c^{-1/2} \omega_v^{-1/2} \, d\omega_c, \qquad (30.11)$$

with

$$\omega = \omega_c + \omega_v + \omega_g + \frac{H}{c}\left[\frac{n_0}{m_v^*} + \frac{n_f}{m_c^*}\right],$$

where ω_c and ω_v are the electron and hole energies (in the conduction and valence bands, respectively) measured from the band extrema. The integrand in Eq. (30.11) is simply the product of the density of states for the possible initial and final states: all transitions are equally probable. Integration of Eq. (30.11) yields for a given set of initial and final subbands (n_0 and n_f):

$$\begin{aligned} \varepsilon_i &= \text{constant} \quad \text{for } \omega > \omega_g + \frac{H}{c}\left[\frac{n_0}{m_v^*} + \frac{n_f}{m_c^*}\right] \\ \varepsilon_i &= 0 \quad \text{for } \omega < \omega_g + \frac{H}{c}\left[\frac{n_0}{m_v^*} + \frac{n_f}{m_c^*}\right]. \end{aligned} \qquad (30.12)$$

The spectrum of indirect transitions between magnetic subbands should be a collection of steps at the energies:

$$\omega = \omega_g + \frac{H}{c}\left[\frac{n_0}{m_v^*} + \frac{n_f}{m_c^*}\right] \qquad n_0, n_f = 0, 1, 2, \ldots. \qquad (30.13)$$

The energies at which the steps occur, once the values of n_0 and n_f have been assigned, yield both masses m_v^* and m_c^* independently. As discussed earlier, direct transitions in a magnetic field give only the reduced mass $1/\mu = 1/m_c^* + 1/m_c^*$ because of the selection rule $n_0 = n_f$. The derivative spectrum corresponding to that of Eq. (30.12) is a series of delta functions at the energies given in Eq. (30.13).

So far we have treated the case of spherical constant energy surfaces. It is possible[330] to find the solutions of Eq. (30.2) even for arbitrary masses (positive or negative). Such solutions are of interest for discussing the effect of a uniform magnetic field on transitions between critical points other than a valence band maximum and a conduction band minimum. Following Baldereschi and Bassani[330] we write the effective mass equation with respect to principal axes:

$$\left[\frac{1}{2m_x}\left(p_x + \frac{1}{c}A_x\right)^2 + \frac{1}{2m_y}\left(p_y + \frac{1}{c}A_y\right)^2 \right.$$
$$\left. + \frac{1}{2m_z}\left(p_z + \frac{1}{c}A_z\right)^2\right]F(\mathbf{r}) = \omega_H F(\mathbf{r}). \quad (30.14)$$

We choose the gauge:

$$\mathbf{A} = \tfrac{1}{2}\mathbf{H} \times \mathbf{r} + \tfrac{1}{2}\,\text{grad}[H_y xz + H_z xy - H_x yx] = H_y z \hat{\mathbf{x}} + (H_z x - H_x z)\hat{\mathbf{y}}, \quad (30.15)$$

where $\hat{\mathbf{x}}, \hat{\mathbf{y}}$ are the unit vectors along x and y. Equation (30.15) has the constants of the motion:

$$p_y; \quad S = \mathbf{H} \cdot \mathbf{p} + \frac{1}{c}(H_z H_y x) = \mathbf{H} \cdot \left(\mathbf{p} + \frac{1}{c}\mathbf{A}\right) = \mathbf{H} \cdot \boldsymbol{\pi}. \quad (30.16)$$

In Eq. (30.16) $\boldsymbol{\pi} = \mathbf{p} + (1/c)\mathbf{A}$ is the generalized momentum. The operator S is a constant of the motion since the component of the velocity along H (proportional to the expectation value of $\mathbf{H} \cdot \boldsymbol{\pi}$ for a given $\mathbf{H}$) is not affected by $\mathbf{H}$. We perform the transformation to the new canonically conjugate variables $\bar{p}$ and $\bar{q}$:

$$\bar{p} = p_z - \frac{m_c^2 H_z}{m_x m_y H^2} S$$
$$\bar{q} = \frac{c}{H_y}\left(p_x + \frac{1}{c}H_y z\right) + \frac{c}{m_y}\frac{H_x H_z}{H_y \beta}\left(p_z - \frac{S}{H_z}\right), \quad (30.17)$$

where

$$\frac{1}{m_c} = \frac{1}{H}\left[\frac{H_x^2}{m_y m_z} + \frac{H_y^2}{m_x m_z} + \frac{H_z^2}{m_x m_y}\right]^{1/2} \quad (30.18)$$

and

$$\beta = \left[\frac{H_x^2}{m_y} + \frac{H_y^2}{m_x}\right]. \quad (30.19)$$

By placing Eqs. (30.17) into Eq. (30.14) we obtain the transformed Hamiltonian:

$$\bar{\mathcal{H}} = \frac{H^2}{m_c^2 \beta}\bar{p}^2 + \frac{1}{c^2}\beta\bar{q}^2 + \frac{m_c^2}{2m_x m_y m_z}\left(\frac{S}{H}\right)^2, \quad (30.20)$$

for $m_c^2 > 0$. The Hamiltonian of Eq. (30.20) is that of a harmonic oscillator with a shift in the origin of energies (there is a trivial change in the sign if $\beta < 0$). Hence, we find the eigenvalue:

$$\omega_H = (n + \tfrac{1}{2})\omega_c + \frac{m_c^2}{2m_x m_y m_z}\left(\frac{S}{H}\right)^2, \tag{30.21}$$

with

$$\omega_c = \frac{\beta}{|\beta|}\frac{H}{m_c c}.$$

Therefore, Landau quantization occurs for any type of M_0 and M_3 critical point. For M_1 and M_2 critical points, Landau quantization occurs only if H is within the cone:

$$\frac{H_x^2}{m_y m_z} + \frac{H_y^2}{m_x m_z} + \frac{H_z^2}{m_x m_y} \geq 0. \tag{30.22}$$

It is easy to see that Eq. (30.22) represents the semiclassical condition of closed orbits, i.e., that the intersection of a plane perpendicular to H with the constant energy surface be *elliptical*. The constant of the motion (S/H) in Eq. (30.21) plays the role of k_z in Eq. (30.6). The matrix element for interband transitions is proportional to [Eq. (29.9)]:

$$\langle u_0 | \mathbf{p} \cdot \hat{\mathbf{n}} | u_f \rangle \int F^*_{n_0, k_{y0}, s_0}(\mathbf{r}) F_{n_f, k_{yf}, s_f}(\mathbf{r})\, dV_r. \tag{30.23}$$

Contrary to the case of isotropic masses discussed above, the wave functions F are proportional to harmonic oscillator wave functions with different orientations and centers. The center of each harmonic oscillator wave function, for instance, is shifted in $\mathbf{p}$-space from $p = 0$ an amount:

$$\frac{m_c^2 H_z}{m_x m_y H^2} S. \tag{30.24}$$

Hence, the wave functions of the valence and the conduction band need not be orthogonal for $n_0 = n_f$. This orthogonality is preserved for a general orientation of $\mathbf{H}$ if the effective mass tensors of the valence and conduction band have proportional components referred to the same axes. If this condition does not hold, but the principal axes of both mass tensors are the same, the selection rule still holds for $\mathbf{H}$ along the principal axes.

For transitions between extrema with completely general effective mass tensors, one should find peaks for any integer values of n_0 and n_f as a result of the breakdown in the selection rule. In this respect the situation is similar to that for indirect transitions and conduction and valence band masses can, in principle, be independently determined from the magnetooptical spectra.

The effect of a modulating parameter which preserves translational invariance, e.g., wavelength or stress on interband transitions for $H \neq 0$ is mainly to yield the derivative of the line shapes discussed above with respect to the energy gap; the change in ω_c caused by stress induced mass changes is usually negligible. Uniaxial stress modulation may introduce splittings and polarization effects which can be calculated as in Chapter VI. It is interesting to discuss also the case of electric field modulation (magnetoelectroreflectance or magnetoelectroabsorption). We shall assume isotropic masses for the purpose of our discussion. Two different situations can arise: parallel modulating electric field and static magnetic field ($\mathscr{E} \parallel \mathbf{H}$) and perpendicular or crossed[334, 335] fields ($\mathscr{E} \perp \mathbf{H}$). For $\mathscr{E} \parallel \mathbf{H}$ the effects of the electric and magnetic fields are decoupled; the magnetic field acts only on the motion perpendicular to it while $\mathscr{E}$ acts on the motion parallel to $\mathbf{H}$. The effective mass wave function can be written as the product of a wave function f_H fulfilling Eq. (30.5) for $k_z = 0$ times the solution f_E of a one-dimensional electrooptic equation. The magnetic selection rules are preserved and the problem of interband transitions between 2 Landau subbands with $n_0 = n_f$ is equivalent to that of transitions between one-dimensional bands (the Landau subbands) in an electric field, already treated in Section 24a. The line shapes of ε_i at each Landau singularity are given by Eq. (24.45) and Table XIV instead of the one-dimensional density of states $[\omega - \omega_g - \omega_c(n + \tfrac{1}{2})]^{1/2}$. If the effect induced by $\mathscr{E}$ is measured, e.g., $\varepsilon_i(\mathscr{E}, \mathbf{H}) - \varepsilon_i(0, \mathbf{H})$, one obtains the line shapes of Fig. 82.

The case of crossed fields ($\mathscr{E} \perp \mathbf{H}$) is somewhat more complicated since the effects of $\mathbf{E}$ and $\mathbf{H}$ are not *a priori* decoupled. The effective mass equation becomes, when the electric field in the x direction is added to Eq. (30.3):

$$-\frac{1}{2m^*}\left\{\frac{\partial^2}{\partial x^2} + \left(\frac{\partial}{\partial y} + i\frac{Hx}{c}\right)^2 - 2m^*\mathscr{E}x + \frac{\partial^2}{\partial z^2}\right\}F(\mathbf{r}) = \omega_{\mathbf{H},\mathscr{E}}F(\mathbf{r}).$$

(30.25)

The Hamiltonian of Eq. (30.25) is translationally invariant along y

and z. Hence, the wave function $F(\mathbf{r})$ can still be written as:

$$F(\mathbf{r}) \sim f(x) \exp[i(k_y y + k_z z)], \qquad (30.26)$$

with $f(x)$ the solution of the equation:

$$-\frac{1}{2m^*}\left\{\frac{\partial^2}{\partial x^2} - \left(k_y + \frac{Hx}{c}\right)^2 - 2m^*\mathscr{E}x\right\}f(x) = \left(\omega_{\mathbf{H},\mathscr{E}} - \frac{k_z^2}{2m^*}\right)f(x). \qquad (30.27)$$

It is possible by means of a shift of the origin of x to transform Eq. (30.27) into that of the harmonic oscillator. We first write Eq. (30.27) as:

$$-\frac{1}{2m^*}\left\{\frac{\partial^2}{\partial x^2} - \left(k_y + \frac{m^*c}{H}\mathscr{E} + \frac{Hx}{c}\right)^2\right\}f(x)$$

$$= \left\{\omega_{\mathbf{H},\mathscr{E}} - \frac{k_z^2}{2m^*} + \frac{m^*}{2}\left(\frac{c\mathscr{E}}{H}\right)^2 + \frac{k_y c\mathscr{E}}{H}\right\}f(x). \qquad (30.28)$$

Equation (30.28) is, except for a change in the origin of energies, that of a harmonic oscillator centered at:

$$x_0 = -\frac{c}{H}\left(k_y + \frac{m^*c}{H}\mathscr{E}\right). \qquad (30.29)$$

The energy, measured from the bottom (top) of the unperturbed band, is:

$$\omega_{\mathbf{H},\mathscr{E}} = \omega_c(n + \tfrac{1}{2}) + \frac{k_z^2}{2m^*} - \frac{k_y c\mathscr{E}}{H} - \frac{m^*}{2}\left(\frac{c\mathscr{E}}{H}\right)^2. \qquad (30.30)$$

The wave functions are harmonic oscillator wave functions with the origin shifted by x_0, as given in Eq. (30.29). In an optical transition, k_y and k_z are conserved. For interband transitions between bands with different effective masses, the harmonic oscillator origins are shifted differently for each band. The difference in shifts is proportional to $\mathscr{E}$: the electric field produces a violation of the $n_0 = n_f$ selection rule. Transitions with $n_0 \neq n_f$, forbidden for $\mathscr{E} = 0$, become allowed for $\mathscr{E} \neq 0$ and, as discussed earlier, an independent determination of m_c^* and m_v^* becomes possible. The electric field produces a linear and a quadratic shift in the energy of each state [Eq. (30.30)]. The linear shift is independent of m^* and, hence, the same for the valence and conduction bands: it is not observed in optical transitions. The quadratic shifts of the valence and conduction bands add: they produce a decrease (without broadening) in the energy at which each Landau peak occurs.[334] This decrease is:

$$\Delta\omega(\mathscr{E}, \mathbf{H}) = \frac{m_v^* + m_c^*}{2}\left(\frac{c\mathscr{E}}{H}\right)^2. \tag{30.31}$$

The imaginary part of the dielectric constant can now be calculated[334] by evaluating the two-center overlap integrals of harmonic oscillator wave functions. One obtains[334]:

$$\varepsilon_i = \frac{2H}{c} e^{-a^2/2} \sum_{n_v n_c} (2^{n_v+n_c} n_v! n_c!)^2$$

$$\times |\mathbf{p}\cdot\hat{\mathbf{n}}|^2 \sum_{m=0} \frac{(-1)^{n_c-m} n_c! n_v! 2^m}{m!(n_c-m)!(n_v-m)!} a^{n_c+n_v-2m}(\omega - \omega_{cv})^{-1/2}, \tag{30.32}$$

where

$$\omega_{cv} = \omega_g + \frac{H}{m_c^* c}(n_c + \tfrac{1}{2}) + \frac{H}{m_v^* c}(n_v + \tfrac{1}{2}) - \frac{m_c^* + m_v^*}{2}\left(\frac{c\mathscr{E}}{H}\right)^2$$

and

$$a = \mathscr{E} L_M/\omega_c, \qquad L_M = (c/H)^{1/2}.$$

The summation over m in Eq. (29.32) is carried out up to either n_v or n_c, whichever is smaller. The summation over n_c and n_v is extended to all values which make $(\omega - \omega_{cv})^{-1/2}$ real. L_M is the Landau radius, i.e., the extent of the harmonic oscillator wave function for $n = 0$. Hence, the parameter a is a measure of the ratio of the electric energy across the cyclotron packet to the cyclotron frequency, i.e., the ratio of the electric to the magnetic perturbation.

We show in Fig. 159a–b the effect of an electric field (in terms of the reduced variable a) on the intensity of several allowed and forbidden transitions, as calculated by Aronov[334] with Eq. (30.32). It is clear that a measurement of this field dependence can yield a rather conclusive identification of the Landau levels involved.

The line shape in a standard modulation measurement of $\varepsilon_i(\mathscr{E}, \mathbf{H}) - \varepsilon_i(0, \mathbf{H})$ (or any other optical property) is composed of 2 contributions. One contribution from the shift in energy of the $\varepsilon_i(0, \mathbf{H})$ line $\Delta\omega$, given in Eq. (30.31); this contribution has the shape $(\omega - \omega_g)^{-3/2}$ (for small $\mathscr{E}$) or its broadened counterpart F' (Fig. 158). The other contribution is caused by the change in oscillator strength with $\mathscr{E}$ (a decrease with increasing $\mathscr{E}$ for small $\mathscr{E}$ of transitions with $n_c = n_v$). This contribution has the shape of $(\omega - \omega_g)^{-1/2}$. It is easy to see that the ratio of the $(\omega - \omega_g)^{-3/2}$ to the $(\omega - \omega_g)^{-1/2}$ contribution blows up as $(\omega - \omega_g)/\omega_c$ tends to zero; hence, if the broadening frequency is much smaller than ω_c, the gap shift contribution to the modulation spec-

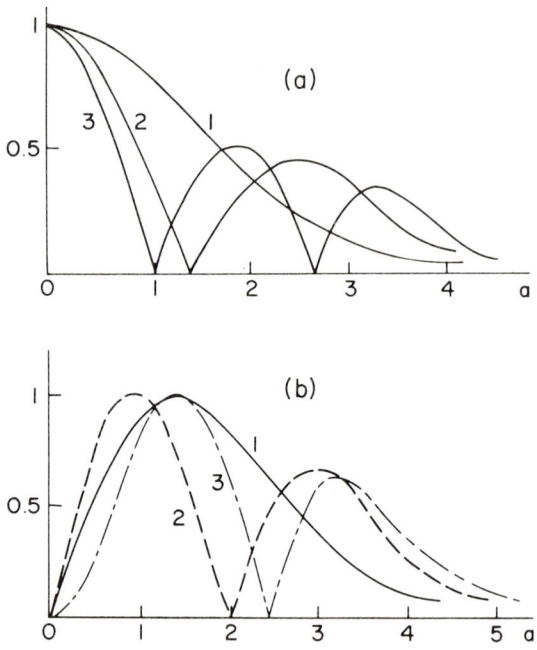

FIG. 159. Relative intensities of transitions in crossed electric and magnetic fields as a function of the parameter a. (a) Allowed transitions: (1) $n_c = n_v = 0$, (2) $n_c = n_v = 1$, (3) $n_c = n_v = 2$. (b) Forbidden transitions: (1) $n_c = 1, n_v = 0$, (2) $n_c = 1, n_v = 2$, (3) $n_c = 1$, $n_v = 3$. From A. G. Aronov, *Soviet Phys.–Solid State* (*English Transl.*) **5**, 402 (1963).

trum becomes dominant near ω_g for allowed transitions. The sign of the signal corresponds to a decrease in ε_i with increasing $\mathscr{E}$. Forbidden transitions yield peaks of the $(\omega - \omega_g)^{-1/2}$ type and opposite sign

Two other calculations of the effect discussed above have been given.[127,335] The treatment of Vrehen[127] is based on a perturbation calculation of the effect of $\mathscr{E}$ on the Landau levels: it is valid only for $a \ll 1$. This treatment can be easily generalized to degenerate bands, for which the treatment given above is not valid. The treatment by Aronov and Pikus[335] is based on the two-band model, in which

k · p interactions between bands other than the lowest conduction and the highest valence band are neglected. Within this approximation, and assuming nondegenerate bands, nonparabolicity effects are included.

In the perturbation treatment for nondegenerate bands, each Landau subband of quantum number n is shifted to second order in the electric field by the interaction with the $n-1$ and $n+1$ subbands: since the electric field Hamiltonian is proportional to the coordinate x, along the direction of $\mathscr{E}$ one has the selection rule $n = \pm 1$ for the matrix elements of the perturbation. The electric field mixing of $|n \pm 1\rangle$ to $|n\rangle$ relaxes the interband optical selection rule and makes the transitions $n_c = n_v \pm 1$ allowed. By this method it is easy to obtain the quadratic energy shift of Eq. (30.31). As shown above, this equation is valid exactly.

The method of Aronov and Pikus[335] is based on the two-band model of the **k · p** interaction between valence and conduction band.[336] The **k · p** Hamiltonian for the interaction between an s-like conduction band and a p-like valence band is:

$$\begin{Vmatrix} & |c\rangle & |v\rangle \\ & \omega_g/2 & kP \\ & kP & -\omega_g/2 \end{Vmatrix}. \tag{30.33}$$

We have neglected the free electron term $k^2/2$, i.e., we assume $m^* \ll 1$. The valence band state in Eq. (30.33) is a linear combination of $|X\rangle, |Y\rangle, |Z\rangle$ with coefficients proportional to the components of **k**. The other 2-valence band states are decoupled in the two-band approximation. We take the origin of energies to be at the center of the gap. P is the corresponding matrix element of **p**. Equation (30.33) has the eigenvalues:

$$\omega = \pm [(\omega_g/2)^2 + k^2 P^2]^{1/2}. \tag{30.34}$$

Equation (30.34) shows a striking formal analogy with the energy of a relativistic electron, $\omega_g/2$ being the equivalent of the rest energy mc^2 and P^2 the equivalent of c^2. The electron and hole bands correspond to electron and positron states. By expanding Eq. (30.34) in power series of k, we find:

$$P^2 = \omega_g/2m^*, \tag{30.35}$$

equivalent to c^2. If uniform crossed electric and magnetic fields are present, it is possible, by applying a Lorentz transformation [with c^2 replaced by Eq. (30.35)], to eliminate either the electric or the magnetic field and to reduce the problem to that of an electron in a uniform electric or magnetic field only. This problem has to be solved including nonparabolicity (relativistic electron in a magnetic field[335]) if the solution of the nonparabolic problem in the presence of $\mathscr{E}$ and H is desired. The electric field can be eliminated if $\beta = \mathscr{E}/(PH/c) < 1$; the magnetic field can be eliminated otherwise. The resulting effective electric and magnetic fields are[335,335a]:

$$\begin{array}{lll} \beta < 1 & \bar{H} = H(1-\beta^2)^{1/2} & \bar{\mathscr{E}} = 0 \\ \beta > 1 & \bar{\mathscr{E}} = \mathscr{E}(1-\beta^{-2})^{1/2} & \bar{H} = 0. \end{array} \quad (30.36)$$

Let us examine the case $\beta < 1$. The transformed frequency $\bar{\omega}$ and crystal momentum $\bar{\mathbf{k}}$ are, with respect to the moving frame:

$$\bar{\omega} = \frac{\omega + P(\mathbf{k} \cdot \hat{\mathbf{u}})}{(1-\beta^2)^{1/2}}; \quad \bar{k}_y = \frac{k_y + \beta\omega/P}{(1-\beta^2)^{1/2}}; \quad \bar{k}_z = k_z; \quad \bar{k}_x = k_x, \quad (30.37)$$

$\hat{\mathbf{u}}$ is the unit vector perpendicular to $\mathscr{E}$ and $\mathbf{H}$ (along y). Equation (30.37) indicates that while the $\mathbf{k}$-vector of the light is practically zero in the rest frame, this is not the case in the moving system because of the Doppler shift. Hence, the transitions occur with a change in k_y, i.e., between harmonic oscillator functions of different center. The selection rule $n_0 = n_f$ does not hold, as already found. The energy of the relativistic Landau levels is in the moving frame[335,335a]:

$$\bar{\omega} = [(\tfrac{1}{2}\omega_g)^2 + 2P^2\bar{H}(n + \tfrac{1}{2} \pm \tfrac{1}{2})]^{1/2}. \quad (30.38)$$

The energies ω (rest frame) at which the absorption peaks occur are:

$$\omega = (1-\beta^2)^{1/2}\{[(\tfrac{1}{2}\omega_g)^2 + 2P^2H(1-\beta^2)^{1/2}(n_c + \tfrac{1}{2} \pm \tfrac{1}{2})]^{1/2} \\ + [(\tfrac{1}{2}\omega_g)^2 + 2P^2H(1-\beta)^{1/2}(n_v + \tfrac{1}{2} \pm \tfrac{1}{2})]^{1/2}\} \quad (30.39)$$

The terms $\pm\tfrac{1}{2}$ in Eq. (30.39) correspond to the electron spin. By expansion of Eq. (30.39) in power series of H we find the results of Eq. (30.31).

The treatments given above obviously break down for the degenerate valence bands of the germanium and zinc blende materials. However, the Landau levels of the valence band in a magnetic field can be obtained, in the effective mass approximation, by solving 4 coupled differential equations.[332] Because of the degeneracy, the levels whose energies are close to the unperturbed edge of the band are not evenly spaced. At higher energies the levels become equally spaced (the degeneracy is lifted by the **k·p** perturbation) and the treatment for nondegenerate bands is valid. The solution of the coupled effective mass equations has the form[332]:

$$|na \pm \rangle = a_{1,n}^{\pm} \phi_{n-2}(x)u_{10} + a_{2,n}^{\pm} \phi_{n}(x)u_{20}$$
$$|nb \pm \rangle = b_{1,n}^{\pm} \phi_{n-2}(x)u_{30} + n_{2,n}^{\pm} \phi_{n}(x)u_{40}, \quad (30.40)$$

where the signs $+$ and $-$ refer to light and heavy holes respectively, ϕ_n and ϕ_{n-2} are harmonic oscillator functions of excitation number n and $n-2$, and u_{i0} are Bloch functions of the band edge. The effect of a crossed electric field ($\mathscr{E} \perp H$) on the energies of the states of Eq. (30.40) can be estimated by perturbation theory.[127,337] It is interesting to note that owing to the smoothness of the perturbation potential over an atomic distance, its matrix elements are proportional to $\langle u_{i0} | u_{j0} \rangle$: since the Bloch functions are orthogonal, no coupling takes place between the a and b ladder. It can be shown[127] that the coupling between the light- and heavy-hole states is small. The second order perturbation repulsion between states caused by $\mathscr{E}$ is expected to produce an equalization of the distance between Landau levels which, as we have mentioned, are unevenly spaced for small n. Figure 160 shows the effect of $\mathscr{E}$ on the Landau levels which has been calculated by Vrehen *et al.*[337] The equalization in level spacings for high a, i.e., high $\mathscr{E}$, is clearly apparent in this figure.

b. Experimental Techniques

All magnetooptical interband modulation experiments which have been reported so far have been performed at the National Magnet Laboratory. Bitter-type solenoid magnets, which provide fields up to about 100 kG at powers of about 5 MW were used. The measurements were performed in the Faraday configuration (direction of propagation of the light parallel to the magnetic field).

The stress modulation experiments were performed using lead zirconate–titanate transducers of the type described in Section 21. For magnetopiezoreflectance measurements the sample was glued to the transducer with Duco cement. For magnetopiezoabsorption measurements[194] the sample was mounted on a yoke-type transducer similar to that shown in Fig. 68; vacuum grease was found to give a good enough bond for low-temperature operation. Aggarwal *et al.*[194]

have reported the use of *web* germanium[338] in their magnetopiezo-absorption measurements. This material grows in the form of thin platelets with as-grown surfaces suitable for optical measurements. Hence it requires neither polishing nor etching.

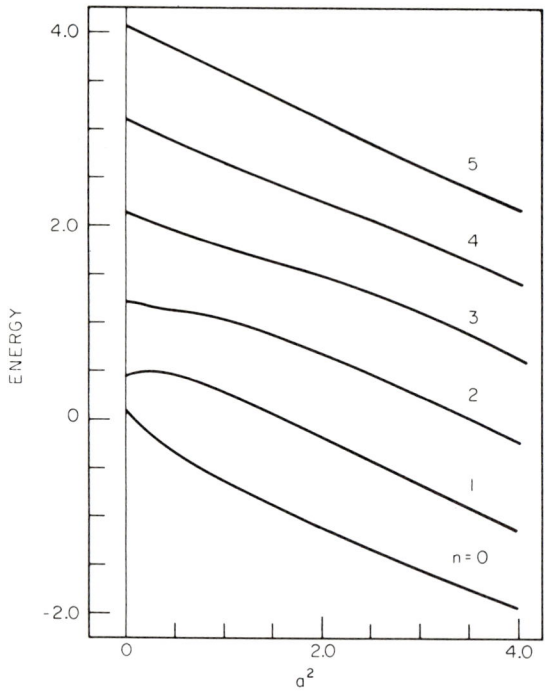

FIG. 160. Calculated energies (in units of the cyclotron energy of the light hole band) of the $a+$ light hole valence band series as a function of the parameter a^2. From Q. H. F. Vrehen, W. Zawadaski, and M. Reine, *Phys. Rev.* **158**, 702 (1967).

Magnetoelectroreflectance measurements have been reported for parallel fields.[70, 271] For these parallel field measurements by Pidgeon *et al.*,[271] one of the space–charge layer methods described in Section 24 was used, the field was applied with a semitransparent nickel electrode and a *photoresist* insulating layer. Measurements at room temperature with the electrolyte technique have also been reported.[70] Magnetoelectroabsorption measurements with crossed fields have been performed[127] on germanium of resistivity high enough to use the method of Fig. 93d. The contacts to the sample were made with

Viking LS232 liquid metal,† alloyed onto the germanium, and gold foil glued to the alloyed areas with Eccobond 57c.‡

c. *Results: Magnetopiezoreflectance*

We have shown in Fig. 161 the magnetopiezoreflectance spectra[227] of InSb near E_0 and $E_0 + \Delta_0$. In these measurements the magnetic field was swept and the photon energy kept constant. This method is preferable to sweeping ω in a constant field for observing detailed structure in a reduced photon energy range: the incident intensity does not change if the field is swept. Sweeping ω in a constant field

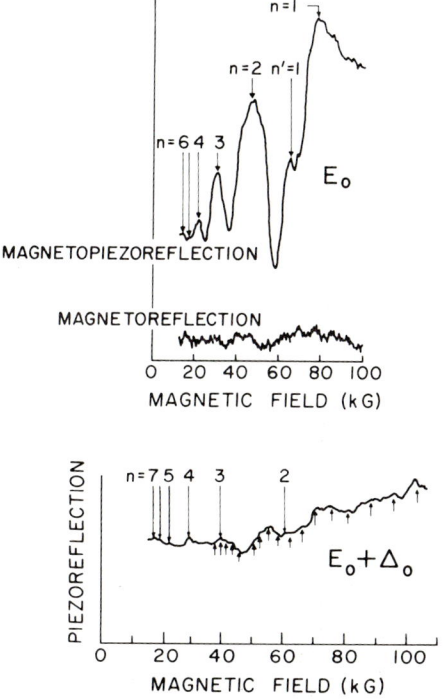

FIG. 161. Magnetopiezoreflection spectra near the E_0 and $E_0 + \Delta_0$ edges of InSb at room temperature. The photon energy has been kept constant in a swept magnetic field. For E_0, $\omega = 0.253$ eV; for $E_0 + \Delta_0$, $\omega = 1.13$ eV. The conventional magnetoreflection spectrum for $\omega = 0.253$ eV is also shown. From J. G. Mavroides, M. S. Dresselhaus, R. L. Aggarwal, and G. F. Dresselhaus, *Proc. Phys. Soc. Japan Suppl.* **21**, 184 (1966).

† Made by the Elmat Corporation, Mount View, California.
‡ Emerson and Cummings, Inc., Canton, Massachusetts.

is more appropriate for surveying a large energy region. Magneto-optical structure connected with the $E_0 + \Delta_0$ edge of germanium and zincblende-type materials is not seen in the conventional absorption or reflection spectra. The E_0 structure is hardly seen in the room temperature reflection spectra: Fig. 161 illustrates rather eloquently the

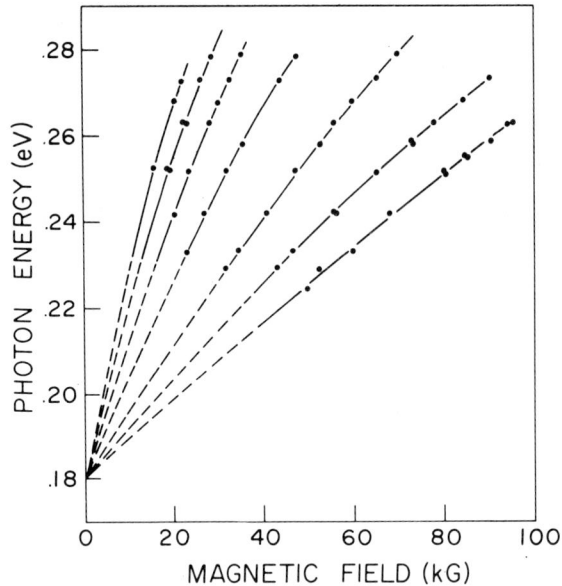

FIG. 162. Energy of the interband magnetopiezoreflectance peaks of InSb as a function of magnetic field. $T = 300°K$. From J. G. Mavroides, M. S. Dresselhaus, R. L. Aggarwal, and G. F. Dresselhaus, *Proc. Phys. Soc. Japan Suppl.* **21**, 184 (1966).

enhancement in sensitivity provided by the modulation techniques.

Figure 162 shows the energies of the various peaks of the E_0 magnetopiezoreflectance spectrum of InSb in the Faraday configuration at room temperature. The various transitions have been extrapolated to the zero-field gap of InSb. The nonlinear dependence of the peak energy on magnetic field H, observed at high H, is caused by the nonparabolic nature of the bands; the bands are expected to be very nonparabolic for energies above the band edge of the order of ω_g. The nonparabolicity is qualitatively similar to that described by Eq. (30.38) but for an accurate description one has to include the valence band degeneracy in the calculations. No detailed analysis of the curves of Fig. 162 has been given. Aggawal[339] has determined the effective mass of the split off valence band of InSb m_{so}^* from data similar to those of Fig. 161, but at 20°K. He also determined by this method the spin orbit splitting of the valence band of this material. These results are listed in Table XXII. Mavroides et al.[227] have reported $m_{so}^* = 0.074$ for germanium as determined with the magnetopiezoreflectance technique. A discussion of these results is given in Section 29e.

d. Results: Magnetopiezoabsorption

It has already been mentioned that the selection rule $n_0 = n_f$ does not hold for indirect transitions and hence that such transitions enable us to determine electron and hole masses separately. Figure 163 shows the magnetopiezoabsorption spectrum of germanium[194] in the neighborhood of the L_1 indirect absorption edge.

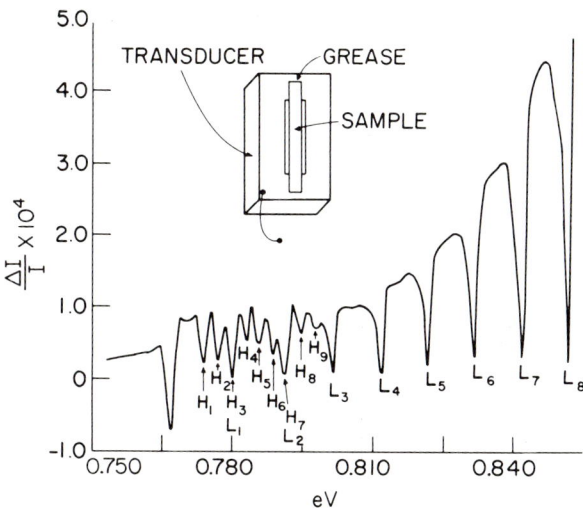

FIG. 163. Magnetopiezoabsorption peaks of the lowest indirect transitions of germanium for $H = 89.9$ kG along [100]. $T = 20°$K. $\mathscr{E} \parallel H \perp$ stress. From R. L. Aggarwal, M. D. Zuteck, and B. Lax, *Phys. Rev. Letters* **19**, 236 (1967).

The peaks of this spectrum have the shape predicted for magnetooptical modulation spectra of indirect transitions: broadened delta functions.

The 4 equivalent L_1 electron valleys fall into 2 equivalent pairs

TABLE XXII. Energy Gaps, Effective Masses, and g-Factors Determined by Magnetooptical Modulation Methods for Ge, InAs, and InSb

Material	E_0	Δ_0	m_{so}^* Expt	m_{so}^* Calc	g_{so}^* Expt	g_{so}^* Calc	$m_{e\perp}^*$ Expt	$m_{e\perp}^*$ Calc	$m_{e\|}^*$ Expt	$m_{e\|}^*$ Calc
Ge	0.795^b (300°K)	0.282^c	0.074^d 0.084^c			-13.2	0.080^e (20°K)	0.0791	1.49^e (20°K)	1.35
InAs	0.181^b (300°K)	0.38^f	0.14^f	0.11	-13.0^f	-13.9		0.13		1.4
InSb		$0.803^{d,f}$	0.10^f	0.14		-9.3		0.13		1.3

[a] The subindex s0 refers to the split off valence band at $\mathbf{k}=0$ while $^l m_{e\perp}^*$ and $m_{e\|}^*$ refer to the transverse and longitudinal masses of the lowest conduction band of germanium. The results are compared with those of $\mathbf{k} \cdot \mathbf{p}$ calculations.
[b] J. G. Mavroides, M. S. Dresselhaus, R. L. Aggarwal, and G. F. Dresselhaus, *Proc. Phys. Soc. Japan. Suppl.* **21**, 184 (1966).
[c] S. H. Groves, C. R. Pidgeon, and J. Feinleib, *Phys. Rev. Letters* **17**, 463 (1966).
[d] R. A. Aggarwal, *Bull. Am. Phys. Soc.* **12**, 100 (1967).
[e] R. L. Aggarwal, M. D. Zuteck, and B. Lax, *Phys. Rev. Letters* **19**, 236 (1967).
[f] C. R. Pidgeon, S. H. Groves, and J. Feinleib, *Solid State Commun.* **5**, 677 (1967).

when a [110] magnetic field is applied: [111]; [11$\bar{1}$] and [$\bar{1}$11]; [1$\bar{1}$1]. This last pair has the cyclotron mass $m_{c2}^* = (m_\perp m_\|)^{1/2}$ while the cyclotron mass of the [111], [11$\bar{1}$] pair is [Eq. (30.18)]:

$$m_{c1}^* = m_\perp \left(\frac{3}{2}\left[1 + \frac{m_\perp}{2m_\|}\right]\right)^{1/2}. \tag{30.41}$$

Aggarwal et al.[194] attributed the peaks of Fig. 163 to transitions from a single Landau level at the top of the valence band to the various conduction band Landau levels. Two series of peaks are evident: series L corresponds to the *light* electrons of mass m_{c1}^* while series H corresponds to the *heavy* electrons of mass m_{c2}^*. The spacing of the peaks decreases slightly with increasing quantum number n_f due to nonparabolicity effects, an extrapolation of this spacing to $n_f = 0$ yields, at 20°K:

$$\left.\begin{array}{ll} m_{c1}^* = (0.0963 \pm 0.001) & m_\|^* = 1.49 \pm 0.06 \\ m_{c2}^* = (0.344 \pm 0.008) & m_\perp^* = 0.0796 \pm 0.001 \end{array}\right\}. \tag{30.42}$$

Figure 164 shows the *light* electron mass m_{c1}^* obtained from data similar to those of Fig. 163, as a function of the energy of the levels

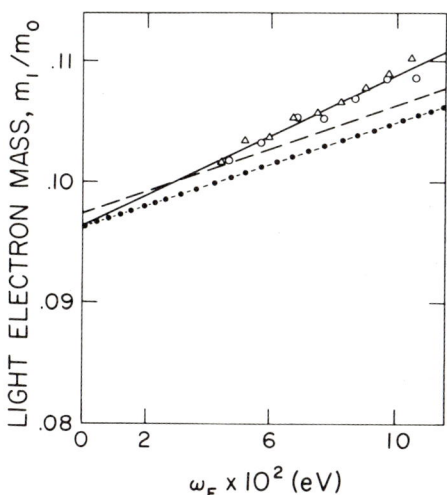

FIG. 164. Light electron mass from Fig. 163 as a function of energy ω_F above bottom of conduction band. From R. L. Aggarwal, M. D. Zuteck, and B. Lax, *Phys. Rev. Letters* **19**, 236 (1967). The solid curve is a least square fit. The dotted curve is obtained from Eq. (30.43), while the dashed curve is the result of a **k·p** calculation [M. Cardona and F. H. Pollak, *Phys. Rev.* **142**, 530 (1966)], which includes the effects of 15 bands.

above the bottom of the conduction band for $\mathbf{H} = 0$. This mass was obtained from the spacing between consecutive levels $\Delta\omega(n_f) = H/m_{c1}^* c$. The nonparabolicity of the conduction band is clearly apparent. This nonparabolicity must come mainly from an increase of $m_\perp^*$ with energy, which can be obtained with the two-band model [Eq. (30.34)]:

$$\frac{1}{m_\perp} = \frac{1}{m_\perp(0)} [(1 - (2\omega_F/\omega_g)], \tag{30.43}$$

to first order in ω_F/ω_g. The gap ω_g in Eq. (30.43) is the $L_{3'} - L_1$ gap of Fig. 29 (~ 2 eV). The dotted curve in Fig. 164 was obtained with Eq. (30.43) while the dashed curve was obtained from the $\mathbf{k} \cdot \mathbf{p}$ band structure by Cardona and Pollak,[103] which includes interactions with 15 bands. The experimental mass increases with $\Delta\omega$ about 30% faster than either calculation indicates. The source of this extra increase is not known.

e. Results: Magnetoelectroreflectance in Parallel Fields

Magnetoelectroreflectance data in parallel fields have been reported for Ge, GaSb, and InSb at room temperature[70] (electrolyte technique) and for InAs and InSb[271] at 1.5°K (semitransparent nickel electrode). Figure 165 shows the energy of the magnetoelectroreflectance peaks of germanium observed in the vicinity of the E_0 and $E_0 + \Delta_0$ edges as a function of the magnetic field. Data obtained with other magnetooptical techniques (magnetoabsorption, crossed fields magnetoelectroabsorption) are also included. The solid lines are the result of a calculation by the method of Pidgeon and Brown.[340] While the degeneracy of the valence band must be explicitly included in the calculation of the peaks associated with E_0, the valence band which takes part in the $E_0 + \Delta_0$ transitions is nondegenerate and a simple calculation of the type described in Section 30a is possible. The slopes of the $E_0 + \Delta_0$ peaks of Fig. 165 yield the sum of the valence and split-off conduction band inverse masses $1/m_e^* + 1/m_{so}^*$. The electron effective mass m_e^* can be obtained from the theoretical fit of the E_0 peaks if the valence band parameters are known: Groves et al.[70] obtained by this procedure $m_e^* = 0.042$ at room temperature. Using this value of m_e^* the slopes of the $E_0 + \Delta_0$ peaks yield $m_{so}^* = 0.084$. These values are compared in Table XXII with the results of theoretical calculations (see Appendix III). As shown by the solid curves in Fig. 165, the $E_0 + \Delta_0$ magnetooptical peaks observed experimentally whould split into doublets according to theory. This splitting is caused by the effective g-factors of the valence and the conduction band: for transitions in the Faraday configuration one must

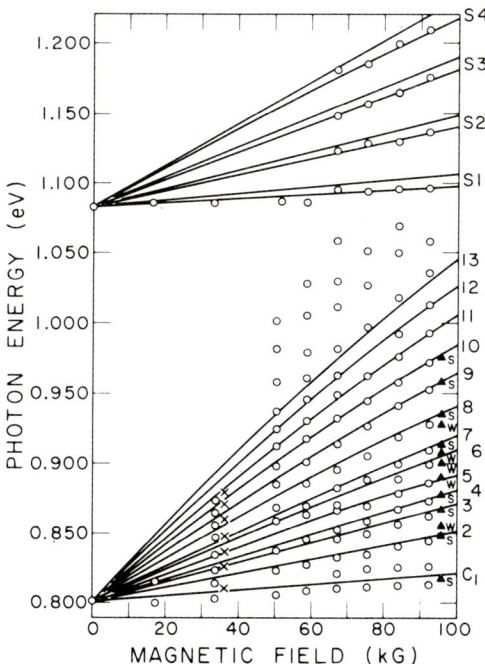

FIG. 165. Photon energies of magnetoelectroreflectance peaks as a function of H (circles) for germanium at room temperature. The crosses are data of Zwerdling *et al.* *Phys. Rev.* **108**, 1402 (1957). The triangles are data of Q. H. F. Vrehen, *Phys. Rev.* **145**, 675 (1966). (w and s indicate weak and strong structure, respectively.) The solid lines are calculated with the theory of C. R. Pidgeon and R. N. Brown, *Phys. Rev.* **146**, 575 (1966). From S. H. Groves, C. R. Pidgeon, and J. Feinleib, *Phys. Rev. Letters* **17**, 463 (1966).

have the selection rule $\Delta J_z = \pm 1$ (we assume the field H is either along [100] or [111]). Since both the conduction and the split off valence bands have $J = \frac{1}{2}$, one can represent their splitting by an effective g-factor (g_e^* and g_{so}^*). The theoretical splitting of the $E_0 + \Delta_0$ lines thus becomes

$$\mu H(g_e^* + g_{so}^*). \tag{30.44}$$

If circular polarization is used, only 1 of the peaks of this doublet appears for each direction of polarization (right σ_R or left σ_L). Equation (30.44) enables us to determine g_{so}^* from the experimental splitting if g_e^* is known. While this splitting is not seen in germanium at room temperature (Fig. 165) it has been seen in the magnetoelectroreflectance spectra of InSb and InAs at 1.5°K.[271] The results for InAs are shown in Fig. 166. The g-factor splitting is clearly re-

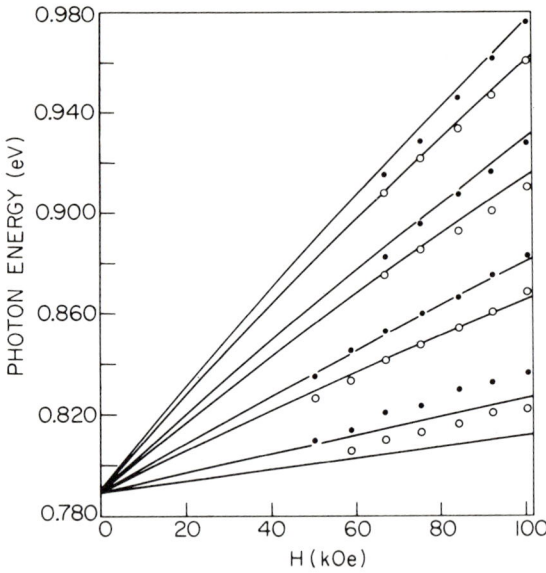

FIG. 166. Energy of the magnetoelectroreflectance peaks associated with the $E_0 + \Delta_0$ transitions of InAs at 1.8°K as a function of magnetic field. The solid lines give the theoretical energy obtained from the effective masses and g-factors including nonparabolicity of the bands, the black circles give the experimental σ_L values, and the white circles give the experimental σ_R values. From C. R. Pidgeon, S. H. Groves, and J. Feinleib, *Solid State Commun.* **5**, 677 (1967).

solved. The solid lines are theoretical lines which include small nonparabolicity effects. (These effects can be approximately described with Eq. (30.38) suitably modified to take into account the fact that $g^* \neq 2$.) Both m_{so}^* and g_{so}^* can be determined from the results of Fig. 166: the values obtained are listed in Table XXII. It has been noted[271] that the discrepancy between theoretical and experimental data for the lowest line of Fig. 162 may be simply related to a question of definition of the peak position.

f. Results: Magnetoelectroabsorption in Crossed Field

Figure 167 shows the magnetoabsorption and the magnetoelectroabsorption spectra of germanium for σ_+ polarization and $H \parallel [110]$ as obtained by Vrehen[127] at 77°K. The number of lines which appears in the modulation spectrum (taken with a dc bias electric field = 1000 V cm^{-1} and a modulation field of 250 V cm^{-1}) should be, according to the calculations, much larger than that obtained in the

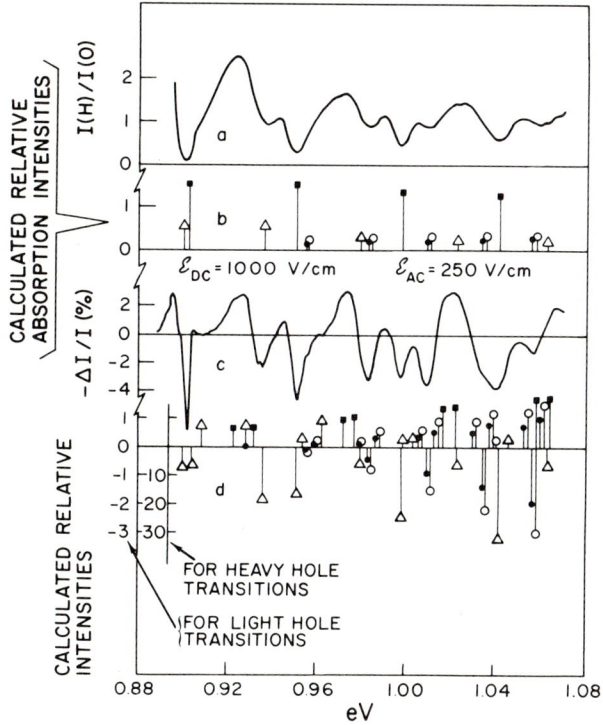

FIG. 167. Magnetoabsorption (a) and crossed fields magnetoelectroabsorption (c) spectrum of germanium as reported by Q. H. F. Vrehen [*Phys. Rev.* **145**, 674 (1966)]. The radiation was circularly polarized (σ_+) and the magnetic field of 96 kOe was parallel to [110]. The temperature was 77°K. The calculated line intensities are given in (b) and (d). Only allowed transitions contribute to (b) while allowed (negative signal) and forbidden (positive signal) contribute to the spectrum in crossed fields (c). The intensity of the heavy hole transitions has been divided by 10 in (d); they are relatively stronger in the modulation spectrum (d) than in the conventional spectrum (b). △ (a^+) ladder, ○ (a^-) ladder, ■ (b^+) ladder, ● (b^-) ladder.

conventional magnetoabsorption. This is a result of the violation of the $n_f = n_0$ selection rule. The large number of lines calculated is not clearly resolved experimentally, although an envelope which corresponds qualitatively to the broadened lines of Fig. 167d can be seen in Fig. 167c. Positive peaks in $-\Delta I/I$ are expected to correspond to a cluster of forbidden transitions. The number of possible transitions can be decreased by application of a uniaxial stress which removes the valence band degeneracy (magnetopiezoabsorption). This procedure enabled Vrehen[127] to resolve 2 forbidden transitions in the magnetoelectroabsorption spectrum of germanium.

Appendix I. A Few Relationships Involving Airy Functions

The purpose of this appendix is to familiarize the reader with the type of calculations with Airy functions performed in Section 24. Many of the relationships derived and the methods used here are found in the work of Aspnes.[76]

A simple substitution and differentiation under the integral sign shows that the differential equation:

$$d^2 F(x)/dx^2 = x F(x). \tag{AI.1}$$

[See Eq. (24.30)] has the 2 independent solutions:

$$\mathrm{Ai}(x) = \frac{1}{2\pi} \int_{-\infty}^{+\infty} ds\, \exp(is^3/3 + ixs)$$

$$\mathrm{Bi}(x) = \frac{1}{\pi} \int_{0}^{\infty} ds [\exp(-s^3/3 + xs) - \sin(\tfrac{1}{3}s^3 + xs)]. \tag{AI.2}$$

The Airy function $\mathrm{Ai}(x)$ is regular while $\mathrm{Bi}(x)$ diverges for $x \to \infty$. By using Eq. (AI.1) we find:

$$\begin{aligned}
\mathrm{Ai}^2(\omega) &= [\mathrm{Ai}^2(\omega) + 2tA(\omega)A'(\omega)] - 2tA(\omega)A'(\omega) \\
&= (d/dt)[\omega\, \mathrm{Ai}^2(\omega)] - 2A''(\omega)A'(\omega) \\
&= (d/dt)[\omega\, \mathrm{Ai}^2(\omega)] - (d/dt)[A'^2(\omega)].
\end{aligned} \tag{AI.3}$$

An integration of Eq. (AI.3) yields:

$$\int_{t_1}^{t_2} \mathrm{Ai}^2(\omega)\, d\omega = [\omega\, \mathrm{Ai}^2(\omega) - A'^2(\omega)]_{t_1}^{t_2},$$

and, in particular, Eq. (24.19):

$$\int_{t}^{\infty} \mathrm{Ai}^2(\omega)\, d\omega = \mathrm{Ai}'^2(t) - t\, \mathrm{Ai}^2(t). \tag{AI.4}$$

In order to derive Eq. (24.24) we shall first obtain an integral representation for $\text{Ai}^2(x)$. Equation (AI.2) enables us to write

$$\text{Ai}^2(x) = \frac{1}{4\pi^2} \int_{-\infty}^{+\infty} \int_{-\infty}^{+\infty} ds\, dt\, \exp\{i[s^3/3 + t^3/3] + ix(s+t)\}. \quad \text{(AI.5)}$$

We perform the change of variables:

$$t + s = \gamma; \quad \tfrac{1}{2}(t - s) = \alpha, \quad \text{(AI.6)}$$

which leads to

$$\text{Ai}^2(x) = \frac{1}{4\pi^2} \int_{-\infty}^{+\infty} \int_{-\infty}^{+\infty} d\gamma\, d\alpha\, \exp(i\gamma^3/12 + i\gamma\alpha^2 + ix\gamma)$$

$$= \frac{1}{4\pi^2} \int_{-\infty}^{+\infty} d\gamma\, \exp(i\gamma^3/12 + ix\gamma) \int_{-\infty}^{+\infty} d\alpha\, \exp i\gamma\alpha^2. \quad \text{(AI.7)}$$

The integration with respect to α in Eq. (AI.7) can be easily performed by replacing the real path of integration $(-\infty, +\infty)$ by the equivalent line $[-(1+i)s, +(1+i)s]$ with $s \to \infty$ (real). We obtain:

$$\int_{-\infty}^{+\infty} d\alpha\, \exp i\gamma\alpha^2 = \exp(\pi i/4) \int_{-\infty}^{+\infty} d\beta\, \exp(-\gamma\beta^2) = (\pi/\gamma)^{1/2} \exp(\pi i/4). \quad \text{(AI.8)}$$

Equations (AI.7) and (AI.8) yield:

$$\text{Ai}^2(x) = \frac{\pi^{1/2}}{4\pi^2} \int_{-\infty}^{+\infty} \gamma^{-1/2}\, d\gamma\, \exp[(\gamma^3/3 + x\gamma + \pi/4)i]$$

$$= \frac{\pi^{1/2}}{2\pi^2} \int_0^{\infty} \gamma^{-1/2}\, d\gamma\, \cos[\gamma^3/3 + x\gamma + \pi/4]. \quad \text{(AI.9)}$$

Similarly one can obtain for $\text{Ai}(x)\,\text{Bi}(x)$ the integral representation[76]:

$$\text{Ai}(x)\,\text{Bi}(x) = \frac{\pi^{1/2}}{2\pi^2} \int_0^{\infty} \gamma^{-1/2}\, d\gamma\, \sin[\gamma^3/12 + x\gamma + \pi/4]. \quad \text{(AI.10)}$$

Equations (AI.9) and (AI.10) yield the integral representation of Eq. (24.24).

We shall now derive the orthonormality relation for 2 Airy functions of arbitrary centers x and y:

$$\int_{-\infty}^{+\infty} dt\, \text{Ai}(t - x)\,\text{Ai}(t - y) = \delta(x - y); \quad \text{(AI.11)}$$

this equation has been used to obtain the normalization constant of Eq. (24.32) in Eq. (24.33). By using Eq. (AI.2) we can write:

$$\int_{-\infty}^{+\infty} dt\, \text{Ai}(t-x)\, \text{Ai}(t-y)$$

$$= \frac{1}{4\pi^2} \int_{-\infty}^{+\infty} \int_{-\infty}^{+\infty} dv\, du\, \exp(i[u^3/3 + v^3/3 - xu - yv])$$

$$\times \int_{-\infty}^{+\infty} \exp[i(u+v)t]\, dt. \qquad (AI.12)$$

By replacing in Eq. (AI.12) the integral over t by its value $2\pi\delta(u+v)$, and performing the integration over v, we obtain:

$$\int_{-\infty}^{+\infty} dt\, \text{Ai}(t-x)\, \text{Ai}(t-y) = \frac{1}{2\pi} \int_{-\infty}^{+\infty} du\, \exp(i[x-y]) = \delta(x-y).$$

Appendix II. Convolution Expression for $\varepsilon(\omega, \mathscr{E})$; Experimental Absorption Edges

Let us derive Eq. (26.2) for direct transitions near an M_0 critical point. We shall start from Eq. (24.39), which is valid for any orientation of $\mathscr{E}$ provided one uses the proper effective mass [Eq. (24.40)] in the expression of θ_1. In order to transform the integral in Eq. (24.39) we make use of the relationship:

$$\frac{4^{-1/3}}{\pi}\int_0^\infty t^{-1/2}\,dt\,\text{Ai}(t+x) = \text{Ai}^2(4^{-1/3}x), \qquad \text{(AII.1)}$$

[Eq. (B17) of Aspnes[76]], which is easily derived from Eq. (AI.2). Equation (24.39) can thus be written:

$$\varepsilon_i(\omega, \mathscr{E}) = \frac{4 \times 4^{-2/3}\pi |\mathbf{p}_{cv}\cdot\hat{\mathbf{n}}|^2 \mu|\mathscr{E}|}{\omega^2 \Omega}\int_{-\infty}^{\omega-\omega_g}\text{Ai}^2\!\left[-4^{-1/3}\frac{W}{\Omega}\right]\frac{dW}{\Omega},$$

$$= \frac{4^{1/3}\pi |\mathbf{p}_{cv}\cdot\hat{\mathbf{n}}|^2 \mu|\mathscr{E}|}{\omega^2 \Omega}\int_{(\omega_g-\omega)/\Omega}^\infty \text{Ai}^2(x)\,dx, \qquad \text{(AII.2)}$$

where we have written $4^{-1/3}\theta = \Omega$ and assumed, for simplicity, spherical energy bands.

By replacing Eq. (AII.1) into Eq. (AII.2) we find:

$$\varepsilon_i(\omega, \mathscr{E}) = \frac{|\mathscr{E}||\mathbf{p}_{cv}\cdot\hat{\mathbf{n}}|^2\mu}{\omega^2\Omega}\int_0^\infty t^{-1/2}\,dt\int_{(\omega_g-\omega)/\Omega}^\infty \text{Ai}(t+x)\,dx. \qquad \text{(AII.3)}$$

An integration by parts transforms Eq. (AII.3) into:

$$\varepsilon_i(\omega, \mathscr{E}) = \frac{2|\mathscr{E}|\mu|\mathbf{p}_{cv}\cdot\hat{\mathbf{n}}|^2}{\omega^2\Omega}\int_0^\infty t^{1/2}\,dt\,\text{Ai}\!\left(t + \frac{\omega_g - \omega}{\Omega}\right). \qquad \text{(AII.4)}$$

Making $t = (\omega' - \omega_g)/\Omega$ and using the expression for $\varepsilon_i(\omega, 0)$ of Eq. (24.21) we obtain:

$$\varepsilon_i(\omega, \mathscr{E}) = \frac{1}{\omega^2}\int_{-\infty}^{+\infty} d\omega'\,\omega'^2\,\varepsilon_i(\omega', 0)\left\{\frac{1}{\Omega}\text{Ai}\!\left(\frac{\omega'-\omega}{\Omega}\right)\right\}. \qquad \text{(AII.5)}$$

While Eq. (AII.5) has been derived for direct transitions near an M_0 critical point, it can be shown to be valid for other types of critical points and for indirect transitions.[231]

In the spirit of Franz's calculation[72] we shall assume that Eq. (AII.6) is valid for an experimental edge of the type given in Eq. (26.1). Since the experimental form is only valid near ω_0, one can take the ω'^2 factor out of the integral sign in Eq. (AII.5). We thus obtain, using Eq. (AI.2):

$$\varepsilon_i(\omega, \mathscr{E}) \propto \int_{-\infty}^{\omega_0} d\omega' \exp[\lambda(\omega' - \omega_0)]\left\{\frac{1}{\Omega} \text{Ai}\left(\frac{\omega' - \omega}{\Omega}\right)\right\}$$

$$= \frac{\exp(-\lambda\omega_0)}{2\pi\Omega} \int_{-\infty}^{+\infty} ds \exp\left(\frac{i}{3}s^3 - \frac{is\omega}{\Omega}\right) \int_{-\infty}^{\omega_0} d\omega' \exp\left(\left[i\frac{s}{\Omega} + \lambda\right]\omega'\right)$$

$$= \frac{1}{2\pi\Omega} \int_{\infty}^{\infty} \frac{\exp\left(\frac{i}{3}s^3 + \frac{is(\omega_0 - \omega)}{\Omega}\right)}{\left[i\frac{s}{\Omega} + \lambda\right]} ds. \qquad \text{(AII.6)}$$

The final integral in Eq. (AII.6) can be evaluated in the limit $(\omega_0 - \omega) \to \infty$ by the method of stationary phase (see Franz[72]). One obtains for this integral the residue of the pole $s = i\lambda\Omega$:

$$\varepsilon_i(\omega, \mathscr{E}) \propto \exp(\lambda^3\Omega^3/3 + \lambda(\omega - \omega_0)), \qquad \text{(AII.7)}$$

which is the expression of Eq. (26.3).

Appendix III. k·p Perturbation Theory, Effective Masses, and Effective g-Factors of Semiconductors[161,162]

The **k · p** perturbation Hamiltonian in the presence of spin orbit interaction [see Eqs. (9.5) and (9.6)] is:

$$\mathcal{H}' = \left(\mathbf{p} + \frac{1}{4c^2}\boldsymbol{\sigma} \times \nabla V\right) \cdot \mathbf{k} = \boldsymbol{\pi} \cdot \mathbf{k}. \qquad \text{(AIII.1)}$$

As already mentioned in Section 9 the spin dependent term $(1/4c^2)\boldsymbol{\sigma} \times \nabla V$ in Eq. (AIII.1), sometimes referred to as the **k**-dependent spin-orbit interaction, produces usually a rather small contribution to the energies[161]: we shall neglect it in our treatment. Application of second order perturbation theory to the **k · p** Hamiltonian in the neighborhood of a nondegenerate band extremum yields:

$$\omega_l(k) = \tfrac{1}{2}\sum_{\mu,\nu}\left(\frac{1}{m}\right)_{\mu\nu} k_\mu k_\nu, \qquad \text{(AIII.2)}$$

with the inverse effective mass tensor $(1/m)_{\mu\nu}$ given by:

$$\left(\frac{1}{m}\right)_{\mu\nu} = \delta_{\mu\nu} + \sum_{i\neq l}\frac{\langle l0|p_\mu|i0\rangle\langle i0|p_\nu|l0\rangle + \langle l0|p_\nu|i0\rangle\langle i0|p_\mu|l0\rangle}{\omega_{l0} - \omega_{i0}}, \qquad \text{(AIII.3)}$$

where the components of $\mathbf{k}(k_\mu)$ are measured with the extremum as origin. Equations (AIII.2) and (AIII.3) yield Eq. (2.9).

A similar treatment for degenerate bands requires the use of degenerate perturbation theory. This is equivalent to the diagonalization of the matrix:

$$\omega^{jh} = \sum_{\mu,\nu} D^{jh}_{\mu\nu} k_\mu k_\nu, \qquad \text{(AIII.4)}$$

where:

$$D^{jh}_{\mu\nu} = \tfrac{1}{2}\delta_{\mu\nu} + \sum_{i\neq l}\frac{\langle l_j 0|p_\mu|i0\rangle\langle i0|p_\nu|l_h 0\rangle}{\omega_{l0} - \omega_{i0}}. \qquad \text{(AIII.5)}$$

The superscripts j and h label the various states of the degenerate multiplet. Equation (AIII.4) can be written as an effective quasi-free electron Hamiltonian:

$$\mathscr{H}' = \sum_{\mu\nu} D_{\mu\nu}^{jh} p_\mu p_\nu. \qquad (AIII.6)$$

It can be shown[342] that Eq. (AIII.6) remains valid in the presence of a magnetic field provided one simply replaces $\mathbf{p}$ by $\mathbf{p} + \mathbf{A}/c$ in the usual manner [Eq. (30.2)]. Let us treat the case of an applied uniform magnetic field for a Kramers–degenerate extremum. The tensor $D_{\mu\nu}$ cannot be symmetrized in the presence of the magnetic field and hence Eq. (AIII.6) cannot be diagonalized to take the form of Eq. (30.14). We can, however, split $D_{\mu\nu}$ into its symmetric and antisymmetric parts. The symmetric part yields the Hamiltonian of Eq. (30.14) with the inverse effective mass tensor of Eq. (AIII.3). The antisymmetric part yields the contribution:

$$\mathscr{H}'' = -(i/c)[D_{xy}^A H_z + D_{yz}^A H_x + D_{zx}^A H_y],$$

where

$$D_{\mu\nu}^A = \frac{1}{2}\sum_{i\neq l} \frac{\langle l0|p_\mu|i0\rangle\langle i0|p_\nu|l0\rangle - \langle l0|p_\nu|i0\rangle\langle i0|p_\mu|l0\rangle}{\omega_{l0} - \omega_{i0}}, \qquad (AIII.7)$$

and $\langle l0|$ must be one of the members of a Kramers doublet. Equation (AIII.7) can be written in terms of an effective g-factor tensor which includes the contribution of the electron spin. The orbital contribution to this effective g-factor is:

$$g_{\mu\nu}^*(\text{orbital}) = -2\sum_{i\neq l} \frac{\langle l0|p_\mu|i0\rangle\langle i0|p_\nu|l0\rangle - \langle l0|p_\nu|i0\rangle\langle i0|p_\mu|l0\rangle}{\omega_{l0} - \omega_{i0}}$$

$$= 4\,\text{Im}\sum_{i\neq l} \frac{\langle l0|p_\mu|i0\rangle\langle i0|p_\nu|l0\rangle}{\omega_{l0} - \omega_{i0}}. \qquad (AIII.8)$$

Let us consider the $\Gamma_{2'}$ minimum of a germanium-type material [See Fig. 29]. Within the energy range of Fig. 29, the only states interacting with $\Gamma_{2'}$ via $\mathbf{k}\cdot\mathbf{p}$ are the $\Gamma_{25'}$ states: the interaction with Γ_{15} is forbidden by parity. We shall only include in Eqs. (AIII.3) and (AIII.8) matrix elements of $\mathbf{p}$ between $\Gamma_{2'}$ and $\Gamma_{25'}$: any other matrix elements are assumed to have a negligible contribution to m* and g* because of the large energy denominators involved. Using the wave functions of the spin orbit split $\Gamma_{25'}$ bands, given in Eq. (9.3), we find for the $\Gamma_{2'}$ the isotropic effective mass and g-factor[343]:

$$\frac{1}{m^*} = 1 + \frac{P^2}{3}\left(\frac{2}{E_0} + \frac{1}{E_0 + \Delta_0}\right)$$

$$g^* = 2\left[1 - \frac{P^2\Delta_0}{3E_0(E_0 + \Delta_0)}\right], \quad \text{(AIII.9)}$$

where $P^2 = 2|\langle\Gamma_{2'}|p_x|x\rangle|^2$, E_0 is the $\Gamma_{25'}(\Gamma_8^+) - \Gamma_{2'}$ gap, and Δ_0 the spin–orbit splitting of $\Gamma_{25'}$. It is known from energy band calculations that P^2 varies little among various materials of a given family[113,116]: for all group IV and group III–V materials $P^2 \simeq 23$ eV.[116]

Let us now calculate the effective mass m_{so}^* and the g-factor g_{so}^* of the Γ_7^+ valence band split from $\Gamma_{25'}$ by spin orbit interaction. Within the range of Fig. 29, **p** couples Γ_7^+ to $\Gamma_{2'}$ and Γ_{15}. It is customary to consider also the interaction to the next conduction band state $\Gamma_{12'}$.[20] Using the wave functions of Eq. (9.3) we find that m_{so}^* and g_{so}^* are isotropic and are given by:

$$\left(\frac{1}{m_{so}^*}\right) = -1 + \frac{1}{3}\left[\frac{P^2}{E_0 + \Delta_0} + \frac{2Q^2}{E_0' + \Delta_0} + \frac{2R^2}{E_0'' + \Delta_0}\right]$$

$$g_{so}^* = 2\left[-\frac{1}{3} - \frac{1}{3}\frac{P^2}{E_0 + \Delta_0} + \frac{1}{3}\frac{Q^2}{E_0' + \Delta_0} + \frac{1}{3}\frac{R^2}{E_0'' + \Delta_0}\right]. \quad \text{(AIII.10)}$$

Q and R are proportional to a matrix element of P between $\Gamma_{25'}$ and Γ_{15} and between $\Gamma_{25'}$ and $\Gamma_{12'}$, respectively. E_0' and E_0'' are the $\Gamma_8^+ - \Gamma_{15}$ and $\Gamma_8^+ - \Gamma_{12'}$ gaps, respectively. In order to calculate the values of m_{so}^* and g_{so}^* given in Table XXII, values of P^2, Q^2, and R^2 obtained from band calculations[103,113] were used. Quite accurate values are nevertheless obtained without the need of detailed band calculations, using $P^2 = 23$, $Q^2 = 15$, and $R^2 = 20$ (gaps in eV) for all group IV and III–V compounds.

Equation (AIII.3) yields for the transverse mass $m_{e\perp}^*$ (see Table XXII) of the L_1 conduction band minimum:

$$\frac{1}{m_{e\perp}^*} = 1 + \frac{P'^2}{2}\left[\frac{1}{E_1} + \frac{1}{E_1 + \Delta_1}\right]. \quad \text{(AIII.11)}$$

The matrix element P'^2 is practically the same as P^2.[103] The gaps at L, E_1, and $E_1 + \Delta_1$ are nearly equal to those at Λ.[103] It is possible to use a similar argument of obtain the nonparabolicity of the L_1

minimum discussed in Section 30d. We assume that only the interaction via P between L_1 and $L_{3'}$ takes place (we neglect the spin orbit splitting of $L_{3'}$). We thus obtain a Hamiltonian such as that of Eq. (30.33), which yields for the energy with the center of the gap as the origin:

$$\omega = \left[\left(\frac{\omega_g}{2}\right)^2 + \frac{k_\perp^2 P'^2}{2}\right]^{1/2} + \frac{k_\perp^2}{2}, \qquad \text{(AIII.12)}$$

where ω_g is equal to $E_1 + \Delta_1/2$. The transverse cyclotron effective mass which is obtained from magnetooptical experiment is[344]:

$$\frac{1}{m^*} = \frac{1}{k}\left(\frac{\partial \omega}{\partial k}\right)_{\omega_F} = 1 + \frac{P'^2}{2[(\tfrac{1}{2}\omega_g)^2 + \tfrac{1}{2}k^2 P'^2]^{1/2}} \simeq \frac{P'^2}{\omega_g(1 + (2\omega_F/\omega_g)}, \qquad \text{(AIII.13)}$$

where ω_F is the energy of the point of observation measured with respect to the bottom of the band. The approximate form at Eq. (AIII.13) is equivalent to Eq. (30.43). The error committed by neglecting the free electron mass (1) in Eq. (AIII.13) is less than 10% for germanium.

References

[1] J. Becquerel, *Le Radium* **4**, 328 (1907).
[2] P. Pringsheim and A. Kronenberger, *Z. Physik* **63**, 493 (1930).
[3] I. W. Obseimov and A. F. Prichotjo, *Phys. Z. Sowjetunion* **1**, 203 (1932).
[4] D. S. McClure, *Solid State Phys.* **8**, 1 (1959).
[5] H. C. Wolf, *Solid State Phys.* **9**, 1 (1959).
[6] R. Hilsch and R. W. Pohl, *Z. Physik* **57**, 145 (1929); **59**, 812 (1930).
[7] H. R. Philipp and H. Ehrenreich, *Phys. Rev.* **131**, 2016 (1963).
[8] J. Franck, H. Kuhn, and G. Rollefson, *Z. Physik* **43**, 11 (1927).
[9] J. Frenkel, *Phys. Rev.* **37**, 17 (1931); **37**, 1276 (1931).
[10] R. E. Peierls, *Ann. Physik* [5] **13**, 905 (1932).
[11] G. H. Wannier, *Phys. Rev.* **52**, 191 (1937).
[12] H. Hayashi and K. Katzuki, *J. Phys. Soc. Japan*, **5**, 380 (1950); **7**, 599 (1952).
[13] E. F. Gross and N. A. Karryev, *Izv. Akad. Sci. USSR*, **84**, 261 (1952).
[14] S. Nikitine, G. P. Perny, and M. Sieskind, *J. Phys. Radium*, **15**, S18 (1954).
[15] J. E. Eby, K. J. Teegarden, and D. B. Dutton, *Phys. Rev.* **116**, 1099 (1959).
[16] J. Bardeen, L. H. Hall, and F. J. Blatt, in "Photoconductivity Conference" (E. Breckeneridge, ed.), p. 146. Wiley, New York, 1954.
[17] R. J. Elliott, *Phys. Rev.* **108**, 1384 (1957).
[18] See J. C. Slater, Electronic Structure of Solids, in "Handbuch der Physik" (S. Flügge, ed.), Vol. 19, Springer, Berlin, 1956, and references therein.
[19] F. Herman, *Proc. Inst. Radio Eng.* **43**, 1073 (1955); *Physica* **20**, 801 (1954).
[20] G. Dresselhaus, A. Kip, and C. Kittel, *Phys. Rev.* **98**, 368 (1955).
[21] G. G. Macfarlane and V. Roberts, *Phys. Rev.* **97**, 1714 (1955).
[22] G. G. Macfarlane, T. P. McLean, J. E. Quarrington, and V. Roberts, *Phys, Rev.* **108**, 1377 (1957).
[23] D. G. Avery and P. L. Clegg, *Proc. Phys. Soc. (London)* **B66**, 512 (1953).
[24] R. J. Archer, *Phys. Rev.* **110**, 354 (1958).
[25] F. C. Jahoda, *Phys. Rev.* **107**, 1261 (1957).
[26] H. R. Philipp and E. A. Taft, *Phys. Rev.* **113**, 1002 (1959).
[27] J. N. Zemel, J. D. Jensen, and R. B. Schoolar, *Phys. Rev.* **140A**, 330 (1965).
[28] J. C. Phillips, *J. Phys. Chem. Solids* **12**, 208 (1960).
[29] L. M. Roth and B. Lax, *Phys. Rev. Letters* **3**, 217 (1959).
[30] See D. Brust, J. C. Phillips, and F. Bassani, *Phys. Rev. Letters* **9**, 94 (1962).
[31] J. Tauc and E. Antončik, *Phys. Rev. Letters* **5**, 253 (1960).
[32] J. Tauc and A. Abraham, *Proc. Intern. Conf. Phys. Semiconductors, Prague, 1960*, p. 375. Czech. Acad. Sci., Prague, 1961.

[33] M. Cardona, *J. Appl. Phys.* **32S**, 2151 (1961).
[34] J. C. Phillips, Solid State Phys. **18**, 1 (1966).
[35] M. Cardona, *Proc. Intern. Conf. Phys. Semiconductors, Paris, 1964*, p. 181. Dunod, Paris, 1965.
[36] See D. Pines, *in* "Elementary Excitations in Solids" p. 209. Benjamin, New York, 1963, and references therein.
[37] B. Segall, *Phys. Rev.* **125**, 109 (1962).
[38] H. Ehrenreich and H. R. Philipp, *Phys. Rev.* **128**, 1622 (1962).
[39] H. Mayer and M. H. El Naby, *Z. Physik* **174**, 289 (1963).
[40] M. H. Cohen and J. C. Phillips, *Phys. Rev. Letters* **12**, 662 (1964).
[41] A. W. Overhauser, *Phys. Rev. Letters* **13**, 190 (1964).
[42] M. S. Dresselhaus, *in* "Optical Properties and Electronic Structure of Metals and Alloys" (F. Abelès, ed.), p. 59. Wiley, New York, 1966.
[43] J. C. Phillips and L. Kleinman, *Phys. Rev.* **116**, 287 (1959).
[44] F. Herman, *Proc. Intern. Conf. Phys. Semiconductors, Prague, 1960*, p. 20. Czech. Acad. Sci., Prague, 1961. F. Herman, R. L. Kortum, C. D. Kuglin, and R. A. Short, *in* "Quantum Theory of Atoms, Molecules, and the Solid State" (P. O. Lowdin, ed.), p. 381. Academic Press, New York, 1968.
[45] D. Brust, *Phys. Rev.* **134**, A1337 (1964).
[46] M. Cardona and G. Harbeke, *J. Appl. Phys.* **34**, 813 (1963).
[47] J. C. Phillips, *Phys. Rev. Letters* **12**, 142 (1964).
[48] B. Velický and J. Sak, *Phys. Status Solidi* **16**, 147 (1966).
[49] C. B. Duke and B. Segall, *Phys. Rev. Letters* **17**, 19 (1966).
[50] Y. Toyozawa, M. Inoue, T. Inui, M. Okazaki, and E. Hanamura, *J. Phys. Soc. Japan Suppl.* **21**, 133 (1967).
[51] J. Hermanson, *Phys, Rev.* **150**, 660 (1966); **166**, 893 (1968).
[52] W. Paul and D. M. Warschauer, *J. Phys. Chem. Solids* **24**, 586 (1963).
[53] W. Paul, *in* "Optical Properties of Solids" (J. Tauc, ed.), p. 257. Academic Press, New York, 1966.
[54] E. F. Gross and B. P. Zakharchenia, *J. Phys. Radium* **18**, 68 (1957) and refs. therein.
[55] J. J. Hopfield and D. G. Thomas, *Phys. Rev.* **122**, 35 (1961).
[56] J. O. Dimmock and R. Wheeler, *Phys. Rev.* **125**, 1805 (1962).
[57] E. Burstein, G. S. Picus, R. F. Wallis, and F. Blatt, *Phys. Rev.* **113**, 15 (1959).
[58] S. Zwerdling, L. M. Roth, and B. Lax, *Phys. Rev.* **109**, 2207 (1958).
[59] W. E. Spicer, *Phys. Rev. Letters* **11**, 243 (1963).
[60] G. W. Gobeli and F. G. Allen, *Phys. Rev.* **137**, A245 (1965) and references therein.
[61] B. O. Seraphin and R. B. Hess, *Phys. Rev. Letters* **14**, 138 (1965).
[62] W. E. Engeler, H. Fritzsche, M. Garfinkel, and J. J. Tiemann, *Phys. Rev. Letters* **14**, 1069 (1965).
[63] G. W. Gobeli and E. O. Kane, *Phys. Rev. Letters* **15**, 142 (1965).
[64] B. Batz, *Solid State Commun.* **4**, 241 (1965).
[65] W. E. Engeler, M. Garfinkel, and J. J. Tiemann, *Phys. Rev.* **155**, 693 (1967).
[66] C. N. Berglund, *J. Appl. Phys.* **37**, 3019 (1966).
[67] I. Balslev, *Phys. Rev.* **143**, 636 (1966).
[68] F. H. Pollak, K. L. Shaklee, and M. Cardona, *Phys. Rev. Letters* **16**, 942 (1966).
[69] R. L. Aggarwal, L. Rubin, and B. Lax, *Phys. Rev. Letters* **17**, 8 (1966).
[70] S. H. Groves, C. R. Pidgeon, and J. Feinleib, *Phys. Rev. Letters* **17**, 463 (1966).
[71] M. Cardona, K. L. Shaklee, and F. H. Pollak, *Phys. Rev.* **154**, 696 (1967).
[72] W. Franz, *Z. Naturforsch.* **13a**, 484 (1958).

REFERENCES

[73] L. V. Keldysh, *Soviet Phy. JETP* (*English Transl.*) **34**, 788 (1958).
[74] J. Callaway, *Phys. Rev.* **130**, 549 (1963).
[75] K. Tharmalingham, *Phys. Rev.* **130**, 2204 (1963).
[76] D. E. Aspnes, *Phys. Rev.* **147**, 554 (1966).
[77] T. S. Moss, *J. Appl. Phys.* **32S**, 2136 (1961).
[78] A. Frova, P. Handler, F. A. Germano, and D. E. Aspnes, *Phys. Rev.* **145**, 575 (1966).
[79] C. B. Duke and M. E. Alferieff, *Phys. Rev.* **145**, 583 (1966).
[80] R. K. Willardson and A. Beer, eds., "Semiconductors and Semimetals," Vol. III. Academic Press, New York, 1967.
[81] J. Tauc, *in* "Progress in Semiconductors" (A. F. Gibson and R. E. Burgess, eds.) Vol. 9. Temple Press, London, 1965.
[82] C. G. Kuper and G. D. Whitfield, eds. "Polarons and Excitons." Plenum Press, New York, 1963.
[83] R. S. Knox, "Theory of Excitons." Academic Press, New York, 1963.
[84] D. Dexter and R. S. Knox, "Excitons," Wiley (Interscience), New York, 1965.
[85] D. L. Greenaway and G. Harbeke, "Optical Properties and Band Structure of Semiconductors." Pergamon Press, Oxford, 1968.
[86] T. S. Moss, "Optical Properties of Semiconductors." Butterworth, London and Washington, D.C., 1959.
[87] *J. Appl. Phys.*, Supplement to Vol. 32, 1961.
[88] D. G. Thomas, ed., "II-IV Semiconducting Compounds." Benjamin, New York, 1967.
[89] L. Onsager, *Phys. Rev.* **37**, 405 (1931); **38**, 2265 (1931).
[90] H. A. Kramers, *Atti del Congr. Intern. dei Fisici*, Como-Pavia-Roma, 1927, **2**, 545 (1928).
[91] M. Cardona, *Proc. NATO Summer Inst. Opt. Properties Solids*, (S. Nudelman and S. S. Mitra, eds.) Freiburg i./Br. 1966, Plenum Press 1969.
[92] F. Stern, *Solid State Phys.* **15**, 300 (1963).
[93] F. Seitz, "Modern Theory of Solids," p. 650. McGraw-Hill, New York, 1940.
[94] H. Ehrenreich, *in* "Optical Properties of Solids" (J. Tauc, ed.), p. 106. Academic Press, New York, 1966.
[95] M. Cardona, *Proc. Latin Am. School Phys. Santiago de Chile*, 1967, Benjamin, New York (1967).
[96] H. Ehrenreich and M. H. Cohen, *Phys. Rev.* **115**, 786 (1959).
[97] N. Wiser, *Phys. Rev.* **129**, 62 (1963); S. Adler, *Ibid* **126**, 413 (1962).
[98] N. Bloembergen, "Non-Linear Optics" p. 21, Benjamin, New York, 1965.
[99] W. G. Spitzer and H. Y. Fan, *Phys. Rev.* **106**, 882 (1957).
[100] J. C. Phillips, *Phys. Rev.* **104**, 1263 (1956).
[101] L. Van Hove, *Phys. Rev.* **89**, 189 (1953).
[102] F. Bassani, *in* "Optical Properties of Solids" (J. Tauc, ed.), p. 33. Academic Press, New York, 1966.
[103] M. Cardona and F. H. Pollak, *Phys. Rev.* **142**, 530 (1966).
[104] R. Sandrock, *Phys. Rev.* **169**, 642 (1968).
[105] L. I. Korovin, *Soviet Phys.-Solid State* (*English Transl.*) **1**, 1202 (1959).
[106] D. T. F. Marple, *J. Appl. Phys.* **33**, 539 (1964).
[107] M. Cardona, *J. Appl. Phys.* **36**, 2181 (1965).
[108] J. O. Dimmock, "Semiconductors and Semimetals" (R. K. Willardson and A. Beer, ed.), Vol. III. Academic Press, New York, 1967.
[109] G. D. Mahan, *Phys. Rev. Letters* **18**, 448 (1967).

[110] W. Kohn, *Solid State Phys.* **5**, (1957).
[111] O. Madelung, "Physics of III-V Compounds" Wiley, New York, 1964.
[112] M. Sturge, *Phys. Rev.* **127**, 768 (1962).
[113] F. H. Pollak, C. W. Higginbotham, and M. Cardona, *J. Phys. Soc. Japan Suppl.* **21**, 20 (1966); C. W. Higginbotham, F. H. Pollak, and M. Cardona, *Proc. Intern. Conf. Phys. Semiconductors, Moscow, 1968*, Vol. 1, p. 57.
[114] R. J. Elliott, *Phys. Rev.* **124**, 340 (1961).
[115] S. Flugge and H. Marshall, "Rechenmethoden der Quantentheorie." Springer, Heidelberg, 1952.
[116] M. Cardona, *J. Phys. Chem. Solids* **24**, 1543 (1963).
[117] M. Cardona and D. L. Greenaway, *Phys. Rev.* **131**, 98 (1963).
[118] D. T. F. Marple and H. Ehrenreich, *Phys. Rev. Letters* **8**, 87 (1962).
[119] M. L. Cohen and T. K. Bergstresser, *Phys. Rev.* **141**, 789 (1966).
[120] G. F. Koster and J. C. Slater, *Phys. Rev.* **96**, 1208 (1954).
[121] See G. Baldini, *Phys. Rev.* **128**, 1562 (1962).
[122] M. Inoue, M. Okazaki, Y. Toyozawa, T. Inui, and E. Hanamura, *Proc. Phys. Soc. Japan* **21**, 1850 (1966).
[123] M. Gershenzon, D. G. Thomas, and R. E. Dietz, *Proc. Intern. Conf. Phys. Semiconductors, Exeter, 1962*, p. 752. Inst. Phys. Physical Soc., London, 1962.
[124] R. J. Elliott and R. Loudon, *J. Phys. Chem. Solids* **15**, 146 (1960).
[125] W. J. Scouler and G. B. Wright, *Phys. Rev.* **133**, A736 (1964).
[126] B. Batz, *Solid State Commun.* **5**, 985 (1967).
[127] Q. H. F. Vrehen, *Phys. Rev.* **145**, 675 (1966).
[128] H. Y. Fan, *Phys. Rev.* **82**, 900 (1961).
[129] W. W. Scanlon, *Solid State Phys.* **9** (1959).
[130] W. Paul, M. Demeis, and L. X. Finegold, *Proc. Intern. Conf. Phys. Semiconductors, Exeter, 1962*, p. 712. Inst. Phys. Physical Soc., London, 1962.
[131] M. Cardona and H. S. Sommers, Jr., *Phys. Rev.* **122**, 1382 (1961).
[132] J. Tauc and A. Abraham, *J. Phys, Chem. Solids* **20**, 190 (1961).
[133] J. C. Woolley, A. G. Thompson, and M. Rubenstein, *Phys. Rev. Letters* **15**, 670 (1965); T. K. Bergstresser, M. L. Cohen, and E. W. Williams, *Ibid.* **15**, 662 (1965).
[134] Y. Toyozawa, *Prog. Theoret. Phys.* (*Kyoto*) **20**, 53 (1958).
[135] W. Martienssen, *J. Phys. Chem. Solids* **2**, 257 (1957).
[136] C. D. Salzberg and J. J. Villa, *J. Opt. Soc. Am.* **47**, 244 (1957).
[137] M. Cardona, *Phys, Rev.* **129**, 69 (1963).
[138] R. J. Zollweg, *Phys. Rev.* **111**, 113 (1958).
[139] M. Cardona and D. L. Greenaway, *Phys. Rev.* **133**, A1685 (1964).
[140] W. C. Dash and R. Newman, *Phys. Rev.* **99**, 1151 (1955).
[141] G. Harbeke, *Z. Naturforsch.* **19a**, 548 (1964).
[142] H. H. Sonpaa, *J. Phys. Chem. Solids* **23**, 407 (1962).
[143] H. R. Philipp and H. Ehrenreich, *Phys. Rev.* **129**, 1550 (1963).
[144] B. Velický, *Czech. J. Phys.* **B11**, 541 (1961).
[145] D. M. Roessler, *Brit. J. Appl. Phys.* **16**, 1119 (1965); **17**, 1313 (1966).
[146] H. C. Gatos and M. C. Lavine, *in* "Progress in Semiconductors" (A. F. Gibson and R. E. Burgess, eds.), Vol. 9 p. 1. Temple Press, London, 1965.
[147] W. J. Mc G. Tegart, "The Electrolytic and Chemical Polishing of Metals." Pergamon Press, Oxford, 1959.
[148] M. Cardona and D. L. Greenaway, *Phys. Rev.* **125**, 1291 (1962).

REFERENCES

[149] B. O. Seraphin and N. Bottka, *Phys. Rev.* **145**, 628 (1966).
[150] M. H. Cohen, *Phys. Rev. Letters* **12**, 664 (1964).
[151] D. G. Avery, *Proc. Phys. Soc.* **65B**, 425 (1952).
[152] D. W. Berreman, *J. Opt. Soc. Am.* **156**, 1784 (1966).
[153] R. Lindquist and A. W. Ewald, *J. Opt. Soc. Am.* **53**, 247 (1963).
[154] R. F. Potter, *Phys. Rev.* **150**, 562 (1966).
[155] R. Braunstein and E. O. Kane, *J. Phys. Chem. Solids* **23**, 1423 (1962).
[156] C. E. Moore, "Atomic Energy Levels" (Natl. Bur. Std. Circ. No. 467), **1**, 1949; **2**, 1952; **3**, 1958.
[157] F. Herman and S. Skillman, "Atomic Structure Calculations." Prentice-Hall, Englewood Cliffs, New Jersey, 1963.
[158] D. G. Thomas, *J. Phys. Chem. Solids* **15**, 86 (1960).
[159] K. Shindo, A. Morita, and H. Kamimura, *Proc. Phys. Soc. Japan* **20**, 2054 (1965).
[160] D. G. Thomas, *J. Appl. Phys.* **32S**, 2298 (1961).
[161] E. O. Kane, *J. Phys. Chem. Solids* **1**, 83 (1956); "Semiconductors and Semimetals" (R. K. Willardson and A. Beer, eds.), Vol. III. Academic Press, New York, 1967.
[162] E. O. Kane, *Phys. Rev.* **146**, 558 (1966).
[163] D. Brust, *Phys. Rev.* **134**, A1337 (1964).
[164] C. W. Higginbotham, F. H. Pollak, and M. Cardona, *Solid State Commun.* **5**, 513 (1967).
[165] G. Dresselhaus and M. S. Dresselhaus, *Phys. Rev.* **160**, 649 (1967).
[166] D. Brust, *Phys. Rev.* **139**, A489 (1965).
[167] G. Gilat and L. J. Raubenheimer, *Phys. Rev.* **144**, 390 (1966).
[168] P. J. Lin and L. Kleinman, *Phys, Rev.* **142**, 478 (1966).
[169] P. J. Lin and J. C. Phillips, *Phys. Rev.* **147**, 469 (1966).
[170] L. M. Falicov and P. J. Lin, *Phys. Rev.* **141**, 564 (1966).
[171] J. L. Shay, W. E. Spicer, and F. Herman, *Phys. Rev. Letters* **18**, 649 (1967).
[172] T. E. Fischer, *Phys. Rev.* **142**, 519 (1966).
[173] E. O. Kane, *Proc. Intern. Conf. Phys. Semiconductors*, Kyoto, 1966, p. 37 (1966).
[174] B. R. Cooper, H. Ehrenreich, and H. R. Philipp, *Phys. Rev.* **138**, A494 (1965).
[175] F. M. Mueller and J. C. Phillips, *Phys. Rev.* **157**, 600 (1967).
[176] G. Bonfiglioli and P. Brovetto, *Appl. Opt.* **3**, 1417 (1964).
[177] R. E. Drews, *Bull. Am. Phys. Soc.* **12**, 384 (1967).
[178] M. Garfinkel, J. J. Tiemann, and W. E. Engeler, *Phys. Rev.* **148**, 698 (1966).
[179] W. J. Scouler, *Phys. Rev. Letters* **18**, 445 (1967).
[180] A. Frova and P. J. Boddy, *Phys. Rev. Letters* **16**, 688 (1966).
[181] P. G. Wilkinson, *J. Opt. Soc. Am.* **45**, 1044 (1950).
[182] R. Haensel, C. Kunz, and B. Sonntag, *Phys. Rev. Letters* **20**, 262 (1968).
[183] See for instance, S. Seely, "Electron Tube Circuit." McGraw-Hill, New York, 1950.
[184] C. H. Anderson, Private communication.
[185] D. E. Aspnes, *Rev. Sci. Instr.* **38**, 1663 (1967).
[186] A. G. Thompson and J. C. Woolley, *Can. J. Phys.* **45**, 2597 (1967).
[187] J. J. Pankove, *J. Phys. Soc. Japan*, Suppl. **21**, 49 (1967).
[188] U. Gerhardt, D. Beaglehole, and R. Sandrock, *Phys. Rev. Letters* **19**, 309 (1967).
[189] I. Balslev, *Solid State Commun.* **5**, 315 (1967).
[190] A. G. Thompson, M. Cardona, K. L. Shaklee, and J. C. Woolley, *Phys, Rev.* **146**, 601 (1966).

[191] B. O. Seraphin, *Phys. Rev.* **140A**, 1716 (1965).
[192] B. Lax and J. G. Mavroides, *in* "Semiconductors and Semimetals" (R. K. Willardson and A. Beer, eds.), Vol. III, p. 321. Academic Press, New York, 1967.
[193] S. E. Schnatterly, *Bull. Am. Phys. Soc.* **13**, 387 (1968).
[194] R. L. Aggarwal, M. D. Zuteck, and B. Lax, *Phys. Rev. Letters* **19**, 236 (1967).
[195] A. Collier and C. Singleton, *J. Appl. Chem.* **6**, 495 (1956).
[196] E. C. Olson and C. D. Alway, *Anal. Chem.* **32**, 370 (1960).
[197] T. P. Pemsler, *Rev. Sci. Inst.* **23**, 274 (1957).
[198] I. Balslev, *J. Phys. Soc. Japan Suppl.* **21**, 101 (1966).
[199] G. Bonfiglioli, P. Brovetto, G. Busca, S. Levialdi, G. Palmieri, and E. Wanke, *Appl. Opt.* **6**, 447 (1967).
[200] A. Gilgore, P. J. Stoller, and A. Fowler, *Rev. Sci. Inst.* **38**, 1535 (1967).
[201] B. Batz, Ph.D. Thesis, Univ. Libre de Bruxelles, 1967.
[202] T. P. McLean "Progress in Semiconductors" (A. F. Gibson, ed.), Vol. 5. Heywood, London, 1960.
[203] M. Cardona and G. Harbeke, *Proc. Intern. Conf. Phys. Semiconductors, Paris, 1964*, p. 217. Dunod, Paris, 1965.
[204] A. Lempicki, J. Birman, H. Samelson, and G. Neumark, *Proc. Intern. Conf. Phys. Semiconductors, Prague*, p. 768, 1961.
[205] J. B. Arthur, A. C. Baynham, W. Fawcett, and E. G. S. Paige, *Phys. Rev.* **152**, 740 (1966).
[206] M. Cardona, *Phys. Rev.* **121**, 752 (1961).
[207] J. Hanus, J. Feinleib, and W. J. Scouler, *Phys. Rev. Letters* **19**, 16 (1967).
[208] A. K. Ghosh, *Solid State Commun.* **4**, 565 (1966).
[209] F. Herman, R. L. Kortum, C. D. Kuglin, and R. A. Short, *in* "Quantum Theory of Atoms and Molecules" (P. O. Lowdin, ed.). Academic Press, New York, 1968.
[210] W. Fawcett, *Proc. Phys. Soc. (London)* **85**, 931 (1965).
[211] H. Ehrenreich, H. R. Philipp, and D. J. Olechna, *Phys. Rev.* **131**, 2469 (1963); J. C. Phillips, *Ibid* **133**, A1020 (1964).
[212] E. Matatagui and M. Cardona, *Solid State Commun.* **6**, 313 (1968).
[213] H. Mayer and B. Hietel, *in* "Optical Properties and Electronic Structure of Metals and Alloys" (F. Abeles, ed.). Wiley, New York, 1966.
[214] J. L. Robins and P. E. Best, *Proc. Phys. Soc. (London)* **79**, 110 (1962).
[215a] C. S. Smith, Solid State Phys. **6**, (1958).
[215b] G. E. Pikus and G. L. Bir, *Soviet Phys.-Solid State (English Transl.)* **1**, 136 (1959); **1**, 1502 (1960).
[216] H. Brooks, *Advan. Electron. Electron Phys.* **7**, 85 (1955).
[217] A. A. Kaplyanskii, *Soviet Phys–Opt. Spectry.* **16**, 557 (1964).
[218] Y. Onodera and Y. Toyozawa, *J. Phys. Soc. Japan* **24**, 341 (1968).
[219] M. Lax and J. J. Hopfield, *Phys. Rev.* **124**, 115 (1961).
[220] G. F. Koster, *Solid State Phys.* **5**, 174 (1957).
[221] E. Erlbach, Phys. Rev. **150**, 767 (1966).
[222] U. Gerhardt and E. Mohler, *Phys. Status Solidi* **18**, K45 (1966).
[223] I. Balslev, *Rev. Sci. Inst.* **38**, 1528 (1967).
[224] A. Feldman, *Phys. Letters* **23**, 627 (1966).
[225] H. B. Huntington, *Solid State Phys.* **7**, 213 (1957).
[226] R. M. Bhagarva and M. Nathan, *Phys. Rev.* **161**, 695 (1967).
[227] J. G. Mavroides, M. S. Dresselhaus, R. L. Aggarwal, and G. F. Dresselhaus, *Proc. Phys. Soc. Japan Suppl.* **21**, 184 (1966).

REFERENCES

[228] J. Callaway, *Phys. Rev.* **134**, A998 (1964).
[229] K. S. Viswanathan and J. Callaway, *Phys. Rev.* **143**, 564 (1966).
[230] F. Aymerich and F. Bassani, *Nuovo Cimento* **48**, 358 (1967).
[231] D. E. Aspnes, P. Handler, and D. F. Blossey, *Phys. Rev.* **166**, 921 (1968).
[232] D. E. Aspnes, *Phys. Rev.* **153**, 972 (1967).
[233a] P. Argyres, *Phys. Rev.* **126**, 1386 (1967).
[233b] V. S. Vavilov, V. B. Stopachinskii, and V. Sh. Chanbarisov, *Soviet Phys–Solid State (English Transl.)* **8**, 2126 (1967).
[233c] Yu. A. Kurskii and V. B. Stopachinskii, *Fiz. i Tech. Poluprov.* **1**, 106 (1967).
[234] H. A. Antonsiewicz, *in* "Handbook of Mathematical Functions" (M. Abramowitz and I. A. Stegun, eds.). U.S. Dept. Commerce, Natl. Bur. Std., Washington, D.C., 1964.
[235] J. C. Phillips and B. O. Seraphin, *Phys. Rev. Letters* **15**, 107 (1965).
[236] J. C. Phillips, *Phys. Rev.* **146**, 584 (1966).
[237] R. Enderlein, *Phys. Status Solidi* **20**, 295 (1967); D. E. Aspnes, *Ibid* **23**, K79 (1967).
[238] D. E. Aspnes, Private communication.
[239] U. Rössler and N. Bottka, *Solid State Commun.* **5**, 939 (1968).
[240] C. M. Penchina, *Phys. Rev.* **138**, A924 (1965).
[241] M. Chester and L. Fritsche, *Phys. Rev.* **139**, A518 (1965).
[242] Y. Yacoby, *Phys. Rev.* **140**, A263 (1965).
[243] V. Rehn, *Bull. Am, Phys. Soc.* **11**, 205 (1966).
[244] L. D. Landau and E. M. Lifschitz, "Quantum Mechanics." Addison–Wesley, Reading, Massachusetts, 1965.
[245] L. J. Slater, *in* "Handbook of Mathematical Functions" (M. Abramowitz and I. A. Stegun, eds.), p. 503. U.S. Dept. Commerce, Natl. Bur. Std., Washington, D.C. 1964.
[246] R. Kern, *J. Phys. Chem. Solids* **23**, 249 (1962); G. Harbeke, *Ibid* **24**, 957 (1963).
[247] A. Frova and P. J. Boddy, *Phys. Rev.* **153**, 606 (1967).
[248] A. Frova, P. J. Boddy, and Y. S. Chen, *Phys. Rev.* **157**, 700 (1967).
[249] C. Gähwiller, *Solid State Commun.* **5**, 65 (1967); *Phys. Kond. Mat.* **6**, 269 (1967).
[250] B. C. Frazer, H. R. Danner, and R. Pepinsky, *Phys. Rev.* **100**, 745 (1955).
[251] A. H. Kahn and A. J. Leyendecker, *Phys. Rev.* **135**, A132 (1964).
[252] M. Cardona, *Phys. Rev.* **140**, A651 (1965).
[253] J. D. Zook and T. N. Casselman, *Phys. Rev. Letters* **17**, 960 (1966).
[254] J. B. Brews, *Phys. Rev. Letters* **18**, 662 (1967).
[255] R. C. Casella, *Phys. Rev.* **154**, 743 (1962).
[256] R. A. Forman and M. Cardona, *in* "II–VI Semiconducting Compounds" (D. G. Thomas, ed.), p. 100. Benjamin, New York, 1967.
[257] Y. Yacoby, *Phys. Rev.* **142**, 445 (1966).
[258] V. Rehn and D. S. Kyser, *Phys. Rev. Letters* **18**, 848 (1967).
[259] R. A. Forman, Private communication.
[260] W. E. Wilhelm, *Z. Angew. Math. Mech.* **45**, 121 (1965); S. Oberlander and W. E. Wilhelm, *Phys. Status Solidi* **12**, 569 (1965).
[261] L. M. Milne-Thomson, *in* "Handbook of Mathematical Functions" (M. Abramowitz and I. A. Stegun, eds.), p. 567. U.S. Dept. Commerce Natl. Bur. Std., Washington, D.C., 1964.
[262] C. G. B. Garrett and W. H. Brattain, *Phys. Rev.* **99**, 376 (1955).
[263] R. H. Kingston and S. Neustadter, *J. Appl. Phys.* **26**, 718 (1955).
[264] R. Seiwatz and M. Green, *J. Appl. Phys.* **29**, 1034 (1958).

[265] A. R. Plummer, in "The Electrochemistry of Semiconductors" (P. J. Holmes, ed.). Academic Press, New York, 1962.
[266] C. A. Mead and W. G. Spitzer, *Phys. Rev. Letters* **10**, 471 (1963); *Phys. Rev.* **134**, A713 (1964).
[267] B. O. Seraphin, *Surface Sci.* **8**, 399 (1967).
[268] W. L. Brown, *Phys. Rev.* **100**, 590 (1955).
[269] B. O. Seraphin, *Proc. Intern. Conf. Phys. Semiconductors, Paris, 1964*, p. 165. Dunod, Paris, 1965.
[270] K. L. Shaklee (to be published).
[271] C. R. Pidgeon, S. H. Groves, and J. Feinleib, *Solid State Commun.* **5**, 677 (1967).
[272] R. Ludeke and W. Paul, in "II–VI Semiconducting Compounds" (D. G. Thomas, ed.), p. 123. Benjamin, New York, 1967.
[273] E. Y. Wang, W. A. Albers, and C. E. Bleil, in "II–VI Semiconducting Compounds" (D. G. Thomas, ed.), p. 136. Benjamin, New York, 1967.
[274] M. Cardona, K. L. Shaklee, and F. H. Pollak, *Proc. Phys. Soc. Japan Suppl.* **21**, 20 (1966).
[275] M. Chester and P. M. Wendland, *Phys. Rev.* **140**, A1384 (1965).
[276] See, for instance, C. G. B. Garrett, in "The Electrochemistry of Semiconductors" (P. J. Holmes, ed.) Academic Press, New York, 1962.
[277] R. Williams, *Phys. Rev.* **126**, 442 (1962); **17**, 1487 (1960).
[278] J. F. Dewald, in "Semiconductors", p. 727. (N. B. Hannay, ed.). Reinhold, New York, 1960.
[279] D. R. Turner, in "The Electrochemistry of Semiconductors" (P. J. Holmes, ed.). Academic Press, New York, 1962.
[280] W. H. Brattain and C. G. B. Garrett, *Bell System Tech. J.* **34**, 129 (1955).
[281] M. Cardona, K. L. Shaklee, and F. H. Pollak, *Phys. Letters* **23**, 37 (1966).
[282] F. Lukeš and E. Schmidt, *Phys. Letters* **23**, 413 (1966).
[283] D. E. Aspnes and M. Cardona, *Phys. Rev.* **173**, 714 (1968).
[284] V. S. Vavilov, A. F. Plotmikov, and G. V. Zakhvathin, *Soviet Phys.–Solid State (English Transl.)* **1**, 894 (1959).
[285] S. Brahms and M. Cardona, *Solid State Commun.* **6**, 733 (1968).
[286] T. S. Moss, *J. Appl. Phys. Suppl.* **32**, 2136 (1961).
[287] H. D. Rees, *Solid State Commun.* **5**, 365 (1967).
[288] E. Gutsche and H. Lange, *Proc. Intern. Conf. Phys. Semiconductors, Paris, 1964*, p. 129. Dunod, Paris, 1965.
[289] R. E. Drews, *Appl. Phys. Letters* **9**, 347 (1966).
[290] Y. Hamakawa, F. A. Germano, and P. Handler, *Phys. Rev.* **167**, 703 (1968).
[291] E. G. S. Paige and H. D. Rees, *Phys. Rev. Letters* **16**, 444 (1966).
[292] S. Ballaro, A. Balzarotti, and V. Grasso, *Phys. Letters* **23**, 405 (1966).
[293] T. O. Poehler and D. Abraham, *Phys. Letters* **23**, 523 (1966).
[294] G. G. Macfarlane, T. P. McLean, J. E. Quarrington, and V. Roberts, *Proc. Phys. Soc.* **71**, 863 (1958).
[295] B. O. Seraphin, *Proc. Phys. Soc. (London)* **87**, 239 (1966); *J. Appl. Phys.* **37**, 721 (1966).
[296] K. L. Shaklee, F. H. Pollak, and M. Cardona, *Phys. Rev. Letters* **15**, 883 (1965).
[297] M. Cardona, *J. Appl. Phys.* **32**, 958 (1961).
[298] M. Cardona, P. McElroy, F. H. Pollak, and K. L. Shaklee, *Solid State Commun.* **4**, 319 (1966).
[299a] S. H. Groves and W. Paul, *Phys. Rev. Letters* **11**, 194 (1963).
[299b] S. Bloom (Private communication).
[300] E. Gutsche and H. Lange, *Phys. Status Solidi* **22**, 229 (1967).

REFERENCES

343

[301] Y. Hamakawa, P. Handler, and F. A. Germano, *Phys. Rev.* **167**, 709 (1968).
[302] J. D. Axe and R. Hammer, *Phys. Rev.* **162**, 700 (1967).
[303] W. Y. Laing and A. D. Joffe, *Phys. Rev. Letters* **20**, 59 (1968).
[304] T. K. Bergstresser and M. L. Cohen, *Phys. Rev.* **164**, 1069 (1967).
[305] M. L. Belle, *Soviet Phys.–Solid State* **5**, 2401 (1964).
[306] D. E. Aspnes, Private communication.
[307] P. J. Lin, W. Saslow, and M. L. Cohen, *Solid State Commun.* **5**, 973 (1967).
[308] F. Vázquez, R. A. Forman, and M. Cardona, *Phys. Rev.* **176**, 905 (1968).
[309] N. O. Folland, *Phys. Rev.* **158**, 764 (1967).
[310] P. M. Lee, *Phys. Rev.* **135**, A1110 (1964).
[311] L. A. Lott and D. W. Lynch, *Phys. Rev.* **141**, 681 (1966).
[312] W. J. Scouler, *Phys. Rev.* **178**, 1353 (1969).
[313] J. Feinleib, *Phys. Rev. Letters* **16**, 1200 (1966).
[314] A. Prostak and W. N. Hansen, *Phys. Rev.* **160**, 600 (1967).
[315] W. N. Hansen, T. Kuwana, and A. Osteryoung, *Anal. Chem.* **38**, 1810 (1966).
[316] H. L. Stadler, *Phys. Rev. Letters* **14**, 979 (1965).
[317] R. A. Forman, D. E. Aspnes, and M. Cardona (to be published).
[318] G. E. Pake and E. M. Purcell, *Phys. Rev.* **74**, 1184 (1948).
[319] N. J. Harrick, *J. Phys. Chem. Solids* **8**, 106 (1958); *Phys. Rev.* **125**, 116 (1962).
[320] G. Samoggia, A. Nuccioti, and G. Chiarotti, *Phys. Rev.* **144**, 749 (1966).
[321] I. Balslev, *Phys. Letters* **24A**, 113 (1967).
[322] F. H. Pollak, M. Cardona, and K. L. Shaklee, *Phys. Rev. Letters* **16**, 942 (1966).
[323] F. H. Pollak and M. Cardona, *Phys. Rev.* **172**, 816 (1968).
[324] A. G. Thompson and J. C. Woolley, *Can. J. Phys.* **45**, 2597 (1967).
[325] J. S. Kline, F. H. Pollak, and M. Cardona, *Helv. Phys. Acta* **41**, 968 (1968).
[326] J. Barber (unpublished).
[327] M. Cuevas and H. Fritzsche, *Phys. Rev.* **137**, A1847 (1965).
[328] H. R Philipp, W. C. Dash, and H. Ehrenreich, *Phys. Rev.* **127**, 762 (1962).
[329] W. Kohn and J. M. Luttinger, *Phys. Rev.* **98**, 915 (1955).
[330] A. Baldereschi and F. Bassani, *Phys. Rev. Letters* **19**, 66 (1967).
[331] U. Gerhardt, *Phys. Rev. Letters* **15**, 401 (1965).
[332] J. M. Luttinger and W. Kohn, *Phys. Rev.* **97**, 869 (1955).
[333] L. M. Roth, B. Lax, and S. Zwerdling, *Phys. Rev.* **114**, 90 (1959).
[334] A. G. Aronov, *Soviet Phys.–Solid State* (English Transl.) **5**, 402 (1963).
[335] A. G. Aronov and G. E. Pikus, *J. Phys. Soc. Japan Suppl.* **21**, 608 (1966).
[335a] S. S. Schweber, "Relativistic Quantum Field Theory," p. 103. Harper, New York, 1962.
[336] E. O. Kane, *in* "Semiconductors and Semimetals" (R. K. Willardson and A. Beer, eds.), Vol. I, p. 75. Academic Press, New York, 1967.
[337] Q. H. F. Vrehen, W. Zawadski, and M. Reine, *Phys. Rev.* **158**, 702 (1967).
[338] S. N. Dermatis, *J. Appl. Phys.* **36**, 3396 (1965).
[339] R. L. Aggarwal, *Bull. Am. Phys. Soc.* **12**, 100 (1967).
[340] C. R. Pidgeon and R. N. Brown, *Phys. Rev.* **146**, 575 (1966).
[341] C. Kittel, "Quantum Theory of Solids," p. 182. Wiley, New York, 1963.
[342] J. M. Luttinger, *Phys. Rev.* **84**, 814 (1951).
[343] C. Kittel, "Quantum Theory of Solids," p. 283. Wiley, New York, 1963.
[344] C. Kittel, "Quantum Theory of Solids," p. 227. Wiley, New York, 1963.

Author Index

Numbers in parentheses are reference numbers and indicate that an author's work is referred to, although his name is not cited in the text. Numbers in italics show the page on which the complete reference is listed.

A

Abraham, A., 4(32), 52(132), *335, 338*
Abraham, D., 234, *342*
Aggarwal, R. L., 7(69), 92(69), 103(69, 194), 148(69, 194), 157(227), 277 (69, 194), 313(194), 315(227), 316, 317(194, 227), 319, *336, 340, 343*
Albers, W. A., 217(273), 274(273), 275 (273), *342*
Alferieff, M. E., 7, 191(79), 192(79), 193, 196(79), *337*
Allen, F. G., 6(60), 81(60), *336*
Alway, C. D., 105(196), *340*
Anderson, C. H., 100(184), *339*
Antončik, E., 4, 65(31), *335*
Antonsiewicz, H. A., 171(234), 172 (234), 175(234), *341*
Archer, R. J., 4(24), 63, 65, *335*
Argyres, P., 169(233a), *341*
Aronov, A. G., 307(334, 335), 309(334), 310(335), 311, 312(335), *343*
Arthur, J. B., 118(205), 128, 129, *340*
Aspnes, D. E., 7(78), 99(78), 101(185), 167(76, 231, 232), 169(231), 172 (76), 174, 176(76), 177(76, 232), 178(76, 232), 180, 185, 190, 217 (78), 219(78), 224(78, 283), 225 (78), 226(78), 228(78), 229(231), 231(78), 259, 269(317), 270(317), 325, 326(76), 329, *337, 339, 341, 342, 343*
Avery, D. G., 4(23), 64(23, 151), *335, 339*
Axe, J. D., 250, *343*
Aymerich, F., 167(230), 180(230), *341*

B

Baldereschi, A., 304(330), 305, *343*
Baldini, G., 39(121), *338*
Ballaro, S., 234, *342*
Balslev, I., 6(67), 7(67), 90(67), 103 (189), 105(67, 198), 108(67), 112, 113, 147(67), 149, 150(67), 277(67, 189), 285, 287, 288(67), 294(189), *336, 339, 340, 343*
Balzarotti, A., 234(292), *342*
Barber, J., 283(326), *343*
Bardeen, J., 3, 24(16), *335*
Bassani, F., 4(30), 15(102), 16(102), 74 (30), 167(230), 180(230), 304 (330), 305, *335, 337, 341, 343*
Batz, B., 6(64), 50(126), 95(64), 111 (126, 201), 125, 127(201), 128, *336, 338, 340*
Baynham, A. C., 118(205), 128(205), 129(205),*340*
Beaglehole, D., 102(188), 143(188), 149 (188), 160(188), *339*
Becquerel, J., 2, *335*
Beer, A., 8(80), *337*
Belle, M. L., 259, *343*
Berglund, C. N., 6(66), 95(66), 122 (66), 127, *336*
Bergstresser, T. K., 39(119), 52(133), 80(119), 257(304), *338, 343*
Berreman, D. W., 64(152), *339*
Best, P. E., 135(214), *340*
Bhagarva, R. M., 152(226), *340*
Bir, G. L., 139(215b), 288(215b), *340*
Birman, J., 115(204), *340*
Blatt, F. J., 3(16), 6(57), 24(16), 92 (57), 103(57), *335, 336*

Bleil, C. E., 217(273), 274(273), 275 (273), *342*
Bloembergen, N., 11(98), *337*
Bloom, S., 242(299b), *342*
Blossey, D. F., 167(231), 169(231), 229 (231), *341*
Boddy, P. J., 97(180), 197(247, 248), 198(248), 202(247), 263, 264(247), 265(248), 266(248), *339, 341*
Bonfiglioli, G., 90(176), 105(99), 108 (176), *339, 340*
Bottka, N., 61(149), 62(149), 103(149), 104(149), 180, 188, *339, 341*
Brahms, S., 228, 234(285), *342*
Brattain, W. H., 209(262), 220(280), *341, 342*
Braunstein, R., 66(155), 68, *339*
Brews, J. B., 200(254), 201, 202, *341*
Brooks, H., 141(216), *340*
Brovetto, P., 90(176), 105(199), 108 (176), *339, 340*
Brown, R. N., 320, *343*
Brown, W. L., 211(268), *342*
Brust, D., 4(30), 5(30), 74(30, 163), 75 (163, 166), 76(166), 237(45), 282 (166), *335, 336, 339*
Burstein, E., 6, 92(57), 103(57), *336*
Busca, G., 105(199), *340*

C

Callaway, J., 7, 167(74, 228, 229), 169 (228), *337, 341*
Cardona, M., 4(33), 5, 7(68, 71), 9(91), 10(95), 12(95), 15(103), 20(107), 21, 29(113), 34(113), 36(35, 46), 37(113), 38(116), 39(113), 49(46, 117), 52(46, 117, 131), 57(46, 137, 139), 59(91), 60(139, 148), 68(33, 137), 71(116), 72(35, 113), 73(35, 103, 113), 74(113, 164), 75(164), 76(113), 78(139, 164), 79(139), 94 (71), 99(71), 102(33), 103(71, 190), 104(71, 91), 115(203), 119 (206), 134, 140(116), 200(252), 203(256), 217(274), 224(283), 228, 230(116), 234(285), 236(71), 240 (296, 297), 241(71, 298), 242(71, 116), 243(71), 245(71), 247(46, 71, 117), 248, 249(274), 255(71),
261(308), 269(256, 317), 270(317), 275(274), 277(322, 323), 278(71, 190, 325), 279(71, 206), 280(71, 325), 282(190), 283(137, 190), 288 (116), 290(323), 291, 293(323), 294(323), 297(103, 323), 298(323), 320, 333(103, 113, 116), *336, 337, 338, 339, 340, 341, 342, 343*
Casella, R. C., 200(255), *341*
Casselman, T. N., 200, 202(253), *341*
Chanbarisov, V. Sh., 169(233b), *341*
Chen, Y. S., 197(248), 198(248), 265 (248), 266(248), *341*
Chester, M., 189(241), 217(275), 224 (275), 228, *341, 342*
Chiarotti, G., 273(320), 274(320), *343*
Clegg, P. L., 4(23), 64(23), *335*
Cohen, M. H., 5(40), 10(96), 64(40, 150), *336, 337, 339*
Cohen, M. L., 39(119), 52(133), 80 (119), 257(304), 259(307), *338, 343*
Collier, A., 105(195), *340*
Cooper, B. R., 83(174), 86(174), 131 (174), 132(174), *339*
Cuevas, M., 284, *343*

D

Danner, H. R., 198(250), *341*
Dash, W. C., 57(140), 285(328), *338, 343*
Demeis, M., 51(130), 117(130), *338*
Dermatis, S. N., 314(338), *343*
Dewald, J. F., 219(278), *342*
Dexter, D., 8, 25(84), *337*
Dietz, R. E., 46(123), *338*
Dimmock, J. O., 6, 25(108), 230(56), *336, 337*
Dresselhaus, G. F., 3(20), 75(165), 77, 157(227), 315(227), 317(227), *335, 339, 340*
Dresselhaus, M. S., 5(42), 8(42), 64 (42), 75(165), 77, 157(227), 315 (227), 317(227), *336, 339, 340*
Drews, R. E., 90(177), 107(177), 108 (177), 113, 230(289), *339, 342*
Duke, C. B., 6, 7, 32, 33(49), 191(79), 192(79), 193, 196(79), *336, 337*
Dutton, D. B., 3(15), 57(15), *335*

E

Eby, J. E., 3, 57(15), *335*
Ehrenreich, H., 2(7), 5, 10(94, 96), 11 (94), 13(94), 14(94), 39(118), 59 (143), 78, 82(38), 83(38, 174), 86 (174), 131(174), 132(174, 211), 285 (328), *335, 336, 337, 338, 339, 340, 343*
El Naby, M. H., 5, 64, 135, 164(39), *336*
Elliott, R. J., 3, 25(17), 27(17), 28(17), 30(17, 114), 44(17), 45(17), 47 (124), *335, 338*
Enderlein, R., 180, *341*
Engeler, W. E., 6(62, 65), 91(62, 178), 100(62), 104(178), 130(178), 139 (65, 178), 148(62, 65), 152, 153, 154(65), 157(65), 160(178), 163 (178), 164(178), *336, 339*
Erlbach, E., 147, 287(221), *340*
Ewald, A. W., 64(153), *339*

F

Falicov, L. M., 80(170), *393*
Fan, H. Y., 14(99), 50(128), *337, 338*
Fawcett, W., 118(205), 128(205), 129 (205), 130, *340*
Feinleib, J., 7(70), 92(70), 103(70), 124(207), 132(207), 217(271), 224 (70), 239(271), 267, 277(271), 314(70, 271), 320(70, 271), 322 (271), *336, 340, 342, 343*
Feldman, A., 149, 152, *340*
Finegold, L. X., 51(130), 117(130), *338*
Fischer, T. E., 81(172), *339*
Flugge, S., 35(115), *338*
Folland, N. O., 261(309), *343*
Forman, R. A., 203(256), 204(259), 261 (308), 269(256, 317), 270(317), *341, 343*
Fowler, A., 107(200), *340*
Franck, J., 2, *335*
Franz, W., 7, 165, 229, 330, *336*
Frazer, B. C., 198(250), *341*
Frenkel, J., 2, *335*
Fritsche, L., 189(241), *341*
Fritzsche, H., 6(62), 91(62), 100(62), 148(62), 152(62), 284, *336, 343*
Frova, A., 7(78), 97(180), 99(78), 197 (247, 248), 198(248), 202(247), 217(78), 219, 224(78), 225, 226, 228, 231, 263, 264(247), 265, 266 (248), *337, 339, 341*

G

Gähwiller, C., 197(249), 198(249), 200 (249), 202, 263(249), 267, *341*
Garfinkel, M., 6(62, 65), 91(62, 178), 100(62, 65), 104(178), 130, 139 (65, 178), 148(62, 65), 152(62), 153(65), 154(65), 157(65), 160, 163 (178), 164, *336, 339*
Garrett, C. G. B., 209(262), 219(276), 220(280), *341, 342*
Gatos, H. C., 60, *338*
Gerhardt, U., 102(188), 143, 149, 158, 160, 297, *339, 340, 343*
Germano, F. A., 7(78), 99(78), 217 (78), 219(78), 224(78), 225(78), 226(78), 228(78), 231(78, 290), 232(290), 236(290), 247(301), *337, 342, 343*
Gershenzon, M., 46, *338*
Ghosh, A. K., 126(208), 247, 248, *340*
Gilat, G., 75(167), *339*
Gilgore, A., 107(200), *340*
Gobeli, G. W., 6(60, 63), 81(60), 91 (63), 139, 148(63), 152, 153, *336*
Grasso, V., 234(292), *342*
Green, M., 209, *341*
Greenaway, D. L., 8, 49(117), 52(117), 57(139), 60(139), 148, 78(139), 79 (139), 247(117), *337, 338*
Gross, E. F., 3, 6, *335, 336*
Groves, S. H., 7(70), 92(70), 103(70), 217(271), 224(70), 239(271), 241 (299a), 277(271), 314(70, 271), 320 (70, 271), 322(271), *336, 342*
Gutsche, E., 230(288), 243(300), 255 (300), *342*

H

Haensel, R., 98(182), *339*
Hall, L. H., 3(16), 24(16), *335*
Hamakawa, Y., 231(290), 232(290), 236, 247(301), *342, 343*

Hammer, R., 250, *343*
Hanamura, E., 6(50), 39(50), 43(122), *336, 338*
Handler, P., 7(78), 99(78), 167(231), 169(231), 217(78), 219(78), 224(78), 225(78), 226(78), 228(78), 229(231), 231(78, 290), 232(290), 236(290), 247(301), *337, 341, 342, 343*
Hansen, W. N., 267, 268(314, 315), *343*
Hanus, J., 124(207), 132, *340*
Harbeke, G., 5, 8, 36(46), 49(46), 52(46), 57(46, 141), 115(203), 247(46), *336, 337, 338, 340*
Harrick, N. J., 272, 273(319), *343*
Hayashi, H., 3, *335*
Hensel, J. C., *291*
Herman, F., 3, 5, 66(157), 67, 81(171), 127, *335, 336, 339, 340*
Hermanson, J., 6, 25(51), 26(51), 39(51), 43(51), *336*
Hess, R. B., 6(61), 94(61), 166(61), 206(61), 216(61), 234(61), *336*
Hietel, B., 135, *340*
Higginbotham, C. W., 29(113), 34(113), 37(113), 39(113), 72(113), 73(113), 74(113, 164), 75(164), 76(113), 78, 333(113), *338, 339*
Hilsch, R., 2, *335*
Hopfield, J. J., 6, 146(219), 147(219), 230(55), *336, 340*
Huntington, H. B., 151(225), *340*

I

Inoue, M., 6(50), 39(50), 43(122), *336, 338*
Inui, T., 6(50), 39(50), 43(122), *336, 338*

J

Jahoda, F. C., 4(25), *335*
Jensen, J. D., 4(27), 55(27), 57(27), 257(27), *335*
Joffe, A. D., 254(303), *343*

K

Kahn, A. H., 198, *341*
Kamimura, H., 68(159), *339*

Kane, E. O., 6(63), 66(155), 68(155), 71(161), 72, 74(162), 76, 77, 82(173), 91(63), 130, 139, 148(63), 152, 153, 282(162), 331(161), *336, 339, 343*
Kaplyanskii, A. A., 144(217), *340*
Karryev, N. A., 3(13), *335*
Katzuki, K., 3, *335*
Keldysh, L. V., 7, 165, *337*
Kern, R., 197(246), *341*
Kingston, R. H., 209(263), *341*
Kip, A., 3(20), 333(20), *335*
Kittel, C., 3(20), 333(20, 343), 334(344), *335, 343*
Kleinman, L., 5, 75(43), 79(168), *336, 339*
Kline, J. S., 278(325), 280(325), *343*
Knox, R. S., 8, 25(83), *337*
Kohn, W., 28(110), 288(329), 301(329, 332), 312(332), 313(332), *338, 343*
Korovin, L. I., 19(105), *337*
Kortum, R. L., 5(44), 127(209), *336, 340*
Koster, G. F., 39(120), 146(220), *338, 340*
Kramers, H. A., 9(90), *337*
Kronenberger, Z., 2(2), *335*
Kuglin, C. D., 5(44), 127(209), *336, 340*
Kuhn, H., 2(8), *335*
Kunz, C., 98(182), *339*
Kuper, C. G., 8(82), 25(82), *337*
Kurskii, Yu. A., 169(233c), *341*
Kuwana, T., 268(315), *343*
Kyser, D. S., 203(258), 269(258), 270(258), *341*

L

Laing, W. Y., 254(303), *343*
Landau, L. D., 192(244), *341*
Lange, H., 230(288), 243(300), 255(300), *342*
Lavine, M. C., 60, *338*
Lax, B., 4, 6(58), 7(69), 92(58, 69), 103(58, 69, 192, 194), 148(69, 194), 277(69, 194), 302(333), 303(333), 313(194), 317(194), 319(194), *335, 336, 340, 343*
Lax, M., 146(219), 147(219), *340*

AUTHOR INDEX

Lee, P. M., 261(310), *343*
Lempicki, A., 115(204), *340*
Levialdi, S., 105(199), *340*
Leyendecker, A. J., 198, *341*
Lifschitz, E. M., 192(244), *341*
Lin, P. J., 79(168), 80(170), 259(307), *339, 343*
Lindquist, R., 64(153), *339*
Lott, L. A., 262, *343*
Loudon, R., 47(124), *338*
Ludeke, R., 217, 234(272), 238, 255(272), *342*
Lukeš, F., 224(282), *342*
Luttinger, J. M., 288(329), 301(329, 332), 312(332), 313(332), 332(342), *343*
Lynch, D. W., 262, *343*

M

McClure, D. S., 2, *335*
McElroy, P., 241(298), *342*
Macfarlane, G. G., 3(21), 25(21), 113, 128(22), 236(294), *335, 342*
McLean, T. P., 3(22), 112(202), 113(22, 202), 128(22), 147(202), 236(294), *335, 340, 342*
Madelung, O., 29(111), 151(111), *338*
Mahan, G. D., 25(109), *337*
Marple, D. T. F., 20, 21, 39(118), *337, 338*
Marshall, H., 35(115), *338*
Martienssen, W., 54(135), *338*
Matatagui, E., 134, *340*
Mavroides, J. G., 103(192), 157(227), 315(227), 317, *340*
Mayer, H., 5, 64, 135, 164(39), *336, 340*
Mead, C. A., 211(266), *342*
Milne-Thomson, L. M., 205(261), *341*
Mohler, E., 148(222), 158, 160, *340*
Moore, C. E., 66(156), *339*
Morita, A., 68(159), *339*
Moss, T. S., 7(77), 8, 229(286), 230(286), *337, 342*
Mueller, F. M., 87, *339*

N

Nathan, M., 152(226), *340*
Neumark, G., 115(204), *340*
Neustadter, S., 209(263), *341*
Newman, R., 57(140), *338*
Nikitine, S., 3, *335*
Nuccioti, A., 273(320), 274(320), *343*

O

Obseimov, I. W., 2(3), *335*
Okazaki, M., 6(50), 39(50), 43(122), *336, 338*
Olechna, D. J., 132(211), *340*
Olson, E. C., 105(196), *340*
Onodera, Y., 144(218), *340*
Onsager, L., 9, *337*
Osteryoung, A., 268(315), *343*
Overhauser, A. W., 5(41), 64(41), *336*

P

Paige, E. G. S., 118(205), 128(205), 129(205), 234(291), *340, 342*
Pake, G. E., 272(318), *343*
Palmieri, G., 105(199), *340*
Pankove, J. J., 102(187), *339*
Paul, W., 6(52, 53), 8(53), 51(130), 102(53), 117(130), 141(53), 151(53), 217, 234(272), 238, 241(299a), 255(272), *336, 338, 342*
Peierls, R. E., 2, *335*
Pemsler, T. P., 105(197), *340*
Penchina, C. M., 189(240), *341*
Pepinsky, R., 198(250), *341*
Perny, G. P., 3(14), *335*
Philipp, H. R., 2(7), 4(26), 5, 59(143), 78, 82(38), 83(38, 174), 131(174), 132(174, 211), 285, *335, 336, 338, 339, 340, 343*
Phillips, J. C., 4(30), 5(30, 40), 6, 8, 15(34, 100), 16(28), 18(100), 64(40), 65(34), 74(30), 75(43), 79, 87, 177(235, 236), *335, 336, 337, 339, 341*
Picus, G. S., 6(57), 92(57), 103(57), *336*
Pidgeon, C. R., 7(70), 92(70), 103(70), 217(271), 224(70), 239(271), 277(271), 314(70, 271), 320(70), 322(271), *336, 343*
Pikus, G. E., 139(215b), 288(215b), *340*

Pikus, G. E., 307(335), 310(335), 311, 312 (335), *343*
Pines, D., 5(36), 14(36), *336*
Plotmikov, A. F., 224(284), 229(284), *342*
Plummer, A. R., 211(265), *342*
Poehler, T. O., 234, *342*
Pohl, W., 2, *335*
Pollak, F. H., 7(68, 71), 15(103), 29 (113), 34(113), 37(113), 39(113), 72(113), 73(103, 113), 74(113, 164), 75(164), 76(113), 78(164), 94(71), 99(71), 103(71), 104(71), 217(274), 236(71), 240(296), 241 (71, 298), 242(71), 243(71), 245 (71), 247(71), 248(71), 249(274), 255(71), 275(274), 277, 278(71, 325), 279(71), 280(71, 325), 290 (323), 291, 293(323), 294(323), 297(103, 323), 298(323), 320, 333 (103, 113), *336, 337, 338, 339, 342, 343*
Potter, R. F., 65, 72, 126(154), 247, *339*
Prichotjo, A. F., 2(3), *335*
Pringsheim, P., 2(2), *335*
Prostak, A., 267, 268(314), *343*
Purcell, E. M., 272(318), *343*

Q

Quarrington, J. E., 3(22), 113(22), 128(22), 236(294), *335, 342*

R

Raubenheimer, L. J., 75(167), *339*
Rees, H. D., 229(287), 231, 234(291), *342*
Rehn, V., 191(243), 203(258), 269(258), 270(258), *341*
Reine, M., 313(337), *343*
Roberts, V., 3(21, 22), 25(21), 113(22), 128(22), 236(294), *335, 342*
Robins, J. L., 135(214), *340*
Roessler, D. M., 60, *338*
Rössler, U., 188, *341*
Rollefson, G., 2(8), *335*
Roth, L. M., 4, 6(58), 92(58), 103(58), 302(333), 303(333), *335, 336, 343*

Rubenstein, M., 52(133), *338*
Rubin, L., 7(69), 92(69), 103(69), 148(69), 277(69), *336*

S

Sak, J., 6, 31(48), 33, 35(48), 36(48), 39(48), 40(48), *336*
Salzberg, C. D., 55(136), *338*
Samelson, H., 115(204), *340*
Samoggia, G., 273(320), 274, *343*
Sandrock, R., 15(104), 75(104), 102(188), 143(188), 149(188), 160(188), *337, 339*
Saslow, W., 259(307), *343*
Scanlon, W. W., 51(129), 55(129), *338*
Schmidt, E., 224(282), *342*
Schnatterly, S. E., 103(193), *340*
Schoolar, R. B., 4(27), 55(27), 57(27), 257(27), *335*
Schweber, S. S., 312(335a), *343*
Scouler, W. J., 49(125), 95(179), 98, 124(179, 207), 130, 132(207), 262, *338, 339, 340, 343*
Seely, S., 99(183), *339*
Segall, B., 5, 6, 32, 33(49), 86, *336*
Seitz, F., 10(93), 12(93), *337*
Seiwatz, R., 209, *341*
Seraphin, B. O., 6, 61(149), 62(149), 94(61), 103(149, 191), 104(149), 166(71), 177(235), 180, 206, 211(267), 212(267), 216, 234(61, 191, 295), 236, 237, 239(295), 243(295), *336, 339, 340, 341, 342*
Shaklee, K. L., 7(68, 71), 94(71), 99(71), 103(71, 190), 104(71), 217 274), 236(71), 240(296), 241(71, 298), 242(71), 243(71), 245(71), 247(71), 248(71), 249(274), 255(71), 275(274), 277(322), 278 (71, 190), 279(71), 280(71), 282(190), 283(190), *336, 339, 342, 343*
Shay, J. L., 81(171), *339*
Shindo, K., 68(159), *339*
Short, R. A., 5(44), 127(209), *336, 340*
Sieskind, M., 3(14), *335*
Singleton, C., 105(195), *340*
Skillman, S., 66(157), 67, *339*
Slater, J. C., 3, 39(120), *335, 338*
Slater, L. J., 193(245), *341*

Smith, C. S., 138(215a), *340*
Sommers, Jr., H. S., 52(131), *338*
Sonntag, B., 98(182), *339*
Sonpaa, H. H., 57(142), *338*
Spicer, W. E., 6, 81(59, 171), *336, 339*
Spitzer, W. G., 14(99), 211(266), *337, 342*
Stadler, H. L., 268, *343*
Stern, F., 9(92), 19(92), *337*
Stoller, P. J., 107(200), *340*
Stopachinskii, V. B., 169(233b), 169(233c), *341*
Sturge, M., 29, 234, *338*
Suzuki, K., *291*

T

Taft, E. A., 4(26), *335*
Tauc, J., 4(32), 8, 52(132), 56(81), 65(31), *335, 337, 338*
Teegarden, K. J., 3(15), 57(15), *335*
Tegart, W. J. McG., 60(147), *338*
Tharmalingham, K., 7, 167(75), 174, 183(75), 185(75), *337*
Thomas, D. G., 6, 8(88), 46(123), 68(158, 160), 102(160), 230(55), 239, 254(158), *336, 337, 338, 339*
Thompson, A. G., 52(133), 102(186), 103(190), 278(190, 324), 282(190, 324), 283(190), *338, 339, 343*
Tiemann, J. J., 6(62, 65), 91(62, 178), 100(62, 65), 104(178), 130(178), 139(65, 178), 148(62, 65), 152(62), 153(65), 154(65), 157(65), 160(178), 163(178), 164(178), *336, 339*
Toyozawa, Y., 6, 39(50), 43(122), 52(134), 53(134), 54(134), 144(218), *336, 338, 340*
Turner, D. R., 219(279), *342*

V

Van Hove, L., 15(101), 16(101), 18(101), *337*
Vavilov, V. S., 169(233b), 224(284), 229(284), *341, 342*
Vázquez, F., 261(308), *343*
Velický, B., 6, 31(48), 33, 35(48), 36(48), 39(48), 40(48), 59(144), *336, 338*

Villa, J. J., 55(136), *338*
Viswanathan, K. S., 167(229), *341*
Vrehen, Q. H. F., 50(127), 103(127), 231(127), 302(127), 310(127), 313(127, 337), 314(127), 323, 324, *338, 343*

W

Wallis, R. F., 6(57), 92(57), 103(57), *336*
Wang, E. Y., 217, 274, 275(273), *342*
Wanke, E., 105(199), *340*
Wannier, G. H., 3, 27(11), *335*
Warschauer, D. M., 6(52), *336*
Wendland, P. M., 217(275), 224(275), 228, *342*
Wheeler, R., 6, 230(56), *336*
Whitfield, G. D., 8(82), 25(82), *337*
Wilhelm, W. E., 205 (260), *341*
Wilkinson, P. G., 98(181), *339*
Willardson, R. K., 8(80), *337*
Williams, E. W., 52(133), *338*
Williams, R., 219, *342*
Wiser, N., 10(97), *337*
Wolf, H. C., 2, *335*
Woolley, J. C., 52(133), 102(186), 103(190), 278(190, 324), 282(190, 324), 283(190), *338, 339, 343*
Wright, G. B., 49(125), 130, *338*

Y

Yacoby, Y., 189(242), 203(257), 224(257), 227, *341*

Z

Zakharchenia, B. P., 6, *336*
Zakhvathin, G. V., 224(284), 229(284), *342*
Zawadski, W., 313(337), *343*
Zemel, J. N., 4(27), 55(27), 57(27), 257(27), *335*
Zollweg, R. J., 57(138), *338*
Zook, J. D., 200(253), 202, *341*
Zuteck, M. D., 103(194), 148(194), 277(194), 313(194), 317(194), 319(194), *340*
Zwerdling, S., 6, 92(58), 103(58), 302(333), 303(333), *336, 343*

Subject Index

Acetonitrile, transmission of, 223
Adiabatic approximation and hyperbolic excitons, 33
Airy function integrals, Ai_1, Gi_1, 180
Airy functions
  asymptotic expansion of, Ai, 172
  of complex argument, 182
  Gi, 180
  important relationships and integral representations, 325
  normalization, 175
  regular at infinity Ai, 171
  singular at infinity Bi, 173
Alkali halides, 2, 3, 54, 57, 68
Alkali metals, 4, 64, 136
Antimony, 79
Augmented plane waves (APW) method, 74

Back reflection, 114
Band structure, 2, 3
  of copper, 87
  of hexagonal CdSe, 257
  of SnTe, 260
  of $SrtiO_3$, 199
Barium titanate, 198, 263, 267
Bismuth sulfide-telluride, 57
Bose-Einstein function, 24
Brewster angle, 65
Broadening
  band-to-band transition, 47
  of excitons, 52
  Gaussian, 54
  by phonon emission and absorption, 47
Burstein-Moss shift in GaAs, 278

Cadmium sulfide, 269, 275
Cadmium telluride, 19, 38, 43, 235, 238
Cesium, 134
Chemical etching, 60, 57
Collision frequency, 13, 14, 77
Combined density of states, 17, 40

Compounds
  I-VII, 57
  II-VI, 57, 72
  III-V, 57, 72
Conductivity tensor, 9
Conformal mapping, 205
Coplanar electrodes, 204
Copper, 82, 86, 132, 160
  band structure, 87
Core corrections to matrix element of **p**, 75
Coulomb interaction, *see* Excitons
Critical points, 15
  at arbitrary points of **k**-space, 16
  M_1, 5
  one- and two-dimensional, 16
  symmetry, 16
  two- and one-dimensional, 21
  types of, 16
Crossed fields, 307
Crystal field splitting, 69
Crystals
  covalent, 2
  ionic, 2
  molecular, 2
Cuprous chloride, 68
Cuprous oxide, forbidden excitons, 29, 228
Cyclotron resonance, 3

Data reducing equipment, 97
Debye screening length, 209
Deformation potentials, 103
  of Ge and GaAs, 293
  of Ge and Si, 289
  intervalley, 141
Density of states, 75
  near critical points, 17
  two-dimensional, 176
Detector, 97
  PbS and PbSe cells, 99
  photomultiplier, 99

Dielectric constant
  calculations, 74
  infrared, 27
  interband contribution, 13
  intraband or free election contribution, 12, 14
  nondispersive, 14
  in presence of crossed electric and magnetic fields, 307
  static, 27
Dielectric tensor, 9
Direct transitions, 15–23
Dispersion relations, 4, 59–62
Dividing circuit ($\Delta R/R$),
  analog, 99
  electronic servo, 100, 101
  mechanical servo, 100
  for PbS and PbSe cells, 101
  strip chart recorder, 101, 102
Double beam system, 107
Dynamic range, 100

Effective g-factor, 318–322, 331–333
Effective mass, 331–334
  conductivity or optical, 14
  from magnetooptical modulation data, 318
Effective mass approximation, 32–37, 175–178, 301–313
Eigenfunctions in presence of electric field, 167
Elastic compliance constants, 140
Electric field modulation, 94–96, *See also* Electroreflectance, Electroabsorption
  allowed M_0 critical point, 170
  broadened critical points, 182
  effective mass approximation, 174–178
  electrolytic method, 219–224
  excitons, 191–196
  experimental techniques
    insulators, 202
    semiconductors, 206
  experimental results, *see* Electroreflectance, Electroabsorption
  forbidden excitons, 196
  forbidden interband transitions, 183–187
  forbidden transitions at M_1 critical points, 187
  indirect transitions, 189–191
  matrix element for direct allowed transitions, 168
  M_1 critical point, 177–179
  M_2 critical points, 178, 179
  M_3 critical point, 177
  nonuniform field, 213–215
  paraelectric and ferroelectric crystals, 197–202
  p–n junction, 218
  theory of, 166
    intraband effects, direct transitions, 166
    two- and one-dimensional bands, 180
    valleys off $\mathbf{k}=0$, 188
Electroabsorption
  direct edge, 229
  direct forbidden edge of Cu_2O, 228
  effective masses and, 230
  indirect edge of Cu_2O, 228
  of Ge and Si, 225
Electrolytic cell, 202
Electrooptic effect, convolution expression, 328–330
Electrooptic functions
  $F(\eta)$, 172
  $G(\eta)$, 173
  one- and two-dimensional F_1, G_1, F_2, G_2, 180
  spatial average, 215
Electropolish, 60
Electroreflectance, 6, 7, *see also* Electric field modulation
  in binary alloys, Ge-Si, 280
  CdS, 271
  CdTe, 246
  electrolyte method: germanium-zinc-blende-type materials, 240–254
  field effect configuration, 234–240
  ferro- and paraelectric materials, 262–267
  GaAs, 270
  Ge, 248
    under uniaxial stress, 291–300
  hexagonal CdSe, 256
  impurity levels, 243
  InP, 250
  insulators, 269–272
  metals, 267

SUBJECT INDEX

Mg_2Ge, 261
plasma resonance, 250
plasmons and phonons in GaAs, 254
potassium tantalate, 265
in pseudobinary alloys
    GaAs-GaP, 279
    GaAs-InAs, 283
rutile, 266
Si, 251, 252
silver and gold, 267, 268
SnTe and GeTe, 260
surface states, 272
total internal reflection, 272
wurtzite, 254–257
ZnO, 255
ZnSe, 271
Electrostriction, 197
Ellipsometry, 4, 62
Elliptic functions, 205
Elliptic integrals, 205
Envelope function, 27
Epitaxial growth, 4
Excitons, 3, 25–47
    forbidden, 29
    hydrogenic or Wannier, 27
    indirect, 3, 44, 112
    at M_3 critical point, 31
    in metals, 25
    nonhydrogenic, 39
    near saddle point singularities, 32, 39
    Wannier, 3, 27
Experimental techniques
    for measuring optical constants, 55
    for measuring refractive index, 55

Ferroelectric Curie temperature, 198
Field effect, 211
    in germanium, 212
Forbidden transitions, 22
    two-dimensional, 23
Fourier components of antisymmetric potential, 74

g-Factors, from magnetooptical modulation data, 318
Gallium antimonide, 320
Gallium arsenide, 29, 151, 235, 239, 269, 289, 298
Gallium phosphide, 46

Gap modulation, 90, 91
Germanium, 52, 62, 65, 72, 75, 112, 125, 128, 129, 151, 225, 234, 237, 275, 289, 291, 298, 320
    phonon spectrum, 145
Germanium telluride, 257
Germanium-zincblende-type compounds, 68, 72, 80
Globar, 98
Gold, 160
Gouy layer, 219

Helmholtz layer, 219
Hydrogen arc, 98
Hydrogen atom
    continuum, 28
    two-dimensional, 34

Impurity levels in electroreflectance, 243
Indirect edge of Ge and Si versus stress, 287
Indirect transitions, 3, 23
    modulation spectra, 92
    in presence of magnetic field, 304
    under static uniaxial stress, 285
Indium antimonide, 49, 36, 52, 78, 239, 320, 322
Indium arsenide, 239, 320, 322
Indium phosphide, 240
Interaction representation, 11
Interband transitions, in magnetic field, 92
Intravalley splitting, 295

$\mathbf{k} \cdot \mathbf{p}$ Hamiltonian, 71, 331
$\mathbf{k} \cdot \mathbf{p}$ Method, 74
$\mathbf{k} \cdot \mathbf{p}$ Sum rule, 12
Koster-Slater interaction, 39, 43
Kramers-Kronig analysis, 59, 103
Kramers-Kronig relations, 9

Lamps, tungsten-iodine, 98
Landau quantization, 103
Landau subbands, 301
Lead chalcogenides, 51, 57
Lead selenide, 55, 79, 257
Lead sulfide, 55, 78, 257
Lead telluride, 55, 79, 257
Light source, 97

Lithium fluoride, 107
Line shapes
  in interband transitions, 111
  in modulation spectrosopy, 109
Local field corrections, 10
Lock-in amplifier, 96, 97
Lorentzian line,
  asymmetrically broadened, 53

Madelung constant, 202
Magnesium germanide, 261, 262
Magnesium silicide, 261, 262
Magnesium stannide, 261, 262
Magnetic field modulation, 95
Magnetoelectroabsorption, 307
  crossed fields, 323
Magnetoelectroreflectance, 103, 277, 307, 314
  parallel fields, 320
Magnetooptical effects, 6
  in crossed electric and magnetic fields, 310
  M_1 critical points, 305
  theory, 301–315
Magnetopiezoabsorption, 277, 313, 314, 317
Magnetopiezoreflectance, 103, 277, 313, 315, 316
Magnetopiezotransmission, 103
Measurement of optical constants by reflection
  at normal incidence, 58
  at oblique incidence, 64
Mechanical polishing, 57
Mercury telluride, 49
Modulating unit, 97
Modulation techniques, 89–104
  dependence of band structure on static parameters, 277
  effects of temperature and doping, 277
  external, 7
  internal, 7
  in magnetic field, experimental methods, 313
  in presence of magnetic field, 300
  in pseudobinary alloys, 278
  under static uniaxial stress, 284
Molecular spectroscopy, 105
Monochromator, 97
Monte Carlo method, 75

Nernst glower, 98
Nickel, 132

One-electron model, 10
Onsager relations, 9
Optical constants, of metals, 4, 82
Orthogonalized plane waves (OPW) method, 5, 74
Oscillator strength, 12, 14
Oxide, cuprous, 3, 228

Parabolic bands, 170
Parabolic coordinates, 192
Parity selection rule, 15
Pervoskite, unit cell and Brillouin zone, 198
Perturbation theory, $\mathbf{k} \cdot \mathbf{p}$, 331
Phase sensitive detection, 6, 90, 96
Phonon occupation numbers, 24
Photoelectric yield, 81
Photoemission, 6, 81
Photon shot noise, 97
Photoreflectance, 217, 275
Piezoabsorption, experimental techniques, 148
Piezoelectricity, 197
Piezoelectroreflectance, 277, 289
  higher direct gaps, 294, 297
  in Si, 298
Piezopiezoreflectance, 277, 289
Piezoreflectance, *see also* Stress modulation
  experimental techniques, 148
Piezoreflectance spectrum
  of copper, 162
  of germanium, 152
  of gold, 163
  of silicon, 152
Piezoreflectance tensor, 138
Plasma resonance
  broadening modulation, 119
  modulation spectrum, 93
  plasma frequency modulation, 118
Potassium, 64, 134
Potassium bromide, 160
Potassium iodide, 54, 158
Potassium tantalate, 263
Propylene carbonate, 257
  transmission of, 223

Pseudobinary alloys
  GaAs-GaP, 278
  GaAs-InAs, 278
  Ge-Si, 278
Pseudo-Brewster angle, 65, 216
Pseudopotential method, 5, 74

Random phase approximation, 10
Reduced mass, 28, 177, 306
Reflection coefficient, 56
Reflectivity of plate, 56
Refractive index, 9
Relativistic electron, 311
Resolvent, 39
Rubidium, 134
Rutile, 198

Saddle points
  three-dimensional, 18
  two-dimensional, 18
Schottky barrier, 209
Selection rules
  for indirect transitions, 47
  in presence of magnetic field, 303
Semiconductor—electrolyte interface, 219
Seraphin method, 206
Signal-to-noise ratio, 97
Silicon, 75, 113, 127, 147, 225, 226, 235, 238
  phonon spectrum, 145
Silver, 160
Sinusoidal versus square wave modulation, 101
Source stability, 99
Space-charge function, 209
Spin orbit interaction, 2
  k-dependent, 72
Spin orbit splittings, 2, 4, 65, 67
  of groups IV and III-V materials, 242
  of Mg_2Ge, 261
  stress dependence, 296
Stark oscillations, 169
Static perturbations
  hydrostatic pressure, 102
  impurities, 102
  magnetic fields, 102
  mixed crystals, 102
  temperature, 102

uniaxial stress, 102
Stress modulation, 94, 137
  direct excitons, 143
  direct transitions, 139
  E_o edge, 140
  experimental techniques, 148
  hydrostatic, 137
  indirect transitions, 144
  in insulators, 158
  metals, 160
  {100}, {111}, and {110} critical points, 142
  results, 151
  uniaxial, 137
Stressing apparatus, 284
Strontium titanate, 198
  band structure, 199
Surface states
  fast, 211
  slow, 211
Synchrotron radiation, 98

Temperature modulation, 95, 117–136
  direct transition, 117
  excitons, 118
  experimental results, 125
  experimental techniques, 123
  indirect transition, 122
  modulation efficiency, 123
  plasma resonance, 118
Thermoreflectance, see also Temperature modulation, 6
Tight binding method, 198
Thermal expansion, effect of, on energy gaps, 51
Tin, gray or alpha, 78, 241
Tin oxide, 202, 216
Tin telluride, 79, 257
Transducer
  lead zirconate-titanate, 148
  quartz, 148
Transmission measurements, 2
Transmissivity of plate, 56
Transparent electrodes, 202, 216
Two-thirds rule, 69, 71

Uniaxial stress modulation, see also Stress modulation, 92
Urbach rule, 229, 263

Van Hove singularities, 89, *see also* Critical points
  broadening of, 48

Wannier functions, 26
Water, transmission of, 223
Wavelength modulated reflectance, 289
Wavelength modulation, experimental results, 112
Wavelength or frequency modulation, 90, 105

WKB approximation, 36, 213
Wurtzite, 4

Xenon, solid, 39, 43
Xenon arcs, 97

Zincblende, 3, 73, 243, 244
Zinc oxide, 68
Zinc selenide, 269
Zinc sulfide, 115, 269
Zinc telluride, 49, 52, 275

QC
176.8
E9C3

JUL 9 1970